FIFTH EDITION

Fundamentals of Case Management Practice

Skills for the Human Services

NANCY SUMMERS

Harrisburg Area Community College

CENGAGE

Australia • Brazil • Canada • Mexico • Singapore • United Kingdom • United States

CENGAGE

Fundamentals of Case Management Practice: Skills for the Human Services, Fifth edition
Nancy Summers

Product Director: Jon-David Hague

Product Manager: Julie Martinez

Content Developer: Lori Bradshaw

Media Developer: Mary Noel

Associate Content Developer: Sean Cronin

Product Assistant: Kyra Kane

Marketing Manager: Shanna Shelton

Art Director: Vernon Boes

Production Management, and Composition: Lumina Datamatics, Inc.

Manufacturing Planner: Judy Inouye

Text Researcher: Kavitha Balasundaram

Cover Designer: Norman Baugher

Cover Image: © Ajn / Dreamstime.com

© 2016, 2012, 2009 Cengage Learning, Inc.

WCN: 01-100-101

ALL RIGHTS RESERVED. No part of this work covered by the copyright herein may be reproduced or distributed in any form or by any means, except as permitted by U.S. copyright law, without the prior written permission of the copyright owner.

For product information and technology assistance, contact us at
Cengage Customer & Sales Support, 1-800-354-9706
or support.cengage.com.

For permission to use material from this text or product, submit all requests online at **www.cengage.com/permissions**.

Library of Congress Control Number: 2014945917

Student Edition:
ISBN: 978-1-305-09476-5

Loose-leaf Edition:
ISBN: 978-1-305-39956-3

Cengage
200 Pier 4 Boulevard
Boston, MA 02210
USA

Cengage is a leading provider of customized learning solutions with employees residing in nearly 40 different countries and sales in more than 125 countries around the world. Find your local representative at: **www.cengage.com**.

To learn more about Cengage platforms and services, register or access your online learning solution, or purchase materials for your course, visit **www.cengage.com**.

Printed in the United States of America
Print Number: 09   Print Year: 2021

To my parents, whose humor and wisdom about people and relationships formed the foundation for my work with others

Contents

Preface   xiii

Section 1 | Foundations for Best Practice in Case Management

Chapter 1 Case Management: Definition and Responsibilities   1

Introduction   1
A History of Case Management   2
Language in Social Services   2
Why We Use Case Management   3
Case Management as a Process   4
Advocacy   13
Service Coordination   13
Levels of Case Management   16
Separating Case Management from Therapy   19
Case Management in Provider Agencies   19
Managed Care and Case Management   21
Caseloads   25
Generic Case Management   26
Summary   26
Exercises I: Case Management   27
Exercises II: Decide on the Best Course of Action   30

Chapter 2 Ethics and Other Professional Responsibilities for Human Service Workers   33

Introduction   33
The Broader Ethical Concept   34
Dual Relationships   35
Boundaries   40
Value Conflicts   40
The Rights of Individuals Receiving Services   44
Confidentiality   47
Privacy   51
Health Insurance Portability and Accountability Act   52
Social Networking   55
Privileged Communication   56
When You Can Give Information   56
Diagnostic Labeling   59

Involuntary Commitment   60
Ethical Responsibilities   61
Protecting a Person's Self-Esteem   62
Stealing from Clients   64
Competence   65
Responsibility to Your Colleagues and the Profession   65
Professional Responsibility   67
Summary   68
Exercises I: Ethics   69
Exercises II: Ethically, What Went Wrong?   71
Exercises III: Decide on the Best Course of Action   76
Exercises IV: What is Wrong Here?   76

Chapter 3   Applying the Ecological Model: A Theoretical Foundation for Human Services   77

Introduction   77
The Three Levels of the Ecological Model   79
The Micro Level: Looking at What the Person Brings   80
Looking at What the Context Brings   80
Why Context Is Important   81
Seeking a Balanced View of the Client   82
Developmental Transitions   86
Developing the Interventions   87
Working with the Generalist Approach   88
Macro Level Interventions Are Advocacy   88
Summary   90
Exercises I: Looking at Florence's Problem on Three Levels   90
Exercises II: Designing Three Levels of Intervention   91

Section 2 | Useful Clarifications and Attitudes

Chapter 4   Cultural Competence   95

Introduction   95
Culture and Communication   95
Your Ethical Responsibility   96
Where Are the Differences?   96
Strangers   98
Anxiety and Uncertainty   99
Thoughtless versus Thoughtful Communication   100
Dimensions of Culture   104
Obstacles to Understanding   109
Competence   111
Summary   112
Exercises I: Testing Your Cultural Competence   113

Chapter 5   Attitudes and Boundaries   117

Introduction   117
Understanding Attitudes   117
Basic Helping Attitudes   118

Reality Check   123
How Clients Are Discouraged   124
A Further Understanding of Boundaries   127
Seeing Yourself and the Client as Completely Separate Individuals   127
Erecting Detrimental Boundaries   129
Transference and Countertransference   129
Summary   130
Exercises I: Demonstrating Warmth, Genuineness, and Empathy   131
Exercises II: Recognizing the Difference—Encouragement or Discouragement   136
Exercises III: Blurred Boundaries   136

Chapter 6   Clarifying Who Owns the Problem   139

Introduction   139
Boundaries and Power   140
If the Client Owns the Problem   141
If You Own the Problem   143
If You Both Own the Problem   144
Summary   145
Exercises I: Who Owns the Problem?   145
Exercises II: Making the Strategic Decision   147

Section 3 | *Effective Communication*

Chapter 7   Identifying Good Responses and Poor Responses   149

Introduction   149
Communication Is a Process   150
Twelve Roadblocks to Communication   151
Useful Responses   156
Summary   164
Exercises: Identifying Roadblocks   165

Chapter 8   Listening and Responding   169

Introduction   169
Defining Reflective Listening   170
Responding to Feelings   170
Responding to Content   174
Positive Reasons for Reflective Listening   176
Points to Remember   177
Summary   178
Exercises I: How Many Feelings Can You Name?   179
Exercises II: Finding the Right Feeling   179
Exercises III: Reflective Listening   180

Chapter 9   Asking Questions   187

Introduction   187
When Questions Are Important   187
Closed Questions   188
Open Questions   189

Questions That Make the Other Person Feel
   Uncomfortable   190
A Formula for Asking Open Questions   192
Summary   195
Exercises I: What Is Wrong with These Questions?   195
Exercises II: Which Question Is Better?   197
Exercises III: Opening Closed Questions   198
Exercises IV: Try Asking Questions   200

Chapter 10 Bringing Up Difficult Issues   203

Introduction   203
Confrontation   203
Exchanging Views   204
When to Initiate an Exchange of Views   204
Using I-Messages to Initiate an Exchange of Views   207
Asking Permission to Share Ideas   213
Advocacy: Confronting Collaterals   214
On Not Becoming Overbearing   215
Follow-up   217
Summary   217
Exercises I: What Is Wrong Here?   217
Exercise II: Constructing a Better Response   219
Exercises III: Expressing Your Concern   219
Exercises IV: Expressing a Stronger Message   222

Chapter 11 Addressing and Disarming Anger   225

Introduction   225
Common Reasons for Anger   225
Why Disarming Anger Is Important   226
Avoiding the Number-One Mistake   227
Erroneous Expectations for Perfect Communication:
   Another Reality Check   228
The Four-Step Process   229
What You Do Not Want to Do   231
Look for Useful Information   233
Safety in the Workplace   233
The Importance of Staff Behavior   234
Summary   235
Exercises I: Initial Responses to Anger   235
Exercises II: Practicing Disarming   236

Chapter 12 Collaborating with People for Change   239

Introduction   239
What Is Change?   239
Stages of Change   240
Understanding Ambivalence and Resistance   244
Encouragement   247
Recovery Tools   250
Communication Skills That Facilitate Change   252

Trapping the Client   258
From Adversarial to Collaborative   258
Summary   262
Exercises: Helping People Change   263

Chapter 13 Case Management Principles: Optional Review   265

Introduction   265
Combining Skills and Attitudes   265
Practice   267
Exercise I   267
Exercise II   268
Exercise III   271
Exercise IV   273
Exercise V   274

Section 4 | *Meeting Clients and Assessing Their Strengths and Needs*

Chapter 14 Documenting Initial Inquiries   277

Introduction   277
Walk-ins   278
Guidelines for Filling Out Forms   278
Steps for Filling Out the New Referral
  or Inquiry Form   278
Evaluating the Client's Motivation
  and Mood   282
Steps for Preparing the Verification of Appointment Form   282
Summary   284
Exercises I: Intake of a Middle-Aged Adult   284
Exercises II: Intake of a Child   284
Exercises III: Intake of an Infirm, Older Person   285

Chapter 15 The First Interview   287

Introduction   287
Your Role   288
The Client's Understanding   288
Preparing for the First Interview   288
Your Office   290
Meeting the Client   290
Summary   295

Chapter 16 Social Histories and Assessment Forms   297

Introduction   297
What Is a Social History?   298
Layout of the Social History   298
How to Ask What You Need to Know   299
Who Took the Social History   306
Social Histories in Other Settings   310
Writing Brief Social Histories   311

Using an Assessment Form   314
Taking Social Histories on a Computer   316
Taking Social Histories in the Home   316
The Next Step   317
Summary   317
Exercises I: Practice with Social Histories   318
Exercises II: Assessment of a Middle-Aged Adult   318
Exercises III: Assessment of a Child   319
Exercises IV: Assessment of an Infirm, Older Person   320
Exercises V: Creating a File   320

Chapter 17   Using the *DSM*   321

Introduction   321
Is *DSM* Only a Mental Health Tool?   322
Cautions   322
Who Makes the Diagnosis?   323
Background Information   323
The *DSM-IV-TR*   327
DSM 5, the Current Diagnostic Manual   328
Making the Code Using *DSM 5*   330
Multiple Diagnoses   331
Other Conditions That May Be a Focus of Clinical Attention   332
When the Diagnosis Does Not Quite Fit   332
When There Is No Number   333
Summary   333
Exercises: Using the *DSM 5*   334

Chapter 18   The Mental Status Examination   337

Introduction   337
Observing the Client   338
Mental Status Examination Outline   339
Summary   356
Exercises: Using the MSE Vocabulary   356

Chapter 19   Receiving and Releasing Information   359

Introduction   359
Sending for Information   359
If You Release Information   359
Directions for Using Release Forms   360
Examples of the Release Forms   362
When the Client Wants You to Release Information   363
When the Material Is Received   363
Other Issues Related to Releasing Information   365
Summary   365
Exercises I: Send for Information Related to a Middle-Aged Adult   366
Exercises II: Send for Information Related to a Child   366
Exercises III: Send for Information Related to a Frail, Older Person   366
Exercises IV: Maintaining Your Charts   366

Section 5 | *Developing a Plan with the Client*

Chapter 20 Developing a Service Plan at the Case Management Unit   367

    Introduction   367
    Involving the Client and the Family   368
    Using the Assessment   369
    Creating the Treatment or Service Plan   372
    How to Identify the Client's Strengths   373
    Individualized Planning   374
    Understanding Barriers   375
    Sample Goal Plan   375
    Summary   376
    Exercises: Broad General Goal Planning   377
    Exercise I: Planning for a Middle-Aged Adult   377
    Exercise II: Planning for a Child   377
    Exercise III: Planning for an Infirm, Older Person   377
    Exercise IV: Maintaining Your Charts   377
    Exercise V: Checking Services   378

Chapter 21 Preparing for a Service Planning Conference or Disposition Planning Meeting   379

    Introduction   379
    What You Will Need to Bring to the Meeting   380
    Goals for the Meeting   380
    Benefits of Conference Planning   381
    Collaboration   382
    Preparing to Present Your Case   383
    Making the Presentation   383
    Sample Presentation   384
    Follow-Up to Meeting   385
    Summary   385
    Exercises: Planning   386
    Exercise I: Developing a Service Directory   386
    Exercise II: A Simulated Planning Meeting   386

Chapter 22 Making the Referral and Assembling the Record   387

    Introduction   387
    Determining Dates   388
    Sample Referral Notification Form   389
    The Face Sheet   390
    Summary   392
    Exercises: Assembling the Record   393

Chapter 23 Documentation and Recording   395

    Introduction   395
    The Importance of Documentation   396
    Writing Contact Notes   396
    Labeling the Contact   398
    Documenting Service Monitoring   398
    Documentation: Best Practice   399
    Government Requirements   402
    Do Not Be Judgmental   402

Distinguish Between Facts and Impressions   403
Give a Balanced Picture of the Person   404
Provide Evidence of Agreement   404
Making Changes to the Plan   404
Summary   404
Exercises: Recording Your Meeting with the Client   405
Exercise I: Recording Client Contacts   405
Exercise II: Using Government Guidelines to Correct Errors   411
Exercise III: Spotting Recording Errors   411

Section 6 | Monitoring Services and Following the Client

Chapter 24   Monitoring the Services or Treatment   413
Introduction   413
What Is Monitoring?   414
The Financial Purpose of Monitoring   414
Follow-Up   416
Collaboration with Other Agencies   416
Advocating   417
Leave the Office   418
Responding to a Crisis   419
Summary   420

Chapter 25   Developing Goals and Objectives at the Provider Agency   421
Introduction   421
Client Participation/Collaboration   422
Make Objectives Manageable   423
Expect Positive Outcomes   423
Objectives   425
Combining Goals and Treatment Objectives   426
Finishing Touches   428
Review Dates   429
Vocabulary   430
Summary   432
Exercises: Developing Goals and Objectives   432
Exercise I   432
Exercise II   433
Exercise III   434
Exercise IV   436
Exercise V   437

Chapter 26   Terminating the Case   439
Introduction   439
A Successful Termination   440
The Discharge Summary   443
Examples   444
Summary   447
Exercises I: Termination of a Middle-Aged Adult   448
Exercises II: Termination of a Child   448
Exercises III: Termination of a Frail, Older Person   448
Exercises IV: Organizing the Record   448

Appendix A Ten Fundamental Components of Recovery   449

Appendix B Vocabulary of Emotions   451

Appendix C Wildwood Case Management Unit Forms   454

Appendix D Prochaska and DiClemente's Stages of Change Model   487

Appendix E Work Samples   490

Appendix F Grading the Final Files   496

Appendix G Information for Understanding *DSM IV TR* Diagnoses   499

Appendix H Case Manager's Toolbox   506

References   515

Index   518

Preface

In a small nonprofit agency handling cases of domestic violence, a woman answers the phone. She assesses the caller's concerns, accurately notes the caller's ambivalence on the inquiry record, and readily connects the caller to the person most able to assist.

In a mental health case management unit a new worker listens with interest to the other case managers, the psychologist, and the psychiatrist discuss the possible diagnosis for a new client of the agency. The worker is able to understand the conversation as the group talks about the *DSM IV TR* diagnosis and the new *DSM 5* diagnosis.

Down the street a young man acting as a case manager in a substance abuse detox center handles intake calls from physicians' offices. He competently notes the main concerns for incoming patients and asks the questions he knows will give him information that doctors and therapists will need later as they work with these new admissions. His notes are clear and useful.

How long did it take these people to acquire these skills? Did they acquire this ability well after being hired in a social service agency, or did they arrive able to handle case management tasks competently?

Purpose

For me and for students, the issue has been how we can teach the social services skills that will promote their walking from the classroom into the social service setting with confidence. How can we be assured that students, often steeped in sound theoretical knowledge, will be able to fill out an inquiry form or make a referral effectively?

It is important to teach these practical skills. In addition, it is important to equip students with the vocabulary and methods used by more advanced professionals in the human service field so that upon entering the field students are prepared to engage in meaningful discussions around client issues. Although entry-level individuals would not usually give a *DSM* diagnosis, it is useful for individuals entering the field to be knowledgeable about what such a diagnosis is and what is meant by an Axis I or Axis II diagnosis or how diagnoses are given using *DSM 5*. In this way, conversations among professionals will not be misunderstood.

Today individuals with a sparse education or with recent college degrees are finding themselves thrust immediately into roles for which they have had little formal training. It is crucial, therefore, to find a method for teaching the actual human service experience at the entry level. *Fundamentals of Case Management Practice: Skills for the Human Services*, fifth edition, seeks to provide that experience in a thorough, step-by-step process that leads the reader from intake through monitoring to termination.

New in the Fifth Edition

New material has been added to this fifth edition to bring the textbook up to date. Added to this edition:

- Current terms are used throughout
- Information on the *DSM 5* and how entry level individuals can use this
- Recent changes to HIPAA
- Expanded Appendices to include material helpful in completing exercises in the text, a safety planning tool and a case manager's tool box with information to assist in assessment and disposition.
- A clear look at case management as a process
- A discussion of how the size of a caseload affects service
- Ethical considerations for those working in the field of substance abuse
- A discussion of the differences among moral, ethical, and legal behavior and how violations are addressed
- More information on the importance of mandated reporting
- Expanded treatment of the ecological model
- Broader section on empathy and more recent findings on empathy
- An extensive feelings list in the appendix for use in various exercises
- Differences between confrontation and an exchange of points of view
- Enhanced discussion of motivational interviewing and why this is useful
- More detail on the significance of the first interview
- There are fewer chapters as some material has been combined in single chapters

In addition, a considerable number of smaller items and changes specifically requested by our reviewers were added to the textbook.

Fundamentals for Practice with High Risk Populations (Summers, 2002) has been published as an adjunct to this text, giving students information and scenarios on populations in which they are interested or with whom they intend to work. Chapters cover topics such as case management with children and their families, survivors of rape and violence, older people, issues with drug and alcohol dependence, and mental illness and developmental disabilities. Each chapter features information about specific populations and provides exercises and intake forms. This textbook also contains a set of forms that can be copied (see Appendix C). These forms can be found on CengageBrain. Taken from actual social service settings, they give the reader an opportunity to practice accuracy and skill in handling social service forms and records and in organizing information.

If you do not wish to cover all of the populations discussed in the text on high-risk populations and instead want to focus on specific populations, you can order individual chapters from *Fundamentals for Practice with High Risk Populations* (Summers, 2002). Please visit http://www.textchoice2.com/ to view chapters online and to build your custom text. You can pick chapters about specific populations and create individualized booklets that you can bundle with this text. If you would like more information about custom options, please contact your local customer service representative. You can locate your representative by using our rep finder at http://custom.cengage.com/.

Format

For each chapter in the textbook, basic information is laid out, followed in most chapters by many exercises that prompt the reader to handle real issues and practice real skills. Each of the chapters on case management describes one of the case management responsibilities followed by exercises to practice applying the information. As readers progress through the text, they gradually assemble files on specific cases. Students can create and monitor believable fictional clients using one of the high-risk populations discussed in *Fundamentals for Practice with High Risk Populations* (Summers, 2002). Classroom discussions about these cases and the best disposition for each of them are not unlike the discussions that occur every day in a variety of social service settings.

Organization of the Textbook

The organization of the textbook follows a logical progression, beginning with the most basic foundation for good practice, moving to discussions on attitudes, followed by how the student will talk to others effectively. The second half of the book follows a similar process, beginning with the person's first contact with the agency and the assessment and planning process through all the case management procedures to termination.

In Part One, "Foundations for Best Practice in Case Management," readers are introduced to important foundation pieces for this field. A definition of case management and how it is central to social services, ethics and ethical issues, and the importance of the ecological model in assessment and planning give readers an introduction to professional basics.

In Part Two, "Useful Clarifications and Attitudes," readers are invited to examine what in their thinking will impede effective helping in the social service setting. Beginning with issues of cultural diversity and moving to the role of personal attitudes and boundaries, this part concludes with information and exercises related to determining who owns the problem. Each chapter in this part contains exercises encouraging readers to examine realistically their own attitudes and judgments.

Part Three, "Effective Communication," begins by introducing the reader to good and poor responses, with exercises that help students see the consequences of poor communication. Chapters on listening and responding, asking questions, bringing up difficult issues, responding to emotions, confronting problematic behavior, and disarming anger are included. Included is a chapter that gives an expanded examination of some of the techniques and ideas related to motivational interviewing. The section ends with a chapter on the effective application of what students have just learned and exercises designed to have students practice all the communication skills in order to smooth out the communication and allow it to become natural and responsive.

In Part Four, "Meeting Clients and Assessing Their Strengths and Needs," readers begin to take inquiries for services. Forms are provided that ask for basic information, teaching the student what is important to find out in that first call. This section also includes a chapter on preparing for the first interview, helping the reader become sensitive to issues that clients might have at a first meeting. A chapter on social histories and assessment forms teaches students how to use these to assemble relevant information. Introductions to the *DSM* and to the mental status examination allow the reader to become familiar with the vocabulary and the information most important to other professionals in the human service field. Students are encouraged to begin noting how a person seems to them at the time of contact. The chapters and classroom discussions will help students pin down what is important to note. In this section, readers also practice completing release of information forms for the clients they have developed in the classroom setting, mastering which records are useful and which are not.

Part Five, "Developing a Plan with the Client," allows readers to further develop a plan for those clients for whom they have created phone inquiries. Here, individually or in planning teams, according to the instructor's process, students develop realistic plans for their clients. A chapter is included instructing students on how to prepare for and participate in team planning. In the final chapters, students refer cases to providers of services and learn about documentation and recording.

Part Six, "Monitoring Services and Following the Client," is the final section, and the section begins with a chapter on monitoring services and treatment. Students switch to the role of a worker in the agency of a provider of service and take the general goals given them by case managers and develop specific goals and objectives to be accomplished within stipulated time lines. Here students learn how to develop attainable goals for their clients. In this part, readers also learn the importance of monitoring cases from a case management perspective and how to terminate the case. Numerous documentation exercises provide opportunities for students to begin writing professional notes and keeping good records.

Supplements Accompanying This Text

Cengage Learning Testing, powered by Cognero Cognero is a flexible, online system that allows you to author, edit, and manage test bank content as well as create multiple test versions in an instant. You can deliver tests from your school's learning management system, your classroom, or wherever you want.

Online Instructor's Manual The instructor's manual contains a variety of resources to aid instructors in preparing and presenting text material in a manner that meets their personal preferences and course needs. It presents chapter-by-chapter suggestions and resources to enhance and facilitate learning.

Online PowerPoint® These vibrant Microsoft PowerPoint® lecture slides for each chapter assist you with your lecture by providing concept coverage using images, figures, and tables directly from the textbook.

MindTap MindTap for Counseling engages and empowers students to produce their best work—consistently. By seamlessly integrating course material with videos, activities, apps, and much more, MindTap creates a unique learning path that fosters increased comprehension and efficiency.

For students:

- MindTap delivers real-world relevance with activities and assignments that help students build critical thinking and analytic skills that will transfer to other courses and their professional lives.
- MindTap helps students stay organized and efficient with a single destination that reflects what's important to the instructor, along with the tools students need to master the content.
- MindTap empowers and motivates students with information that shows where they stand at all times—both individually and compared to the highest performers in class.

Additionally, for instructors, MindTap allows you to:

- Control what content students see and when they see it with a learning path that can be used as-is or matched to your syllabus exactly.
- Create a unique learning path of relevant readings and multimedia activities that move students up the learning taxonomy from basic knowledge and comprehension to analysis, application, and critical thinking.
- Integrate your own content into the MindTap Reader using your own documents or pulling from sources like RSS feeds, YouTube videos, websites, GoogleDocs, and more.
- Use powerful analytics and reports that provide a snapshot of class progress, time in course, engagement, and completion.

In addition to the benefits of the platform, MindTap for Counseling offers:

- Video clips tied to the learning outcomes and content of specific chapters.
- Activities to introduce and engage students with each chapter's key concepts.
- Interactive exercises and in-platform discussion questions to provide direct, hands-on experiences for students of various learning styles.
- Review and reflection activities to demonstrate growth and a mastering of skills as students progress through the course.

Helping Professions Learning Center Designed to help you bridge the gap between coursework and practice, the Helping Professions Learning Center offers a centralized online resource that allows you to build your skills and gain even more confidence and familiarity with the principles that govern the life of the helping professional. The interactive site consists of five learning components: video activities organized by curriculum area and accompanied by critical thinking questions; ethics, diversity, and theory-based case studies; flashcards and practice quizzes; a professional development center; and a research and writing center.

To the Students

It is always a challenge to know what skills and information you will need on the first day of your first job. Even when you are already working in the field and managing many of the tasks well, you often do not know for certain why agencies choose to do things one way as opposed to another. This textbook seeks to empower you to function competently and to know why you are proceeding or should be proceeding with clients in a particular way.

In *Fundamentals of Case Management Practice*, you will follow a specific series of steps, beginning with what you are thinking and how to incorporate ethics into your thinking in client–worker relationships, continuing through your communication with clients, and ending with your putting together hypothetical case files and managing those hypothetical cases.

Throughout the course you will find yourself in discussions with others about possible treatment or service plans or the dynamics of a person's situation. Use these discussions to learn more about collaboration and to increase your ability to participate in the same sort of discussions in the agency where you will work.

Many students have taken this textbook to work with them and have found it both useful and realistic. Students have contributed their experiences on the job to make this textbook replicate as nearly as possible the issues and concerns you will encounter in your work with other people.

Further, in developing your hypothetical clients, you may want to refer to *Fundamentals for Practice with High Risk Populations* (Summers, 2002). In that textbook, six populations commonly served by social services, such as those associated with domestic violence, substance abuse, or mental health issues, are detailed so that you will be very familiar with their issues and likely problems. It is also possible to purchase individual chapters from that textbook on the population or populations that interest you. Each chapter will give you information on common problems, diagnoses, medications, treatments, and other considerations such as legal issues or common medical problems each specific population often experiences. See the instructions on how to order specific chapters in the earlier section of this Preface titled "New in the Fifth Edition."

To the Instructor: Suggestions for Using This Text

This text can be used to take students step-by-step through the case management process outside of the often harried and pressured atmosphere of a real social service agency. When the student is ultimately confronted with the actual situation, the routine and expectations will not be new. Chapters are broken down into each step in the case management process. Readers progress according to their skill levels, finally creating cases and caseloads with you acting as the supervisor, much as a supervisor would act in an actual agency. Without the urgency, you will have time to let students look up information, discuss possible diagnoses, and develop sound interventions under your guidance. For example, exercises on the *DSM* and on the mental status examination have a number of possible answers. Your discussion with your students, similar to the discussions that take place in agencies about these possibilities, is more important than the actual answers that are chosen.

Most chapters include exercises to help students practice their skills. Often several versions of the same exercise are provided. It is useful to students to begin in small groups to address the issues posed in the exercises. Their discussions and the ideas and concerns they bring back to the larger class are consistent with discussions held in social service agencies. Later, versions of the exercises can be used as tests, or you can go back to them at a later time to make sure students continue to practice their skills.

It is extremely worthwhile for students to apply the skills described in this book to specific populations. To do this, you can use this book in conjunction with my other book, *Fundamentals for Practice with High Risk Populations* (Summers, 2002). After students have read the chapters on the specific populations you have assigned or on those that are most interesting to them, they can create a fictional "typical" client that they can then walk through all the exercises from intake to termination. Case notes would reflect the common problems encountered by the population, and intake would describe a common reason for seeking services among people in this population. This gives students a good beginning look at how cases come in and unfold while clients are receiving services.

Details on six high-risk populations are provided in *Fundamentals for Practice with High Risk Populations* (Summers, 2002). A detailed chapter on children and their families gives students information on how to include others involved in the child's life and how to coordinate all the various entities with whom the family interacts. Another chapter focuses on domestic violence and rape, including how these issues affect children.

A third chapter looks at substance abuse and includes the common social and medical issues that arise for this population. This chapter also includes the common challenges this population presents to case managers and gives tips for how to handle these. Mental health and intellectual disabilities each are featured in chapters, giving common problems and issues, diagnoses, and treatments. Finally, there is a chapter focusing on aging that includes both medical and social issues for this population. All the chapters include an assessment form for that population taken from actual agencies that work with that population, and all the chapters give the most typical diagnoses and medications used with each population. Where a population has special considerations the student should know, these are included as well. For example, in the chapter dealing with issues most likely to affect women, there is a discussion of how women's programs and agencies differ in their approach to clients from other social service agencies. To order specific chapters related to specific populations, see instructions in the earlier section of this Preface titled "New in the Fifth Edition."

Benefits and Advantages

This material has been used in my own classroom for 30 years and has been updated to meet current social service trends and changes. Students have commented that using this text is like walking from the classroom into the social service setting with very little lost time in learning the actual process. Instructors teaching the practicum course have used the word *empowered* when describing what this text has done to give students confidence and skill in their first encounter with a social service position. Employers as well often contact me to say how well prepared students are who have used this textbook.

Three positive features of this textbook make it especially useful in preparing students to work in this field:

1. The text gives very basic information a person needs to handle each of the tasks described. Theoretical information can be found in many other places, and thus the concentration and focus are on what is important to note, think about, document, and pass on in each step of the human service process.
2. Numerous exercises create very real situations for students to consider and handle. These exercises are based on real experiences taken from my 23 years of practice in human services and from the experiences of many others who graciously contributed to this book. Doing the exercises and participating in the classroom discussions that follow will expose students to an extremely broad range of possible circumstances and difficulties in the field.
3. The book contains forms that give students an opportunity to practice compiling information at various times throughout the management of the case. These forms can be copied and used to create files on clients developed by the students. Using each form a number of times gives students practice in preparation for real clients in real social service settings.

These features, when taken together, create a nearly realistic social service setting in the classroom, giving the instructor many opportunities to strengthen student skills and sensitivity.

In addition, *Fundamentals for Practice with High Risk Populations* (Summers, 2002) supports students with applicable details and considerable information on various at-risk populations. This textbook acts as a reference so that the hypothetical clients students develop are real with entirely likely problems. Students can use the material found in this supplemental textbook to develop realistic clients, create useful service plans, and make appropriate referrals.

Acknowledgments

As with each edition of this textbook, I could not write such a realistic work without the wonderful help of the staff at the Dauphin County Case Management Unit. Always ready to give their time and support, they have answered important questions, clarified new national policies, and brought enthusiasm to the writing of this book.

I particularly want to acknowledge the help of Mathew Kopechny, former Executive Director who went out of his way to see that I had access to what I needed to make this textbook current. I am grateful for the time and useful examples provided by Michelle Beahm, who allowed me to shadow her for the preparation for this edition. In addition, I want to express my gratitude to Kim Castle, clerical supervisor, who has worked with me on each edition of this book to clarify issues and give me details I might otherwise have missed. Just knowing she was there to help made writing this book much easier. Thanks go as well to Joel Smith, intake case manager, who graciously allowed me to observe an intake with a client during my time observing at the Dauphin County Case Management Unit.

I am particularly indebted to Charles Curie, MA, ACSW who has always given time to discuss issues related to the textbook. He headed the Pennsylvania Office of Mental Health and Substance Abuse Services, making Pennsylvania a leader in innovative services and procedures. Appointed by President Bush to head the Substance Abuse and Mental Health Services Administration (SAMHSA) he instituted state-of-the art prevention and treatment ideas nationally. His support of this textbook has ensured the teaching of best practices.

At the Dauphin County Executive Commission on Drugs, Alcohol and Tobacco, I want to thank John Sponeybarger for his help in formulating realistic plans and services. I am grateful to the late Ruby Porr for her ideas based on her work as a service provider and to Aimee Bollinger Smith, Karen Polite, Wendy Bratina, and Carol Reinertsen who use this textbook and had teaching suggestions for additions to the text.

I deeply appreciate the support and information my husband, Martin Yespy, contributed to this work. His unfailing assistance and encouragement of these textbooks and the useful material and information he brought from the field of crisis intervention have enhanced this work.

I especially want to thank my editor, Julie Martinez, who has given me support, guidance, and a good dose of humor when needed. She has always been there when I needed help and her advice greatly enhanced this work.

The two students, Danica Zirkle and Keyanna Watkins, who organized and then participated in the videos deserve considerable gratitude for all the work they did to keep everyone on track. The students who participated in the vignettes, writing and rehearsing their work also deserve my gratitude for all their hard work. They are Catherine Wrighstone, Tom Moulfair, Sean Taney, Michele Anthony, and Alison Kilgore. Many thanks to Michelle Beahm again for participating in the videos and giving her ideas to make the vignettes more realistic. Brian Peterson and his crew from Motion Masters provided considerable direction and made the filming smooth and effortless. My thanks to all of them for the time they spent with us.

I would also like to thank the reviewers of this textbook for their very helpful comments:

Susan M. Scully-Hill,
Assumption College

Barry Yvonne,
John Tyler Community College

Paula Gelber Dromi,
California State University, Los Angeles

Alyssa Forcehimes,
University of New Mexico

Monte Gray,
Bronx Community College

Karen Guerrieri,
Kent State University, Salem

Richard Jenks,
Tillamook Bay Community College

Lee Ann Rawlins,
University of Tennessee

Their ideas and suggestions greatly strengthened this work.

CHAPTER 1

Case Management: Definition and Responsibilities

Introduction

Case management is one of the primary places in human service systems where the whole person is taken into account. Unlike specific services, case management does not focus on just one problem but rather on the many strengths, needs, and personal concerns a person brings.

For example, an elderly person may be referred to Help Ministries for a voucher for fuel oil because it has been unusually cold and the elderly person has been unable to pay for the additional oil needed to warm his home adequately. In this case, Help Ministries is concerned with his fuel oil need and the warmth he will need to stay in his home during the winter. That is their only concern with regard to this man.

The **case manager**, on the other hand, is concerned with the person's need for fuel oil, with his desire to move into public housing for the elderly in the spring, with what resources he has among his children, with his recent slurred speech indicating a possible stroke, and with his need for meals-on-wheels. The case manager is aware that there is a neighbor who can look in on him daily, that the man has ties to a church, and that he receives Social Security but little other income. She knows he has a sense of humor, goes to bingo once a month, and should be fitted for a cane.

Case management is a process for assessing the individual's total situation and addressing the needs and problems found in that **assessment**. As a part of this process, the person's strengths and interests are used to improve the overall situation wherever possible. The primary purpose for case management is to improve the quality of life for your client. This might mean more comfortable or safer living arrangements, or it might require psychiatric care or medication for diabetes. Another major purpose of this activity is to prevent problems from growing worse and costing more to remedy in the future. In the situation of the elderly man just described, we find that the meals-on-wheels program will deliver a certain standard of good nutrition, preventing malnutrition and costly medical bills in the future. By getting the man a cane, we may be preventing falls that would shorten his life and cost much more in medical bills to repair his injuries. If we enlist the neighbor to look in on our client every day, we have provided a link between the man and his neighborhood. In addition, the neighbor can alert us to small problems that require our attention. In this way we have foreseen possible difficulties and taken steps to prevent them.

A History of Case Management

In the late 1800s, a formal attempt was made to organize the delivery of services to people in need. Initially the Charity Organization Society took control of this approach, making the collecting of information and the delivery of services more systematic. In the course of its work, the society developed casework as a useful method for tracking needs, progress, and changes in each case. As people had more needs and problems beyond poverty, the need to coordinate these services became important to prevent duplication. Casework also was employed as a means of tracking and using scarce resources to the best advantage. In the 1960s, the process of deinstitutionalization meant that individuals once housed in institutions were now placed in communities where they needed considerable support to live more independent lives; as a result, casework became ever more important for a larger number of people.

In the 1980s, the term *caseworker* evolved into the term *case manager*, and these managers took on greater responsibility for managing resources, finding innovative supports, and coordinating services. Agencies began to use case management as a procedure to assess needs, to find ways to meet those needs, and to follow people as they used those services. In addition to keeping an eye on how scarce resources were spent, case managers were charged with taking a more holistic approach to their clients, looking at all their needs rather than addressing only those that brought the person in for assistance. As part of this charge came the directive to develop **individualized plans**, plans constructed specifically for that person and not a cookie-cutter approach to supplying services.

Today case managers are seen as a significant service in almost all social service settings and are viewed as the most important way to prevent relapse, track clients' needs, and support progress toward good health.

Language in Social Services

Language in the social services is a funny thing. After a word is adopted to describe people who use a service, that word becomes pejorative over time and a new word meaning the same thing is sought. In social services, we have gone from labeling people *patient* (which implied people seeking services were all sick in some way) to *client* and finally to *consumer*.

Client was meant to denote that the person was being served by a case manager in a relationship much like a lawyer–client relationship. This originally conferred an obligation on the part of the case manager to give good service to someone paying, in some manner, for that service. However, as with all words describing people who use social services, the word *client* developed a negative connotation and the word *consumer* was increasingly used instead. Consumer also implied the person was paying for good services from the case manager.

With the **Recovery Model** (Appendix A) developed by the Substance Abuse and Mental Health Services Administration (SAMHSA) and the emphasis on partnerships between case managers and the people seeking services, those words are no longer considered appropriate. The concern is that these words denote a difference in status between case manager and those they serve. Thus, in recent years, the terms *client* and *consumer* have given way to *person* or *individual*, and in many cases no term is used but rather the person's name is used instead.

In this textbook, we subscribe to the idea that case managers and the people they serve are in a partnership to which each brings a certain degree of expertise. In your work, we strongly encourage you to drop the use of the words *consumer* and *client* and adopt what is seen as the more respectful terms of *individual* and *person*.

However, having said that, there are places in this textbook where using *person* and *individual* alters the meaning of the sentence and the point that is being made. For that reason, in this textbook, we need to use *client* to denote a person seeking professional services from a professional case manager in order for the point to make sense. This is in no way meant to diminish the person who does seek service, but rather to make our points more coherent.

Why We Use Case Management

Case management serves two purposes. First, it is a method for determining an individualized service plan for each person and **monitoring** that plan to be sure it is effective. Second, it is a process used to ensure that the money being spent for the person's services is being spent wisely and in the most efficient manner.

The money you oversee in consumer care may be public money, such as the money that comes from the state to a county to administer mental health services or substance abuse treatment. It may be money that is provided by insurance companies for services to a policyholder. It may be money provided directly to an agency from either of these sources for the care of a client. Sometimes organizations, such as United Way, divide the money they have raised among various community agencies. These organizations then employ case managers to make certain the most effective use is made of that money. It is therefore the case manager who determines what is needed and how to prevent needs and problems from escalating. It is the case manager who, in collaboration with the consumer, determines what services should be authorized with the existing money. It is the case manager who then follows the consumer and the consumer's services and treatment to keep the plan on track.

Case management is more than looking out for another entity's money. It is also the most efficient way to make certain a person receives the most individualized plan for service and treatment possible. To ensure that this will be done, case management responsibilities have been broken into four basic categories of service: assessment, **planning, linking,** and monitoring. These four activities constitute the case management process. Let's look at these categories in the order in which they are usually accomplished when working with a person.

Case Management as a Process

Assessment

The first case management task is assessment. A good assessment is the foundation for understanding the problem and informing and guiding the treatment and services. Therefore, it must be done with care. This is an initial assessment, meant to be comprehensive and thorough. For that reason, it covers many different aspects of the person's life in an attempt to develop an accurate profile of the individual and the individual's problem.

There are several kinds of assessments. In some cases you will be asked to do a social history (see Chapter 16). Here you ask a series of questions, and as the person answers, you construct a written narrative. Social histories usually have a number of elements that you are to assess, and each is given a subheading within the narrative. For example, current medical condition, living arrangements, relationships, and work experience are all important. In another kind of assessment, you may be given an intake assessment form that lists all the questions you are to ask and gives you a place to note the answer. Each of these assessment procedures attempts to be comprehensive. Each seeks to assemble a considerable amount of material about the person and his or her problem.

The first thing the case manager does is assess the initial or presenting problem. Why did this person come into the agency, and what is the person asking for? Here case managers look at the extent to which problems have interfered with clients' abilities to function and care for themselves. Does this problem interfere with work or with relationships? It is especially important to note the background of the problem, how long it has gone on, and how it started. In addition, the reasons the person is seeking help now are important.

Case managers include an opinion about what possible problems might arise for this person in the future and what plan might be put into effect with the person to prevent these problems. Your opinions about potential future problems are formed as you listen to consumers describe their situations. Will the individual be likely to be around people who encourage him to drink? Does she have a medical problem that needs attention because it exacerbates her depression?

A discussion of the problem uncovers the person's needs and how he or she views those needs. Case managers look at the overall situation and consider what that person needs to bring stability and resolution to his or her life and problem. Are there needs that can be addressed that will relieve the problem, or at least alleviate it to some extent?

In every assessment with an individual, you will begin to learn what strengths the person has that you and he may draw upon to resolve the current problem. Does your client have an advanced degree, a particularly supportive family, a number of friends, a sympathetic boss, a particular skill? An assessment should never be just about the person's problems, but should also include the strengths the person brings to the problems and the strengths you see in the person's environment.

As you take the information from the consumer, you are, through your observations, also evaluating the person's ability to think clearly and to understand options, and the person's general mood. You are seeking to understand the extent to which the person understands the origin of their problems. Chapter 18 discusses in more detail something called the *mental status examination*. This is not a series of questions but rather your astute observations of the individual during the interview.

At the end of your assessment document, you will be asked to express your assessment and recommendations. Here you will summarize briefly the problem and the person's ability to handle the problem, noting the person's strengths and needs. Then you will give your own recommendations for service or treatment. Recommendations are generally worked out with the consumer as you learn what it is the person is seeking and share with that person what you have to offer.

To summarize, in an assessment you are exploring and evaluating the following:

1. The initial problem and the background to that problem
2. The person's current situation
3. The person's background in areas such as education, relationships, work history, legal history
4. What the person needs to make life more stable and to resolve the current problem
5. The strengths, including those the person brings to the problem and those in the person's environment that would be useful in resolving the situation
6. Observations about how well the person functions cognitively and any seeming mental problems you have noted
7. Recommendations for a service or treatment plan for the person

A good assessment is the foundation for the development of an individual plan for service or treatment. It delineates what essential services should be provided for individualized treatment. The assessment also establishes a baseline detailing where the person was when he or she entered services and against which you can measure progress.

Planning

After the assessment, you will be expected to develop an initial plan with the client that is comprehensive and addresses all the issues raised in your assessment. This plan should show incremental steps toward improvement and expected outcomes.

As a case manager, you cannot plan well with the person unless you are thoroughly aware of the services, social activities, and resources in your community.

Formal Agencies Every community has social service agencies that serve specific needs. The best case managers seem to know all the good places to send people for the services those people need. Some communities and counties have more services than others, but in most locations, agencies are serving children and their families, older adults, individuals with substance abuse problems, individuals on probation, women in abusive or rape situations, and individuals with mental illness or intellectual disabilities. Generally, case managers need to learn about other services as well, and the information and phone numbers for these services should be readily available to you when you practice. You will want to gradually develop contacts in these places so that your referrals are smooth and problems are quickly handled.

Begin by knowing what formal agencies are out there to help with a particular issue. For instance, if your client has a mental health problem, you might refer him to an agency that specializes in mental health treatment. The staff at that agency is familiar with medications, diagnoses, and treatment alternatives for mental health problems. Another individual may be elderly and in need of protective services because you suspect she is being physically abused by her family. You would refer her to a specific agency that offers protective services to older people. A third person may have intermittent problems with substance abuse and need services from an office where there is an intensive outpatient treatment program in the evenings. Knowing the agencies in your community and the **formal services** they offer is a good foundation.

Generic Resources Good planning is not limited to formal agencies, however. Learn about resources that are available for common problems we all have. Not every problem a person with an intellectual disability has will need to be treated by agencies set up exclusively for individuals with intellectual disabilities. For example, a woman with an intellectual disability, grieving the death of her mother, was welcomed into a grief support group at the local church and given much support. In another example, a child with academic problems in school was referred to the free tutoring at a local church. An older person who needs more social contacts might be referred to a senior center where many older people go for social and recreational opportunities. In the previous examples, the older person suffering abuse may also need the services of your local district attorney, and the person with a substance abuse issue might need medical care from a general practitioner and a public defender for pending charges of disorderly conduct. These are all services anyone can use. Knowing how to access them for your clients is important.

Support Groups and Educational Seminars Other resources often overlooked are support groups and educational seminars. For example, you may have referred the family of a child you are working with to formal family therapy sessions. In addition, you would look at support groups where parents dealing with similar problems can get together to support each other. Further, you might find a workshop on parenting skills that would greatly benefit this family, and you would tell them about the workshop and strongly encourage them to attend. A man on probation might benefit from a workshop for job readiness or a support group for ex-offenders attempting to make significant life changes. A woman who is struggling with years of sexual and physical abuse might benefit from a support group of other women facing similar issues.

These resources augment your own efforts and those of formal services and give consumers additional support and information. Often they are free or at very little cost. What your clients gain from their experiences in such groups reinforces the other services you are arranging.

Peer Support A relatively recent trend is to use peer support wherein a former client who is doing well is hired by an agency to support others on the road to recovery and life changes. It might be individuals who were able to turn their lives around after a period in juvenile detention and now are supporting others coming out of juvenile detention to do the same thing. It might be someone who has had a mental illness and is now helping others who are recovering from their own mental illness. And, of course, in substance abuse, Alcoholics Anonymous (AA) has always used that model of one person in AA helping another. The idea is based on the fact that not all professionals know what it is like to experience some problems firsthand. The peer support person is able to say he has been there and can show another how to resolve the issues with firsthand practical information.

Individuals whose functioning is impaired might benefit from a peer support person who can help them function better educationally, socially, or vocationally and may even become involved in helping them with issues of self-care. Much like case management, the peer support person ascertains that the consumer will accept peer support and then works with that person to set realistic and meaningful goals the two can approach together. Good peer support helps people formulate the small action steps needed to move toward the goals the two have identified together, and the peer support person can be there with advice and ideas if the action step doesn't work very well. As a case manager, you will use peer support when a person needs more sustained time than you can give, and the support will significantly help the person move toward recovery.

Informal Resources and Social Support Systems You will also want to be aware of social activities your clients might enjoy that would keep them involved in their communities. Perhaps one person likes to work on models and could become a member of the model railroaders club. Perhaps another genuinely likes people and enjoys being with them. This person might do well as a member of the Jaycees.

People do better living in a community in which they have healthy social support systems. A social support system refers to the kinds of supports most of us have in our communities such as Lions Club, a church, or volunteering on specific community projects. All of us need to feel we are a part of the place where we live, but many people do not have the skills to interact with others and find useful activities on their own. As a case manager, it is your responsibility to integrate your consumer into the community if this is a need. Find social clubs, churches, and groups that pursue similar interests, and help your client make contact with those people. The more contacts the person has and the more useful activities the person engages in, the more support the community can give.

A particularly touching example of the use of informal social supports occurred in a small town in which the firehouse was located just around the corner from a group home for five older men with mental health problems. They had been institutionalized for most of their lives, spent years on medication, and had the common long-term side effects that can develop. One of the men, Nick, wanted to be a fireman, so the case manager connected this man to the fire company around the corner. The men at the firehouse made Nick a part of their everyday routine. Nick helped roll hoses, swept floors, and took his meals with the men. Nick was included in meetings and made decisions about the dinner menu. He became such a part of the fire company that when he died suddenly of cardiac complications the men were deeply saddened. As a tribute to Nick on the day of his funeral, the procession from the funeral home to the cemetery was led by a number of fire trucks, beginning with the trucks from Nick's home station and including some from neighboring communities. This was an excellent example of using social supports to give a person a valued place in his community and a sense of doing something worthwhile.

Case managers often fail to use these valuable informal resources for several reasons. They may feel that their client cannot handle being with ordinary people in ordinary settings. This is often based on the case manager's attitude about the person's disability and is often quite erroneous. Having consumers in small numbers in social activities or organizations that give them an opportunity to practice strengths is an invaluable experience for everyone concerned. Another reason a case manager might be reluctant to place a consumer in a community social group might stem from the case manager's perception that people in such groups do not want to be bothered with people who have disabilities. In some cases, this assessment is correct, but in others it is quite the opposite. Many organizations are set up to provide service and perceive this as an opportunity to grow and serve the community.

Doing your homework pays off. You cannot rely on suppositions and speculations. Know what is available in your community and have places in mind that would serve your consumers as the need arises. Meet people and talk to them about what you would like to have available for your clients. Gradually, you will develop a list of people and places that welcome your clients and provide the specific experiences and support you are seeking. Your task is to have many resources you can use at your fingertips when developing plans for your clients and to continually be developing new ones in your community.

Creating an Individualized Plan After you have worked with people to determine where the problems are and what areas need attention, you will also know about the supports and other resources people have in the community and among their family members and friends. You will know what they do well and what interests them most. Each person will be different.

As you go about designing the plan with the person, you will place in that plan elements that take advantage of the client's strengths and supports. In addition, you will address those problems most outstanding or immediate for that client. Each person has a different set of strengths, life circumstances, immediate problems, and personal goals. No two people view their situations in exactly the same way, so no two plans will be exactly alike. Each plan will be developed specifically for that individual client.

At one time, a small program for homeless women employed a part-time case manager for the children. Homeless women were given 2 years' residence in apartments belonging to the program to work hard on getting an education or training, and a stable source of income. Many of them were distracted from this by concerns about their children. Still others had little time to think about what their children needed as they went about restructuring their lives. The county social services department gave the shelter a small stipend to hire a children's case manager. The shelter hired a young woman who had just graduated from college. This seemed like an ideal choice. She was energetic, related well to the children, and was genuinely concerned about each of them. In the next year, the program monitor from the county noticed two things. First, there seemed to be very little material on the children in any records. No individual plans could be found, and no assessments on each child appeared to have been done. Second, the children were all following much the same plan. All the girls attended gymnastics; all the boys were enrolled in Little League. On certain weekends, all the children, regardless of their age or interests, went to the zoo or to the circus.

After receiving repeated requests for individualized plans for each child and some guidance about how to create them, the case manager quit. She said, on departing, that she did not have time to sit and write up records, that the children had been "having fun," and that the county was unreasonable. The county became more involved in hiring the second case manager, and this person was well aware of the importance of individualized planning.

In the first 6 months, two children began to get orthodontic work done, one received a scholarship to a private school, four boys went to Little League, one took violin lessons, and a third joined the swim team at the YMCA. Most of the younger children went to the circus and to the zoo. Most of the older children went on a bus trip to Washington, D.C., and half of them went to two symphony orchestra performances that winter. No child's plan was the same as that of another child. Each child's needs had been documented and addressed in some way, and each child's strengths and interests were brought into play as the plans were developed.

In developing these plans, the case manager called all her contacts in the community. She asked two dentists to donate their time. She prevailed upon the symphony to give her the tickets for two performances. She went to a private school and talked to them about this particularly gifted child until a plan for financing the child's education was worked out. She found a violin teacher and asked for 15 free lessons as a gift to the shelter. In churches, mosques, and synagogues, she got people enthused about helping the children whose mothers were working so hard to put a stable life together for their families. She looked at scout troops, church youth groups, and organized sports for possible answers to the children's needs. In any number of cases, the plan simply involved the case manager helping an older child choose from among school activities and arranging transportation.

This is what is meant by individualized planning. When it is done well and done creatively, your clients can grow and thrive.

Continued Assessment and Continued Planning In continued assessment and planning, as you follow the case, you will take into account changes the person may face. An example will illustrate this kind of planning, which you may be called upon to provide. Mary Beth has an intellectual disability and was assigned to you when she left a state-run institution for individuals with intellectual disabilities. When you did the intake assessment and planning, you determined that she would do better initially in a sheltered living arrangement for about a year. Because the goal is for her to move to an apartment of her own at the end of the year, your planning should start well in advance of this move. This planning makes the transition easier for her and for you. There are no shocks and sudden surprises that might necessitate her need for hospitalization or a regression back to greater dependence on the agency.

You might begin by setting up services and activities with Mary Beth that involve her in her community. Mary Beth told you when you first talked to her of her interest in singing. The people at the institution said she loved music and sang well, although she could not read music. At the time Mary Beth came out of the institution, you could not find a good place for her to use her musical interests, but you noted this as a strength and kept your eyes open for an appropriate link. Now you have found a choir director at a small church who is willing to have Mary Beth sing with her choir. The church has numerous activities, and there are members who see to it that Mary Beth is included. In this way, you begin to prepare her for a move to more independent living. You seek and find a place for her to live not too far from the church, and you work with interested members to ensure that Mary Beth will have their continued help with transportation and inclusion in church activities.

You may think it best that Mary Beth has other social ties to her community as well. There is the Aurora Club, created by professionals just for people with some intellectual disabilities. This club is a place to go and meet others; and the club takes trips, goes bowling, and goes out to dinner together. You could refer her there; however, you might decide that Mary Beth's mild disability does not warrant her being limited to social activities only for individuals with intellectual disabilities. Instead, you might develop a relationship with a local women's club, getting them to take Mary Beth as a member.

As Mary Beth makes an adjustment to being outside the institution, you look for a job placement. You make a referral to Goodwill, where she is able to develop her social skills, and soon she is hired by a local Wal-Mart as a greeter. All the while Mary Beth is away from the institution, you are meeting with her, assessing her progress toward independence, and planning for her to take a little more personal responsibility.

By the time Mary Beth moves into an apartment of her own, she has gained new confidence and many friends who connect her to the community. Her success is due in large measure to both your wise initial plan and your modifications of the plan as Mary Beth grew more independent.

Linking

Linking is the general term we use to mean connecting clients to people or agencies where they will receive the help or service they need. Once the plan is drawn up, the case manager links the person to the service, activity, organization or club, or people who will carry out the plan. When we connect the person to a formal agency we make a referral (see chapter 22) but, as noted before, we will link people to more than just formal agencies. Linking a client to a specific service requires care and skill on your part. You need to know the best service that will meet the individual issues and needs of your client.

Linking your client to a social service agency that provides a specific service—such as day treatment, drug rehabilitation, or groups for victims of violent crime—will require a written referral. You will state why you are making the referral, indicating the problem for which the referral is being made, and the goal that you expect as a result of your client's contact with the agency. The referral will also indicate the amount of time you estimate it will take for the agency to reach this goal. The time limit is very important. It keeps treatment from becoming endless and unstructured. With a goal and a set amount of time in which to attain that goal, both the agency and the client are more likely to make the most of their time together.

Sometimes people can take advantage of services on their own. You might tell a client about the Aurora Club, for example, and the next week he may take a bus there and begin going to the club regularly, participating in activities and social events. At other times, you may have clients who are unable to take the first step and who will need you to accompany them or to arrange transportation for them.

In a formal social service agency, personnel at that agency will be able to support your clients in their programs and implement the goals and work on the issues you and your clients have identified as important. Some agencies with very fine programs or specialized services are small, which may require you to give more support to your client. For example, at New Start, a staff of three focuses on second-stage groups for victims of rape and domestic violence, and much of the work is done by volunteers. The success rate is excellent, and clients report a high degree of satisfaction with the agency. However, the small staff is not equipped to handle other problems that might develop while your client is in their group. If you refer a client to a group at New Start and your client has landlord problems between group sessions, the staff at New Start may not be aware of it in time to prevent an eviction notice. Even if they become aware of the notice, they will need to refer the client to you to resolve the matter because of the limited staff time available to clients. On the other hand, at Riverview, a day-treatment program, nurses are aware of medication problems, social workers monitor progress toward goals, and staff can work to prevent eviction of a client, if that appears imminent.

On occasion you may find a service for your client at an agency that does not seem interested in serving her. Perhaps they are reluctant because your client has been ill recently or because the agency is not interested in her type of problem. The agency may accept the client into service with them to fill all their slots and draw down payment for services, but in reality they may give poor or no services. In such instances, linking becomes **advocacy** as you advocate for your client or on behalf of your client. In a situation like this, advocacy means you will attempt to seek the best services for your client, and you will insist that your client be treated fairly and with respect.

Beyond the formal agencies, however, we might choose **generic services** that require no formal referral or we might make connections with community social supports where it is often completely unnecessary for the people there to be aware of your client's status with your case management unit. Linking is about choosing all the best options that will support your client toward her goals and the stability she is seeking.

ASSIGNMENT

Begin now to put together a resource book listing agencies and what they do, support groups, and places where educational seminars and workshops are held for the general public. Collect them from the community where you expect to practice and keep the latest copy of social services agencies found in most telephone books.

Monitoring

After the plan has been made and implemented (meaning the referrals and links indicated in your plan have been accomplished), it becomes your responsibility to monitor the services given to your client. When a formal agency is holding a planning or treatment conference about one of your clients, you should be invited to attend. You should also receive written reports about your client's progress and about the services given to him or her. If you do not receive reports at specified intervals from the agency, you need to contact them yourself on a regular basis.

Talking with another agency about the service they are giving your client is done for two reasons:

1. To be certain that the treatment or service you authorized for this person is in fact the treatment or service that is being given
2. To keep track of your client's progress toward the goals you developed with the client and be aware of times when modifications and revisions in either the goals or the plan need to take place. Again, your assessment is on-going.

Less formal groups or institutions that are part of your plan should get a call or visit from you occasionally to monitor how the plan is working. Suppose that the neighbor offers to take your client, Bill, to church with her family every Sunday. In August, the family goes away for a month and does not make arrangements with anyone else to take him to church. He begins to feel lonely, and one day he goes to another church closer to his apartment. There he is extremely friendly to everyone, which seems to bother the minister and several members of the church. They decide he is "inappropriate" and call crisis intervention, whose team gets tied up going to the church and sorting out what happened. All of this could have been avoided if you had been able to have regular contact with the family who took your client to church. In that case, you would have known of the vacation and could have requested that they find a substitute or could have found a substitute yourself.

If you have linked your client to a choir, the Lions club, or courses at the community college it would not be appropriate to check in with these entities, but it would be appropriate for you to check in with your client to see how these connections are working out.

Figure 1.1 outlines the knowledge base and skills needed for case management and offers some useful guidelines for you to follow in practicing case management.

Advocacy

Nearly everything you do in relation to your client is a form of advocacy. When you plan with your clients, you advocate for their voices and opinions to be heard. When you link your clients to specific services or activities, you advocate for the best placements and treatments for your clients. When you monitor your cases, you advocate for the goals you and your client have determined should be met.

There are other instances where you need to be able to stand up for your client or find the leverage in your community where clients' rights or best interests will be supported. For example, suppose that your client has just left a drug rehab facility and is living on her own for the first time. She is in a small public housing apartment and is told she is being evicted along with a number of other clients because the building has been deemed unsafe. However, the city seems unable or reluctant to find other housing and your client can only afford subsidized housing. If you have met with your client to look at the options, and if you have met with your client and the public housing officials and find them unwilling or disinterested in relocating your client, you may need to go with your client to see her state representative.

True advocacy in this case might involve your accompanying your client to a hearing, testifying on her behalf at a hearing, insisting that she receive a fair hearing, assembling the facts and putting them before a particular board, going to meetings with others whose clients will be affected, and even seeking legal aid.

In the chapters on communication, you will learn ways to communicate your concerns so that you do not come across as petulant and demanding. Nevertheless, many clients are not able to organize on their own behalf, defend themselves effectively, or know when they are being exploited or abused. Case managers have an obligation to monitor when their clients are at an extreme disadvantage and to advocate for these people in whatever way is appropriate.

Service Coordination

Service coordination refers to working with other agencies or systems in a person's life. Many of your clients will be involved in other programs and systems, and each of these programs or systems may have a different plan for the person. Each of these plans may be headed more or less toward the same broad general goal, but their specifics for the person are all different. Often the major and most significant role for the

FIGURE 1.1 Knowledge base, skills, and guidelines for case management

Knowledge Base for Case Management

In order to do case management, you need knowledge of the following:

1. *Individual and family dynamics* (which you find in courses such as Human Development, Introduction to Psychology, Marriage and the Family, and Abnormal Psychology)
2. The relationship between and among social, psychological, physiological, and economic factors (as found in the *ecological model*, a theoretical basis for evaluating a person's situation and needs)
3. The *focus and policies of your agency*
4. *State and federal laws and regulations* that affect your agency's delivery of service
5. The vast array of *community services and resources* where you practice

Skills You Need to Be an Effective Case Manager

To be an effective case manager, you need the ability to:

1. *Work effectively with people* to promote their growth
2. *Work collaboratively with people* of various professions, paraprofessionals, the public, and clients and their families
3. *Identify what your client needs*
4. *Keep accurate and well-organized records*
5. *Allow the client to take leadership* in planning services
6. *Develop creative resources* within your community to meet client needs

Guidelines for Case Management

Here are some useful guidelines in practicing case management:

1. *Plan ahead.* Plan before there is a crisis. Develop a plan that will prevent crises based upon what you learned about your clients in their assessments and what you can foresee happening in their situations if the issues are not addressed. Alleviate crisis-provoking situations.
2. *Be accountable* to your client and to the community. Do what you say you will do. Do it *promptly*. And carefully *document* what you have done.
3. *Be optimistic* about your client. Expect improvement and some degree of independence, and that is what you are most likely to get. Reinforce success, and never miss an opportunity to give positive feedback. Set up situations in which your client is likely to succeed rather than situations that are complex and tricky.
4. *Involve your clients* in all phases of planning. Let your people decide what issues and problems in their lives take priority. Get their opinions and feedback about services and about their plans.
5. *Go where your clients are.* Do not stay shut up in your office. Go out and see where your clients are interacting with others, and teach your clients new skills in the field where they will need them.
6. *Promote independence.* Show pride in the independence your clients demonstrate regardless of how small it is. Model independence, encourage independence, and teach as often as it takes the skills to maintain independence.
7. *Develop a large number of resources* and know how to find good ones for your clients. Know what formal agencies exist in your community and their focus. Look for and develop good social support systems on which you can rely. As you move about the community, look for new resources you can add to your list.

Digital Download Download from CengageBrain.com

case manager is to bring representatives of these different systems together, forming a team that collaborates with one another in supporting the client's goal.

This is not as easy as it sounds. Communities and counties have numerous services, such as the school and other educational systems, mental health and the intellectual disabilities systems, the criminal justice system, a substance abuse system, and health care systems. These organizations often operate as though they are the only program with which the client is involved. Case managers who attempt to get everyone to work together are sometimes not welcome, and the program may be closed to outside input and collaboration with other agencies. However, coordinating the different services can enhance clients' movement toward their goals. When coordination is not possible, a person's goals can be impeded tremendously.

For example, Norita was a student at a community college and also a mother on welfare with one child. Because of her mental health problems in the past, her case manager at the mental health unit had facilitated Norita's receiving welfare to support herself while in school, and the case manager had worked with Norita to get her into school where she was an excellent student. Then the welfare worker insisted that Norita drop out of school and take a job readiness course as all single mothers on welfare were required to do. This demand came in the middle of a semester, and time and money would have been wasted if Norita was forced to drop out of school. The case manager worked with both the school and the welfare office to form a team working to support Norita in her movement toward financial independence.

At first, the welfare worker was not happy about working with the case manager. She was curt and unpleasant and stipulated that the rules for remaining on welfare meant that Norita would have to drop out of school and take a 7-week job readiness course. The case manager made an appointment to meet and brought an academic counselor from the college to the meeting. In this face-to-face context, the welfare worker began to soften and see advantages to Norita's current plan. Norita was only one semester away from graduation after she completed the current semester. The college counselor stated that the counseling department was available to help with resumes and job searches. In fact, it was likely that Norita would be hired from her internship as she was going into a field with a high demand for workers.

Gradually, a team was formed, and collaboration around helping Norita become independent took place. In the end, each party felt the outcome was beneficial to its system. This last element is crucial when coordinating with other agencies and systems. To support the overall plan, each party has to feel that what is being agreed to will have a satisfactory outcome for the system he or she represents. In this case, both the welfare worker and the case manager met the goal for Norita to become independent and self-supporting. In addition, the case manager avoided duplication of services. If the college was showing Norita how to get a job and the welfare system was as well, there would have been duplication of services. What could have been competing systems and ill will became, instead, complementary services integrated around a specific goal and working together on Norita's behalf.

You will not always be coordinating with other social services agencies. In one instance Meredith's client, Phillip, believed he was employed by the Fresh 'N Save grocery store near his home. This was a delusion Phillip had held consistently for over a year. Phillip had gone to the store on numerous occasions, rearranging things on the shelves, helping shoppers with their bags and carts, and generally impeding some of the daily tasks at the store. On a number of occasions, Phillip was arrested for defiant trespass, and often he was escorted off the property by the local police. In one instance he was given a short jail sentence, but 6 hours after his release from jail he was back at the Fresh 'N Save. The case manager pulled together a team of people who previously had been working on their own to try to solve this problem. Present at the first meeting was the probation officer, the county mental health representative, a person from the police department, and the manager of the Fresh 'N Save. The question before the team was, "What resources do we need to resolve this problem and move Phillip to more constructive activities?"

Everyone on the team recognized that there was no treatment in the jail for Phillip, so the team looked at what other resources would be needed. It was agreed Phillip needed some level of supervision. A commitment to a partial hospitalization program would be obtained. Phillip would go there 5 days a week. In addition, supervised housing was arranged for Phillip. In this way, someone would know where Phillip was or should be at all times. The police and the store manager agreed to call the partial hospitalization program or the supervised housing unit if Phillip returned to the store. The case manager agreed to work with Phillip and staff in the partial hospitalization unit to seek other goals Phillip might have for himself. The county mental health representative agreed that the crisis intervention team would intervene when necessary if the case manager was off on a weekend or in the evenings. In this way individuals representing a number of different systems or agencies went from feeling frustrated and exasperated to leaving with a plan and some assurance that Phillip would get the assistance he truly needed.

Levels of Case Management

Some agencies have several levels of case management, and clients receive a level of case management commensurate with their need and ability to function. In this text, we look at three levels of case management: administrative, resource coordination, and intensive. In addition, "targeted" or "blended case management" is discussed. In some agencies, these categories may have other names or there may be more than three levels of case management. The following sections provide examples of how case management services might be organized.

Administrative Case Management

This level of case management is assigned to high-functioning individuals who need very little assistance navigating the system. On occasion, they might need a prescription refilled, an emergency appointment, or a return to outpatient substance abuse treatment, but for the most part they are capable of handling these details themselves. These people are placed in a pool with other clients who require little service or follow-up beyond the original referral. For the most part, these people function independently, using well the services to which they were referred. When something does come up for a person in this caseload, an available caseworker handles it. This means that an individual on this caseload does not always see the same case manager.

Resource Coordination

This next level of case management is reserved for individuals who have some trouble handling the details of their treatment or plan. They usually need help and may have more involved or chronic difficulties that require more assistance. They do not, however, pose a risk to themselves or to others. In addition, with good support, they are unlikely to experience repeated hospitalizations or other crises. Here caseloads are larger, and clients are often in need of services and assistance on issues such as housing, medication, and therapy, but generally the clients do well with the services offered. A person going through a particularly difficult time might be moved up to intensive case management and then return to resource coordination after the stressful circumstances have been addressed.

Intensive Case Management

Individuals receiving intensive case management require considerable supervision, support, and assistance in order to remain in the community and in circumstances that do not exacerbate their problems. Generally, the caseloads of intensive case managers are smaller, allowing for more individual attention. Those in this caseload would be people at high risk for repeated emergencies and hospitalization or at risk of deteriorating to the point that they pose a danger to themselves or others. Intensive case management is usually available 24 hours a day and requires intense involvement to ensure that the person has a support network available and is not in high-risk situations, such as running out of medications or living in a housing situation likely to trigger stress and relapse.

Targeted or Blended Case Management

Some agencies are moving toward a type of case management called targeted or, in some cases, blended case management. This is a different way of delivering case management services. Instead of dividing clients according to their level of need, individuals of varying needs are given to a case manager who carries a smaller caseload as a result. In this method, the person has the same case manager through stable times and times of crisis, so that there is good continuity of care and rapport that might support that client when things become unstable.

Margery, a 36-year-old single mother of a daughter, was stable and working when she was transferred to a blended case management caseload. She had needed little contact with the agency in the past year, and the contact she did have was mainly for medication checks and prescription refills. In March, the man she had been dating and talking about marrying was killed in a single-car accident. In the car with him at the time was another woman, who survived the accident and insisted she was actually the man's girlfriend. Margery was devastated. She began to miss work, jeopardizing her place in the career program she had chosen. She took her medication erratically and gradually became in need of emergency care. Her case manager was a person she had known for some time, and it was her case manager who came at once to the emergency room when Margery was brought in by family members.

Liz, her case manager, was shocked at the changes in Margery. Margery was haggard, thin, and unkempt. She looked past Liz and said little. Liz made certain there were arrangements for Margery's daughter with family members and stayed in the emergency room until a bed was found for the patient on the psychiatric unit. During Margery's 6-day stay, Liz visited several times, worked with Margery to make contact with her work and brought her daughter in for a visit after school one afternoon. When they talked, Liz worked on some of the supports Margery would need to return home. Of paramount importance was that Margery stay on her medication and call Liz when she felt distressed about something.

Gradually, with the 9 weeks of counseling arranged by Liz and a return to work the next month, Liz saw positive changes in Margery. Her home became cleaner and brighter and so did Margery's appearance, indicating that she was taking an interest in herself and her surroundings. Liz continued to have contact with Margery regarding renewal of prescriptions and regular contact just to see how Margery was doing.

This is an example of blended case management in which one case manager provides services to the same person regardless of the level of need at any particular time. When Margery needed an emergency intervention, she had a case manager at her side with whom she was familiar, someone she knew and trusted. Figure 1.2 illustrates the general case management process.

FIGURE 1.2 Walking through the case management process

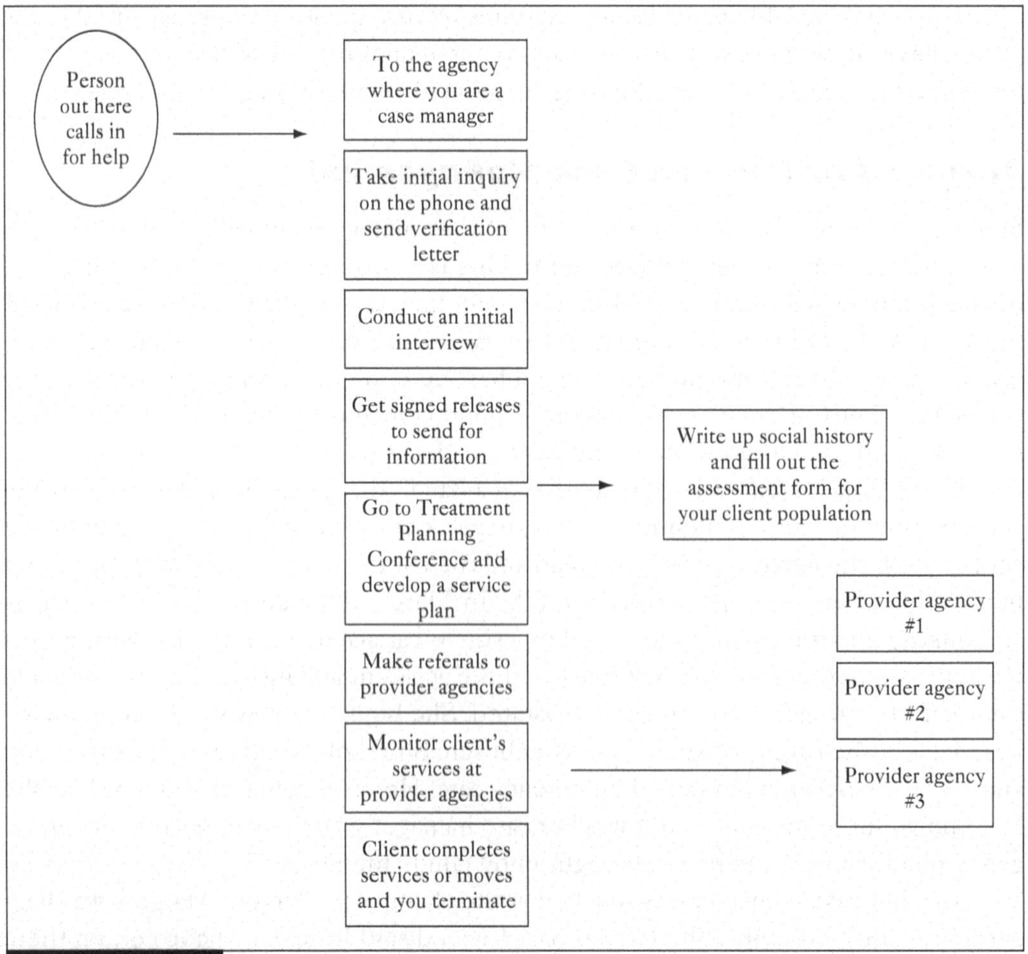

© Cengage Learning®

Digital Download **Download from CengageBrain.com**

Separating Case Management from Therapy

Case management is not therapy. Often beginning case managers believe that they are to do therapy; that is, that they are to provide weekly talking sessions in which deep-seated conflicts and concerns are exposed and resolved. That is not the purpose of case management, and, indeed, most case managers are not prepared to handle this type of work.

In the course of case management with a client, you may uncover deep-seated problems and issues. These become the basis of a piece of the plan developed to resolve these issues. The client is referred to a person or agency that can do that work expertly. As the case manager, you may be the one your clients call when they are having a crisis and their therapist is unavailable. Good listening skills and helping such people develop a way to handle things until they are seen in therapy is the case manager's role. It is not your role to intervene with a therapy session.

Finally, you will find plenty of other problems that do call for innovative interventions on your part. Learning to be independent, adopting useful and appropriate work habits, practicing good interpersonal skills, and behaving appropriately are all areas that you may address with your client in the course of case management. Although the individual may be referred to a specific agency for exactly those skills and that information, you would then support that intervention in your contacts with the person.

Many clients do not require therapy. Perhaps they have had considerable therapy in the past and are not able to benefit from it now or were never able to benefit from it. Perhaps their interpersonal problems are more a result of the chemical imbalance they suffer than psychological dynamics. For instance, research shows that individuals who are depressed or suffer from a bipolar disorder do well when they receive both medication and therapy. Many, however, have had considerable therapy in the past and are now maintained on effective medications.

Other individuals may have intellectual disabilities and only need skills to attain as much independence as they are capable of handling. Some people might have suffered a crisis such as rape or domestic violence and need a plan that focuses on protection and independence. Others may be out of fuel and need a plan that resolves the problem of a cold home with small children in it. A few clients may have completed a therapy program for substance abuse some time ago and need only an AA or Narcotics Anonymous (NA) meeting and occasional supportive contact with their case manager.

What case managers do is therapeutic in the sense that it benefits the client. Conducting clinical therapy, however—where a person comes in about long-standing emotional problems or pervasive affective disorders—takes years of study and training and should never be attempted by a person not specifically trained to conduct therapy.

Case Management in Provider Agencies

A provider agency is an agency that "provides" specific services to clients. A case manager might refer clients to such an agency for a specific service. There are positions in provider agencies, however, that are sometimes titled "case manager."

Case managers in provider agencies have oversight responsibility for the service or treatment given by the provider agency to the client. These case managers generally make sure the reason for the referral from the general case management unit is actually addressed, and they communicate with the client's general case manager on progress, goals for the client, and any changes that may need to be made in the service or treatment plan. Because the client is being seen regularly at the provider agency, the provider agency case manager there may also handle some of the personal issues and problems that arise for the client while in the care of this provider agency.

When Jennine came to the Wildwood Case Management Unit, she was suffering from depression and was not able to go to work. After she and the case manager discussed her mood and a social history was developed, Jennine agreed to go to a partial hospitalization program at Marshall River Center where her medications could be monitored, she would be attending groups, and she would get a lunch every day. Her husband agreed to the plan, feeling that this would prevent Jennine from sleeping all day and going without lunch. Jennine, her husband, and the case manager decided on the goal together. That goal was to alleviate Jennine's depression. They further decided together on the best place for Jennine to go to meet this goal.

At Marshall River, Jennine had a case manager who set up groups and activities to meet the goal. This was important because the Wildwood Case Management Unit generally authorizes payment for the service given to the client and expects that the goals for the client's recovery will be addressed in return.

The goal for Jennine was to alleviate her depression. Jennine and her general case manager decided this was a top priority for her. The case manager at the provider agency addressed this goal by instituting a regular lunch and two healthy snack breaks, physical exercise, and group discussions on ways to handle or relieve depression. During that time, the provider agency case manager set appointments so that Jennine saw the psychiatrist once a week for medication checks and adjustments.

During the 4 weeks Jennine was in the care of the Marshall River Center, her husband lost his job. This threw her into a panic and exacerbated her depression. The case manager was in touch with the Wildwood Case Management Unit about these developments. A reassessment was done and there was agreement that Jennine would need 2 more weeks at Marshall River. In this case, the Marshall River case manager saw Jennine daily to talk to her about her husband's job search and how Jennine was viewing the loss of his job.

In this textbook, you will learn how to do general case management with clients in which you work to address many of the problematic aspects of their lives and make strategic referrals to places where help for them is provided. For example, if Jennine and her husband further stated that part of Jennine's depression came from her inability to discipline her 8-year-old son, the general case manager might have made a separate referral to parent education classes. Once you understand how general case management is done, you will be able to take on case management responsibilities in agencies that provide services and treatment to clients.

Managed Care and Case Management

Managed care is a phrase you will hear often when you go to work. Managed care is a financial system developed to contain the soaring costs of health care. It works like this: A managed care organization (MCO) receives a pool of money, allocated on the basis of a specific number of patients or clients who will be served by this MCO. The MCO hires case managers who oversee the care given in order to prevent the costs of caring for these patients or clients from running over the amount allotted. An MCO can be either a private insurance company or a company that handles public money.

In situations where the insurance covers medical and physical problems, the case manager is generally a nurse who has specific training in the managed care field. In MCOs set up to deal with behavioral health issues, a variety of social service professionals may be employed and trained as case managers.

You are most likely to deal with an MCO when working with clients with intellectual disabilities, those with substance abuse problems, and individuals who have mental health problems. Because managed care is an economic system to control costs, you may find yourself at odds with the decisions of the MCO case manager. For example, you may be required to receive permission from the MCO case manager before you can implement a service or treatment plan. This is called a preauthorization. You may have a person sitting in your office who seems clearly suicidal to you, but the MCO case manager is denying hospitalization and suggesting partial hospitalization instead. There is little room for individual variations or innovative treatment plans in managed care where an MCO is paying for the services because MCOs generally have a cookbook approach to various health problems. There are specific protocols or decision trees to help MCO case managers decide what treatment or services are appropriate, and these protocols do not take into account individual differences in clients' personalities and circumstances.

If you are not an MCO case manager, but rather the case manager seeking payment from the MCO for services you really believe your client needs, it is your role to advocate for your client and the services you feel are the best ones for the current situation. Sometimes you will be able to obtain a better decision for the client and sometimes you won't be able to do that. However, if you are convinced that what the MCO is proposing is not in your client's best interest, you owe it to your client to advocate on his behalf.

Figure 1.3 explains some recent trends in case management in managed care organizations.

The Recovery Model

Recovery is a model put forth by the U.S. Department of Health and Human Services and the office of SAMHSA. They are changing the view of mental illness and substance abuse from one of gradual deterioration or a lifetime of chronic illness to one of recovery and a productive life. Spurred by the self-reports of people who have recovered from their addictions, mental illness, or emotional problems and went on to live

FIGURE 1.3 Recent trends in managed care case management

1. Some MCOs want to make the MCO case managers the only case managers for clients. These MCO case managers see the clients directly, make home and site visits, and manage the case as a general case manager might do. In most instances, these case managers are trained social service professionals and would be likely to follow the steps outlined in this book for case management while keeping a strict eye on the financial bottom line.
2. In some places, MCOs are paying for the contact case managers have with their clients. In this way, the agency draws down funds for case management services while providing considerable support to clients. In other words, the more contact given to clients the more the agency is paid and the better service the client receives. This appears to keep individuals in a more stable situation and acts as prevention for crises which are more time consuming and expensive to resolve.
3. Research tells us that individuals with chronic mental illness or substance abuse problems do not live as long as the general population. These people are less likely to seek help, to be able to use the help that is given, and may have had long-term medical issues. Case managers are urged to integrate the physical health needs and the substance abuse or mental health needs of their clients to facilitate a longer and healthier life span.

productive lives, this model is being applied to mental health and substance abuse, but it works well in many other fields. Professionals who did research on how individuals recover concluded that it is possible to recover and lead a normal life *provided the services are in place that will cause outcomes beneficial to this process.*

In this model, case management is extremely important because the research showed that many times people did not recover because they either did not have access to services that would promote their recovery or these services were improperly coordinated. Now the movement is toward services that are diverse enough to meet peoples' needs, well coordinated, and easily accessible. A person's hopeful personal vision for his life and respect for that person's self-determination are always foremost in planning, reclaiming the role of "healthy person" rather than "sick person." When appropriate, the client's family is included.

Recovery happens when individuals are able to assume control of their lives, working to establish goals and develop a sense of purpose. When people work on recovery, they are encouraged to take the long-range view, looking toward a future that is more productive, more meaningful, and less stressful to them in ways it has not been before. In addition, people are asked to take a broader view beyond the issues that caused them to seek help, looking at many different aspects of a person's life. Starting with where the person is at the time, the client is encouraged to think about where she wants to be in a year or in 5 years. How will she get there? What elements does she want in her life? What would she put in her life if she could? What goals would she like to pursue? What issues does she believe need to be cleared up? What treatments and services does she want to use? In answering these questions, we introduce the elements for positive changes and recovery.

The *recovery* mission is respectful because it seeks to give clients more self-determination or empowerment and improved role functioning. This is done through a set of identified services combined to result in the stated desired outcomes clients have articulated for themselves, such as improved role functioning, self-development, services that support recovery, and symptom relief. In this model, relapse is seen as similar to relapse in any other illness (for example, diabetes or diverticulitis). Most of the time, the person will be healthy and functioning.

Charles Curie, former administrator of SAMHSA in the U.S. Department of Health and Human Services put it this way: "Recovery must be the common, recognized outcome of the services we support." While SAMHSA developed a list of 10 components of recovery for people with substance abuse problems or mental health diagnoses, these components can support any effort to create plans with our clients For the Recovery Model as put forth by the SAMHSA, see Appendix A.

For Children

At the U.S. Department of Health and Human Services, the *Child and Adolescent Service System Program* (CASSP) promotes these concepts. They expect case management to be "child-centered and family focused." This means that case managers need to have a respectful focus on each child as an individual and need to include the child's family as a valued partner in planning for the child. The CASSP emphasizes community-based services that keep children at home, or at least in their own community. Further, the CASSP expectation is that case managers will respect and be competent when dealing with diverse cultures.

Also coming from SAMHSA is the *resiliency model* for children, again based on research. The findings showed that at-risk children often "bounce back" when they have someone in their lives who makes it clear that they are important regardless of the past; with this kind of support, many at-risk children go on to live healthier lives. A child's case manager certainly should be conveying this important sentiment, but the case manager can also find those people in a child's life who will do this as well: family members, teachers, youth counselors or workers, or neighbors.

The *resiliency model* adopts a positive view (the glass is half full) of the child's life and circumstances, actively seeking strengths on which to build a better future. The belief is that children, like all people, have a "self-righting mechanism" that will help them bounce back from problems using their own power and ideas. The case manager facilitates this self-righting by naming the child's strengths and teaching the child to acknowledge them. In addition, the case manager either provides or coordinates with others who provide the following: opportunities for the child to participate in meaningful activities, including those that help others; communicating high standards for the child and equally high expectations; providing consistent support and care; increasing prosocial bonding; setting clear behavioral boundaries and consistently observing these; and teaching life skills for success. Combined successfully, these factors allow a child to bounce back successfully.

Further research has shown that the inclusion of several activities not commonly considered when planning formal services is extremely important. A good children's case manager will develop contacts for these activities. First is reading practice so that at-risk children become competent readers. This is extremely important to later cognitive functioning, success in school, and other endeavors. Another activity involves opportunities for children to give something to others, to be helpful, to be in a situation where they exercise concern for others.

Finally, the Resiliency Model seeks case managers who are culturally competent, respectful of diverse religions, and able to respect and include the children's families.

For Those with Intellectual Disabilities

In the field of intellectual disabilities, the concept is called *self-determination* or, in some places, it is referred to as *Everyday Lives*. The emphasis is on clients' personal choices for the life that is most meaningful to them. Case managers look for services and activities that validate those choices and they collaborate with the family to consider how best to spend money allocated for the care of a person with severe disabilities. Rather than the case manager deciding what the client needs, the family is able to request what would be most helpful to them. Some families might ask to use the money for a ramp and remodeled bathroom to accommodate a wheelchair. Another family might request special speech or physical therapy. What is important is that the case manager takes an active role in helping clients and their families conceive of a hopeful future and looks with them at the supports that could be accessed to make that happen.

Physical Health Is Part of Wellness

Working within a holistic framework involves looking at all areas of a person's life. In the past, case managers and other providers focused on the immediate problems and on relieving immediate symptoms. Now case managers, particularly, are looking at all aspects of a person's life. A study by the National Association of State Mental Health Program Directors in 2006 produced some alarming findings. According to the study, "People served by the public mental health system had a higher relative risk of death" NASMHPD, 2006, p. 11. The study went on to report, "Deceased mental health clients had died at much younger ages and lost decades of potential life" p. 11. On average, those with a mental health diagnosis were likely to die 1 to 10 years earlier than a person with no mental health diagnosis. While case managers and other service providers were focused on the mental illness, other health issues had been overlooked. In addition, many who needed health care or even needed to receive disability payments were unable to successfully navigate the system.

This study alerted people in the fields of mental health and other social services to the importance of looking at a person's health as a whole and to the importance of moving people receiving social services toward wellness on many levels.

Practicing case management will mean making sure people on your caseload have regular physical checkups and chronic conditions are being monitored by a health care professional.

For All Clients of Social Service Agencies

In another move in this direction, director Elizabeth J. Clark of the National Association of Social Workers (NASW) testified before the President's New Freedom Commission on Mental Health in July 2002. She stated that as a profession, the NASW has made a commitment to mental health care that is driven first by the client and the client's family rather than by the preferences of the professional or the limits of the formal service system. She called for more accessibility to services, early mental health screening, and a national campaign to reduce the stigma of mental illness.

From these efforts, you can see that case management is moving rapidly away from coordination of existing formal services for individuals who are seen as chronically handicapped and therefore unable to lead productive, rewarding lives. Instead, in all fields of service, whether mentioned here or not, the emphasis is on respect for the wishes and the hopeful vision of the person seeking services and then undertaking activities that will move that person toward achieving that vision.

For case managers, it now becomes imperative to know what your client wants, not just in terms of services but also in terms of a productive, useful life. It involves careful listening and encouraging people to think about what else might make their lives fuller. It entails respect for the person, their family, and the diversity of human experience. Funding sources are building into their reviews a surveillance of the capacity of case management to support these aspirations not only with standard formal services and, in some cases, medication, but also with the use of community resources and community social supports to assist the individual toward a healthy, productive life.

Caseloads

We have looked at what constitutes good case management. The size of a case manager's caseload can have a considerable impact on how well case management services are provided. Large caseloads can diminish the quality of care and reduce the contacts between the case manager and the client. In such circumstances, case managers are doing little more than reacting to crises. Clients do not have the opportunity to benefit from close, stable relationships with their case manager and case managers can come to feel burdened and overwhelmed. There are a number of factors that go into developing the size of a case manager's caseload. Two of the most obvious relate to financial considerations as to how many case managers the agency can afford to hire and the complexity of the cases the agency handles. Often caseloads are set arbitrarily without evidence that this is the ideal number of cases for case managers in an agency and clearly more work is needed to determine what constitutes an effective caseload for various populations handled by case management.

Generic Case Management

The skills you learn to perform here can be used in any social service setting in which clients' needs and situations are evaluated, addressed, and monitored. Every agency does things in its own way, uses different forms, and often has a specific focus, but the tasks of case management are the same. Once you have learned how take a social history, make a referral, and document contacts, you will be able to take that knowledge to any agency and quickly adapt to that agency's methods and way of doing things.

Summary

Case management is not therapy, but it requires a set of skills that is nonetheless therapeutic to clients and to their well-being. The practice of case management requires an ability to listen to people and accurately assess their problems, offer a range of diverse and innovative interventions, and follow their progress toward the goals they have set for themselves.

New models coming from the federal government and professional organizations stress the importance of giving people ample opportunity to plan for the kind of future they wish to have for themselves. These models have introduced the concept of recovery from the problems and issues that occur throughout the course of life and recovery from substance abuse and mental illness as well. With a belief in the capacity of people to move beyond illness or current difficulties, case managers plan with their clients for more than just immediate treatment and service needs.

In the long run, it is the case manager who takes the long-term view of the individual. In addition, it is the case manager who develops a comprehensive picture of the person, identifying the specific interventions that each particular client will need.

Think of case management as the foundation. Once you understand the practice of case management, regardless of the career you choose in human services, you will have a deeper understanding of how social services work for people, what social services can and cannot do, the obstacles people face when seeking help, and the numerous ways to circumvent those obstacles. You will be better equipped to work within the social services delivery system or to access it on behalf of your client. Finally, you will have a much broader view of the many interventions and avenues available to help others.

Video Examples

To view the videos that accompany this book, go to CengageBrain.com.

- "The First Interview"—Watch "The First Interview" to see how Keyanna begins an assessment with Michelle, a new client.
- "Monitoring: Making a Site Visit"—Keyanna visits with Michelle and her worker at the program where Michelle is receiving services. Here you can see the kinds of concerns that are raised in such a meeting.

Exercises

These exercises can also be filled out online at CengageBrain.com.

Exercises I: Case Management

Instructions: In each of the following situations, develop a tentative plan for the client. List the various services you believe each person needs initially. Include in your plan for each client both formal and generic services, and where appropriate, use social supports and support groups. Suggest other services the person might use later once the case is stabilized. Think about how you can involve others close to the person and how you will involve the client in planning.

1. You are called by the daughter of an elderly woman who lives alone. The daughter lives in another city and is concerned because her mother does not drive and has seemed unhappy and listless on the phone. The daughter expresses concern that her mother seems lonely and is perhaps depressed. The daughter does not know her mother's neighbors and calls you instead at the Office of Aging. She has told her mother she is going to call your agency for help, and the mother had no objection to that.

2. A man with an intellectual disability lives alone with his widowed mother. She has fallen and broken her hip and will be at the rehabilitation hospital for about 6 weeks. He cannot stay alone. He has a job at Goodwill Industries. County transportation takes him there every morning at 8:30 A.M. and brings him home at 5:00 P.M.

3. A woman and her two children are waiting to receive their welfare check. They came to your state from another to escape an abusive husband and father. The woman is frail and appears sick. They have no place to go and have not eaten in several days. The children smell as if they need a bath and are listless.

4. A mother of two preteens has brought her son in for services. The woman is a widow. She confides that she has been having trouble controlling the boy, who is the oldest, and that the girl is disgusted with her brother's behavior and does not want to be involved in helping him. Lately the boy has become involved with teens his age and older. They have been drinking and coming home when they feel like it. The mother allowed them to smoke pot in the garage in hopes that she could keep the boy at home, but now she feels that backfired. The boy

makes it clear that he thinks coming for help is ridiculous and says the one thing he will not do is give up his friends.

5. A man has been referred by his family physician for help. The man seems extremely inebriated. His wife brings him in and says she is worried that he may go into delirium tremens if he withdraws from alcohol too quickly. His family physician did not see him but sent the couple straight to your office.

6. A father brings in his 14-year-old daughter who is running the streets, refusing to listen, and failing in school. He is at his wits' end, saying he must work and cannot be home when the girl returns from school. Her mother died 4 years ago, and the trouble started when the daughter was about 12. The father feels that he and his daughter have a difficult time communicating with one another.

7. A police officer asks you to come to the home of an older man he has been concerned about for several weeks now. The man is delighted to see you and tells you that he is having pains in his legs and is unable to walk. During your visit, he asks you to get things for him that are nearby, but obviously it is too painful for him to get up. He says he does not go to the kitchen often to prepare meals, but the police officer has stopped by several times with sandwiches, and Mrs. Jones from up the street, an old friend of the man's late wife, has brought a casserole on occasion. He is adamant that he wants to stay in his home as long as he can.

8. A woman comes in complaining of depression. She says it started when her husband left with a younger woman and she has not been "right since." She reports having difficulty falling asleep and complains of no appetite. She says she has missed more than 3 weeks of work since he left last month. There are no children, but she tells you she has neglected the dog and cannot remember if she fed him last night or not. She appears listless and very sad, weeping off and on during the interview.

9. A man in his 60s comes in on the recommendation of his doctor. He had a back injury some years ago and was placed on codeine at the time. After the back injury, other things went wrong. The plant where he worked closed down and his

mother died. He found himself feeling very alone and uncertain about finances. "It was then that I started to drink too," he tells you. When you ask what he means by "too," he says his doctor believes he has become addicted to codeine. "I don't know," he says. "I've gotten to the point that I can't get through a day without a lot of help."

10. A single mother brings her 12-year-old son in because they are "not getting along." She reports that he does not listen and comes and goes as he pleases. His homework has fallen off and his grades have slipped, but he is still doing well in math and likes his math teacher. The boy's father was killed in a railroad accident 2 years ago. The mother tells you that the boy and his father enjoyed a close and warm relationship, and that she has felt her influence on him slipping away since the accident.

11. A woman comes to your agency on a referral by the courts after she was arrested for selling various prescription medications on the street. She tells you she currently has some amphetamines, Xanax, and a popular addictive pain medication in her bag. The court is ordering her to show within the next week that she has enrolled in a program that will get her help with her own addiction to some of the medications she sells. "I have regular customers," she tells you. "And they are going to crap when I stop coming around." Asked where she gets her medications, she smiles mysteriously and says, "The police are looking into that—you don't need to." She denies she is addicted to anything, but court records, including an evaluation by a psychologist, which she has brought with her, indicate that she is addicted to several different medications.

12. A woman who was recently placed in the community after 3 years in a state mental hospital is having trouble adjusting to the living arrangement made for her by the hospital. She is not going out and does not participate in any activities. She is friendly when you talk to her and seems glad to have your company, but she does not seem to know how to take care of the details of everyday living. She has a roommate who is more competent and independent. The two get along well.

13. A woman with two small children is referred to you because she recently lost her apartment. She has a meager income from a part-time job as a clerk in a

convenience store and was unable to pay the rent and take care of other bills. She seems unaware that she might be eligible for financial assistance. She is not sure where her children's father is at the moment. All her belongings are packed in five bulging garbage bags. She and her children seem malnourished and thin.

14. A man comes in who was referred by his job for possible crack use. The man admits he uses crack, but rarely and certainly not to the extent that it would interfere with his work. He will be given 4 weeks off if he enrolls in a legitimate program for detoxification. He seems reluctant and torn. Eventually, he agrees to work with you on a plan.

15. An older woman comes to you for help after a particularly abusive incident with her husband. She admits that she is becoming increasingly afraid of him. She has no friends and no job, "because he wouldn't let me out of the house—or out of his sight." She has a number of old bruises and lost two teeth in this last incident. She is asking for help and says, "I don't know. I really fear for my life. It's just gotten out of hand and I don't know where it will end."

Exercises II: Decide on the Best Course of Action

Instructions: Sit with a small group of other students and decide how you will handle each situation. There are many areas both ethically and legally that are not clear, so the discussion you have with your colleagues about these cases is much like a discussion you might have in a real agency. There are no "correct answers."

1. You are the case manager for a man who has only recently had a first manic episode. He had submitted to treatment, responded well, and returned to work. However, he is currently experiencing another episode, and this one seems more severe. He is not sleeping or eating, is sending cryptic messages on the Internet, and believes everything he reads there is directed at him. He believes the government is "monitoring" him, and that he has an important job lined up with a record company in New York. You have determined that he is not leaving his house, and that he is preoccupied with what is happening on message boards and chat rooms on the Internet. You have checked in with him several times by phone. Each time he assures you that he is fine and does not need help.

What is the best course of action? Do you risk seeking an involuntary commitment, knowing that he seems just well enough to convince the emergency room physician he is not in need of hospitalization? Will this alienate him and make it impossible for you to work with him? Do you wait for things to get worse? If so, is there a chance he may leave the house and get into trouble? What might happen if you wait it out?

2. A 72-year-old woman on your caseload has been in mental health services for a number of years, ever since the year after she graduated from college. She has had severe episodes of schizophrenia over the years with countless hospitalizations and prescriptions for numerous antipsychotics and antidepressants. All her life she has been a client of social services and hotlines, reaching out frequently for any support she could get. She has always been very dependent on her case managers, calling them often for support. When support was withdrawn, as it often was when workers tried to get her to be more independent, the woman would deteriorate.

   Sometime after she turned 60, these severe, acute episodes diminished, and she has been living in comfortable subsidized living arrangements for senior citizens. As she has grown older she has developed breathing problems that will eventually take her life. In an effort to be helpful and without your knowledge, her family doctor referred her to hospice for support. In his view she was facing the end of her life. Hospice entered the picture with considerable support, seeing her often, providing meals and companionship. This level of attention met the woman's needs for support and she became extremely dependent on the hospice workers. She seemed happier and began to participate more in the group activities at the apartment building where she lived. This level of support is more than you could give and you have been relieved that she is doing so well with the hospice workers. However, hospice has decided to withdraw their services. The woman's death does not appear to be imminent. In fact she appears to be improving slightly. The woman has called you, her case manager, in a complete panic at the prospect of losing this support. She is not eating and is distraught.

   What are the ramifications here? What does continued hospice support mean for hospice, which relies on public and charitable funding? What are the consequences if the services are withdrawn for this woman and for you? Are there solutions to this problem? Are there ethical considerations here?

CHAPTER 2

Ethics and Other Professional Responsibilities for Human Service Workers

Introduction

Ethical principles are the foundation of good human service practice. In fact, workers who do not practice within ethical parameters cannot be called professional. True professionals understand their ethical obligations and seek guidance when they do not. Each social welfare profession, from psychologists to social workers to human service workers, develops a set of ethical principles appropriate to the practice. Most professions monitor the behavior of their members with regard to these principles, singling out those who violate **ethics codes** for disciplinary measures.

Ethical principles are generally created in order to protect and prevent the exploitation of the individuals who come to us for service. In the work we do, there is considerable opportunity to exploit vulnerable people because the people who seek our help are dependent upon us for the aid they need. Any violation of their trust on our part will only compound the person's problems. Ethical principles provide guidelines to protect individuals from exploitation. However, when professionals practice within the parameters of ethical principles, the public can feel confident that their interests will be respected and protected. Thus, ethical principles inform the decisions we make that affect clients, and they provide guidance in choosing the approaches we take with clients.

In this chapter, we will look at some ethical guidelines common to all the helping professions. Failure to know and follow these guidelines in your future practice can result in dismissal from an agency or, worse yet, in a civil suit brought against you for a violation of the ethical code wherein the violation caused damage to the person.

If you have earned a license to practice in the human services you can potentially lose your license. Although violations of ethical principles may have negative consequences for you and your career, they are always extremely destructive for the individual, who is already vulnerable.

The Broader Ethical Concept

Ethical behavior can be viewed from three different aspects: moral, ethical, and legal. We examine these three aspects below.

1. *Moral.* Moral choices have to do with culture and socialization. We choose what we believe is moral behavior based on what we feel is right, what we have been socialized to view as morally correct, and what the culture in which we grew up would view as morally right. Often, morality is tied to the religion we practice. We internalize certain moral principles and choose to apply these principles as we believe they should be applied. Usually, when we violate what we feel is moral behavior we wrestle with our own guilt, but we usually do not face public condemnation or punitive action unless the behavior is also illegal or violates an ethical code.
2. *Ethical.* Ethical behavior is usually represented in ethical codes. These codes are generally developed as guidelines stipulating the behavior of members in particular groups or professions. They are not applied at the discretion of the individual but are required to maintain professional status. Violations of these codes will often result in sanctions applied by the group or profession against the individual. Suspensions and loss of privileges often follow violations of ethical codes.
3. *Legal.* When we refer to what is legal we are referring to what is the law. Laws will often embody moral and ethical principles, but not always. Governments and the courts determine what is legal and illegal behavior, what behavior is a criminal offense. The courts make decisions about the consequences for illegal behavior.

In all three of these concepts there is certainly overlap. For example, it may be illegal to have sex with your client. Prohibitions against this behavior will almost certainly be embodied in your ethical code as well and clearly sexual encounters with clients would be viewed as immoral behavior by professionals.

However, because some practitioners apparently did not see sex with clients as immoral, professional groups saw fit to place sanctions against such behavior in ethical codes for social service professions and eventually states enacted laws prohibiting this type of behavior and making it a crime.

We can look at a case involving a principal at a school in Colorado. She objected to the fact that children who could not pay for their school lunches had their hands stamped, embarrassing some children to the point that they skipped lunch. Clearly, she found this humiliating to children and therefore an immoral practice. However, there is no legal basis for stopping this and it would take some time to find and apply an ethical prohibition against such behavior. Sometimes we find that what we believe to be moral behavior is included in ethical codes or in the law. Sometimes it is not. We may find that ethical codes involve behaviors we never considered particularly immoral, and laws may or may not coincide with ethical codes or common beliefs about morality.

Our personal responsibility is to respect and follow the ethical codes that apply to our profession and the laws that govern our work. In addition, we have an ethical obligation to know and be aware when we practice what we believe to be moral behavior with regard to our relationship with clients. Many of the situations we encounter that require us to choose the right action are gray areas that call for discussions with supervisors, a rereading of our ethical codes, and sometimes consultation with a lawyer to better understand the law. Codes, whether they are legal, ethical, or our personal morality, are guidelines to help us construct the best response in difficult circumstances. You will, however, encounter circumstances that are not explicitly covered anywhere and these will require you to thoughtfully construct the best legal and ethical decisions based on careful consideration and consultation with others.

ASSIGNMENT

Find the ethical code for the profession you are interested in pursuing. If you want to become a social worker, look for the code of ethics for social workers. If you are interested in becoming a psychologist, find the code of ethics for psychologists. There are codes of ethics for human service workers, for therapists, family therapists, and group therapists. Read the code of ethics that applies to your chosen profession and be able to discuss it in class.

Dual Relationships

A dual relationship occurs when you and a person to whom you are giving services have more than one relationship. You may be this person's case manager as well as her cousin, her boyfriend, or her customer at her beauty salon. Or you may be a person's case manager and also his employer for your yard work, his Sunday school teacher, or his Little League coach. In other words, a dual relationship occurs when you are in two different relationships with a person, one related to your position as the person's case manager and the other unrelated to that role.

The first rule is to avoid all dual relationships. Your practice gives you a position of power. People tend to look up to you as someone who can provide real assistance. Furthermore, you might be the one who will determine when a person can return to work, or you may be the person who reports an individual's attendance in your program—attendance that keeps that person out of jail. It is possible that you could exploit or give the appearance of exploiting this power. In addition, there is enormous potential for a conflict of interest.

Suppose, for example, that your supervisor tells you on Thursday afternoon that you have been chosen to represent the agency at a big dinner being given to honor a county official at the Hilton Hotel on Saturday night. This gives you little time to prepare. You need to get your hair cut and styled. You call a man who receives services from you who is a hair stylist, and prevail on him to work you in at the last moment. He does you a favor and sees that you get a good appointment. You are very grateful, go to the wonderful dinner, and think little more about it.

Several months later this man calls you. He has a need for a prescription refill from his psychiatrist. It is Friday afternoon, and the psychiatrist will not be back in the agency until the following Wednesday. He feels you should be able to do this favor for him because of the favor he performed for you. He does not have time to see the psychiatrist regularly, he tells you. When you refuse to call in a prescription for him without the doctor's prior knowledge, he cannot understand why you are being "so rigid." He indicates that he thought the two of you were friends who helped each other out when needed.

Whatever you do in this situation, you will lose. If you call in the prescription, you will have violated an agency rule that a person must be seen at regular intervals by his psychiatrist before medications can be refilled. This could cost you your position or result in a disciplinary action. On top of that, you could start down a very slippery slope with this individual. He may come to expect special favors from you and offer you special, very tempting favors related to his business in return. On the other hand, if you do not call in the prescription, you have alienated someone who needs the services of your agency. You have created a barrier to his feeling comfortable with you and getting the help he needs in the future. The individual is harmed. A relationship that was or could have been useful to him in resolving problems is now something else. The opportunity for real progress is diluted with issues of friendship and favoritism.

From a shortsighted point of view, you and this person may see the convenience of exchanging favors as trivial and unrelated to the therapeutic relationship. In the long run, however, when scenarios such as the one just described occur, the relationship can never return to a professional one; and if in the future this individual is in acute need, you may no longer be able to provide the professional intervention needed.

When working in the field of substance abuse, opportunities for dual relationships are many and require special vigilance. For many practitioners, addiction is a firsthand experience and helping others cope comes from deep personal trials and challenges. The field of substance abuse emphasizes peer support and diminishes hierarchical patient-therapist relationships. Nevertheless, friendships and therapeutic relationships are not the same thing, and it is not helpful if people come to think of the two as similar.

Perhaps you are still attending AA meetings and your clients are also attending those meetings with you. Issues about how much to reveal to others and to what extent we socialize with clients who are also friends from AA are challenges that must be faced. Lester worked in a halfway house and attended AA every Tuesday evening. Two of his residents went with him. Lester felt that revealing a lot of his past in these meetings undermined his relationship with his clients. For one thing they often asked for more information about things he had revealed in AA when he saw them privately, seeking a level of detail Lester did not feel comfortable discussing with them.

Socializing as friends with people who share your AA meetings can be tricky if you are also the person's counselor. Lester tried to avoid friendships with the members of his AA group because they knew where he worked and often asked him for advice or interventions on their behalf based on the assumption that "we are all friends." One client asked Lester to be his sponsor, pointing out that Lester was in such close proximity that it would be easier for them to communicate. Lester suggested the person find another sponsor, but it seemed to leave a bad feeling between them. It is wise to be aware of the issues Lester faced in his work and know how you will address these.

In some very small, rural communities, it is not possible to avoid dual relationships entirely. In those situations, after doing all that you can to make other arrangements, you must talk with the person about the possible problems that could arise and how each of you must avoid these problems together. Then the person has the choice to continue the relationship, find other arrangements, or discontinue services altogether.

Gifts from People You Serve

Although gift giving by those whom you serve does not pose a dual relationship, people who bring gifts for you do pose a particular conflict of interest. It is usually best to avoid accepting gifts to keep the relationship professional. Often, though not always, gifts are the person's way of manipulating the situation. "I'll give you this item and you accept it. Next time you owe me something," or "I gave you this lovely thing. I am such a nice person. Even you think so or you would not have taken my gift. How can you then refuse to give me what I want?" People need to learn to express their desires clearly rather than by using gifts.

Gifts are not always manipulative, however. For example, a case manager working in a fuel assistance program worked closely with a family. The husband was injured when an automobile he was working on at a garage slipped on the lift. Unable to work, the family's meager resources began to dry up. The wife managed to find work in a greenhouse, but as winter approached, she was laid off and the expenses, particularly for fuel, increased. The couple had two children, both in elementary school, and they struggled to clothe and feed them as the wife sought another job.

The case manager saw this family through their difficulties by getting them welfare checks, seeing that the husband enrolled in the community college for courses in high-tech auto repair while his injuries healed, and finding school clothes for the children. The husband did well in school that winter and set a good example for his children, who seemed to do better in school than they had the previous year. The wife returned to the greenhouse in the spring and found that not only was she needed as a manager, but there was also a strong possibility she would have a year-round position there.

Elated by how well things were going and how much better the future looked, the couple came to see the case manager one day in the early summer and brought her a pot of black-eyed Susans from the greenhouse. "We just wanted you to have these for all you have done for our family," the husband said, smiling expansively. The husband and wife looked pleased and happy. Obviously, the couple was proud to now be in the position to be able to give something too. It was important to them not to see themselves as the recipients of handouts all the time, but to be able to also give something to someone who had been helpful to them.

Refusing a gift in such circumstances can be interpreted as rejection. If the worker had said, "Oh, I can't accept that. You'll have to give it to someone else," a person might have heard a different message: "You are the client and I am the benevolent worker. I help you, but you can never get to the position where you could possibly do anything for me. I don't need anything you could give me; but you, on the other hand, are a poor soul in need of my help."

If your agency has a policy against your personally accepting gifts, try to find a way to accept a gift of this sort on behalf of the agency. In this case, the worker planted the flowers in a planter near the door of the agency. This was a better solution than outright rejection of the gift.

The rule is to be very careful about accepting gifts from people you serve. Whenever an individual offers a gift, make a note in the person's record of the offer as well as whether the gift was accepted or rejected and why.

Sexual or Romantic Relationships

Individuals who come to human service workers for help often feel isolated, discouraged, and misunderstood. The relationship they form with a respectful, concerned worker may make them feel understood and appreciated for the first time. This relationship may be so comforting that people attempt to turn it into something more permanent, more personally meaningful.

It is not uncommon for people to fall in love with their workers in what we call "transference." In a sense, such individuals transfer to the workers the attributes they are seeking in another person. They may assume the love and affection they are seeking will be forthcoming from their workers because the workers have been so kind and helpful. These people fall in love with their workers because of an erroneous perception: They see concern and encouragement as gestures of love and affection—as an invitation to create more than a professional relationship.

Countertransference also can occur. It is not unusual for workers who are harried and overworked, and possibly coping with difficulties in their personal lives, to find the willing ear of a person to whom we give services very supportive. The people we serve are often attractive, sensitive people who can convey warmth and support when case managers are most vulnerable.

Take, for example, Kent, a case manager. Kent's wife left in the middle of a Tuesday morning, and Kent was to be at work that afternoon at 3:00. Though he fought hard to dissuade her from going, she left. For the rest of the day before work, Kent tried to get some money from their joint account, tried to find out where his wife was going, and tried to make some decisions. He arrived at work feeling exhausted and bitterly betrayed.

That evening, Kent made a home visit to Lucy's house. Lucy had first come to the agency with extreme depression, but she was doing so well now that Kent was considering terminating these follow-up visits. Lucy, an artist, greeted Kent warmly. She had put on a pot of tea and made some banana bread for his visit. Gratefully, Kent sank down on her sofa. Instead of asking Lucy how things were going for her, whether she had enough medication, and whether she had any medication questions, Kent found himself talking about his upsetting day.

In response to Lucy's first remark, "You don't look very well tonight, Mr. Paulman," Kent heard himself pour out the day's events; then he went on to talk at length about how his marriage had unraveled. He felt comforted by Lucy's interest in him as she listened intently. Here was a person who appeared to respect him as a professional and as a person. Here was a woman willing to listen to his problems. Here was a warm retreat from the job and the problems of the day where Kent could feel safe and supported.

Kent never intended for a real relationship to develop between Lucy and himself. In fact, as he left that night, he told himself that he might have crossed a dangerous line and that he should avoid further contact of this sort with Lucy. Nevertheless, based on that evening, Lucy called; and because Kent was lonely, his life was uncertain, and he was filled with anger and bitterness about his situation, he continued to see Lucy, finding in her a warm, supportive person, someone who could reassure him by her presence that he was attractive and interesting.

The relationship moved from his visiting in her home after work to his staying overnight at her house to his moving in his belongings and beginning to live there. They went out on dates. The furtiveness of these activities only made the relationship seem more romantic and important. Finally, a supervisor discovered the relationship, and Kent lost his job. After 3 years, he and Lucy have separated, and Kent is not working in the human service field anymore because he violated such an essential ethic. Instead, he sells appliances in a local store. Lucy became depressed when the relationship ended and has entered treatment again.

The person responsible for maintaining a professional relationship regardless of personal feelings is the case manager. Regardless of how you feel about the person on your caseload or how that person apparently feels about you, you are the responsible party. You will be penalized if the relationship crosses from professional to intimate. It is always assumed that the individual you are serving is the vulnerable party.

Figure 2.1 lists some warning signs that indicate when a worker or an individual receiving services might be moving away from a professional relationship and toward a personal one. Make certain you are familiar with these signs.

PLEASE NOTE!

It is a violation of all ethical codes, and in most states against the law, to engage in a sexual or romantic relationship with a person receiving services from you. This is *clearly* exploitation. It is never tolerated. You must be aware that, although attractions can occur between those receiving service and those who provide service, acting on those attractions is entirely unethical and generally illegal in professional practice.

Boundaries

Kent's situation with Lucy demonstrates a problem with boundaries. Karst-Ashman and Hull (2006, p. 402) define boundaries in the human services as "invisible barriers that separate various roles and limit the type of interaction expected and considered ethically appropriate for each role." Many ethical problems arise for social service workers when they have a vague understanding of boundaries or they cross those boundaries. It is the worker and not the client who is responsible for being aware of where the boundaries lie. Clients may attempt to cross boundaries but the worker is the one responsible for making sure that these lines are observed.

Let's look at examples where boundaries are crossed. Obviously, Kent's relationship with Lucy crossed a boundary, but there are many less obvious ways to violate boundaries. Perhaps you have had a similar problem and want to get feedback from the group you are leading about your problem. Maybe you want your daughter to play with the client's daughter and the two of you become friends. Perhaps you listen to a problem your client is having and because you know the people who caused her difficulties you stop speaking to these people and let others know that they have been the cause of your client's problems. In another instance, you might lend something personal to a client, have a client stop by the house on occasion, or find a client so much like yourself that you make him a friend and see him in social settings as well. There are countless, often seemingly harmless ways, that you can turn a professional relationship into a friendship or an exploitive relationship.

A good part of understanding your ethical obligations is knowing where the professional boundaries are for the population with which you work. Some populations have a closer relationship with their workers than others. In women's programs and in some substance abuse programs, sharing common problems is not unusual. In programs for individuals with developmental disabilities, socializing with the clients of the program is not always frowned upon. Enter your work aware of the fact that there are boundaries that govern your behavior toward clients and become familiar with these.

Value Conflicts

Generally, the person's values and your values have little to do with why the individual is seeking services from you. Sometimes, however, religious, moral, and political values play a pivotal role in the problems people bring to agencies. It is rare for case managers to get deeply involved in such primary problems, but it can happen.

First, you can be prepared by consciously knowing yourself and your feelings about certain value-laden issues. Then, if a conflict of values occurs between you and an individual you are serving, you should be able to tell that person that the conflict exists and may interfere with services. You can begin to inventory some of your own attitudes and strong feelings by completing the self-assessment exercise in Figure 2.2.

FIGURE 2.1 Warning signs that the worker–client relationship may become too personal

Warning Signs from the Client

- The client shows overt sexual interest in the worker either through conduct or verbally.
- The client describes dreams that are increasingly sexual in which the worker is prominently involved.
- The client is excessively interested in the worker's private life.
- The client inquires about the worker's relationship with his or her spouse and children.
- The client attempts to give the worker romantic gifts. (Be careful about accepting gifts from a client. Note every such offer in the record, along with whether the gift was accepted or rejected and why.)
- The client wants to see the worker outside the office in places such as restaurants or movie theaters.
- The client gives the worker romantic poetry or brings in romantic articles and books.
- The client dresses in seductive attire.
- The client interprets the worker's statements of concern for the client to mean the worker has a romantic interest in the client.
- The client repeatedly hugs and touches the worker.
- The client indicates a desire to be special to the worker.

Warning Signs from the Worker

- The client is prominent in the worker's dreams.
- The worker looks forward to seeing the client, more so than other clients.
- The worker begins to see the client as more understanding than others in the worker's life.
- The worker inquires about the client's sexual life and fantasies when these are not relevant to case management.
- The worker is more interested in this client's attire than in the attire of other clients.
- The worker is more concerned with his own attire on days when he will see the client.
- The worker begins to see the client as a person without issues or problems or minimizes these so that the client seems more acceptable as a partner or friend.
- The worker takes many innocuous actions the client might interpret to mean the worker has a special interest in him or her.

Source: Based on a list created by attorney O. Brandt Caudill (1996). Used with permission.

Second, if a severe conflict of values exists, you might need to make arrangements to transfer the person to another case manager. You would not do this because of a simple value conflict, but you should try to make such a transfer if you find you can no longer be objective, you are extremely uncomfortable with the person because of her or his values, or you feel compelled to counteract the individual's values by imposing your own.

FIGURE 2.2 Self-assessment exercise

Self-Assessment Exercise: Possible Values Conflicts When Helping Others

Look at each description of a person or group of people, and assign a number to each.
1. Give the description a 1 if you think you could work with the person or group.
2. Give the description a 2 if you think you could work with the person or group, but would find it uncomfortable or difficult.
3. Give the description a 3 if you could not work with the person at all.

_____ 1. A woman who wants you to help her feel comfortable with her decision to have an abortion
_____ 2. A man who frequently brings up his fundamentalist religious beliefs
_____ 3. A homosexual couple who want help in improving their relationship and resolving their interpersonal conflicts
_____ 4. An interracial couple seeking premarital counseling
_____ 5. A man from Iran who strongly opposes the equality of women and talks about women working in denigrating tones
_____ 6. A man who has for years been getting more welfare than he is entitled to receive by using certain tricks he developed to beat the system
_____ 7. A man and woman who say they want to improve their marriage, but the man will not end his affair with a second woman
_____ 8. A white couple seeking help for behavior problems with their adopted son, who is African American
_____ 9. A man who makes it clear he often disciplines his children by using corporal punishment
_____ 10. A person who refuses to discuss feelings and says that all that matters are facts and logic
_____ 11. A woman who has chosen prostitution as a way to support herself and her children
_____ 12. A gay man dying of AIDS who comes in with his lover to resolve conflicts around how he contracted the disease
_____ 13. A man seeking help to curb his extreme abuse of his wife
_____ 14. A woman who sexually molested her son
_____ 15. A lesbian couple seeking to adopt a child
_____ 16. A woman dying of breast cancer who wants to take her own life
_____ 17. A person who relies heavily on cocaine to get through the day
_____ 18. A couple who are openly anti-Semitic
_____ 19. A vocal member of the Ku Klux Klan
_____ 20. An orthodox Muslim who cannot always see you because his appointments often interfere with his times of prayer

Digital Download Download from CengageBrain.com

For example, a human service worker who did not personally believe in birth control (including tubal ligation) was a case manager for individuals with developmental disabilities. When a young couple on her caseload decided to marry, she became actively involved in discouraging them from the idea, particularly when she learned that the woman planned to have a tubal ligation so that they would not have children. The families of the two individuals supported the marriage.

The people were high functioning, each one had a job, and each had the support of other community agencies.

In the months before the wedding, the case manager did not attempt to transfer the cases to another case manager. Instead, she harangued the couple about the sins of birth control and of marriage without children, and about the unwise decision to marry at all, given their "mental impairment." The families complained to the agency, asking that she stop pressuring these vulnerable individuals. Twice the supervisor disciplined the worker. When the worker persisted—visiting the couple's minister who would perform the ceremony, the supervisor where the man worked, and the woman's parents—she was fired from her position.

The worker's behavior was harmful to this couple. For those two people, who had always relied on a case manager who had seemed to be wise, the worker introduced uncertainty and fear. Her constant negative warnings damaged their fragile self-confidence and self-esteem. For them, what should have been a happy decision, supported by family and friends, became a decision fraught with anxiety. Family and friends had to work long and hard to restore their confidence in their original decision.

This is an example of the worst possible way to handle a **values conflict**. In this situation, the worker attempted to impose her own point of view, her own personal values, on the couple. She denied them the right to choose for themselves and interfered in what was largely a personal and family issue unrelated to case management.

Avoiding Value Conflicts

Following are some rules for avoiding value conflicts and ensuring that individuals get professional service from you:

1. Be respectful of attitudes and lifestyles that differ from your own.
2. Never practice prejudice toward minorities, those with disabilities, or those differing in sexual preference.
3. Always give your best service to a person, even when you disagree with the person.
4. Never attempt to change the individual's values to coincide with your own.

Using Values to Motivate People

When people come in seeking assistance, they are usually hoping to make things better in their lives than the way they are right now. In the course of your time with them, it is important to explore the values that caused them to seek help. Find out what goals people have for themselves if their situations were better, and what values those goals reflect. Here we are looking at the things that people value for themselves: being a good parent, living an independent life, living free of the symptoms of schizophrenia, or being free of cocaine addiction. The person is envisioning something up ahead that involves a goal or value dear to that person.

You often learn about people's values when you ask them where they would like to be in 5 years. What you hear when they answer will point to what they hold important for themselves and those around them. When people are having trouble making changes that will move them toward their personal goals and visions, it is helpful for the worker to know what the individual's values are and to look at the person's situation with those values on the table.

The Rights of Individuals Receiving Services

Anyone who gives service in one of the helping professions must be familiar with the rights of the person and make a particular effort to see that people understand they have rights when they seek help. Often people mistakenly assume that they have few or no rights when they come in for services. In addition, professionals may fail to inform individuals of their rights because it is easier to work with people who are vulnerable, dependent, and uninformed. This, of course, sets up a situation in which it is easy to exploit the person. The purpose of educating people about their rights is to allow them the opportunity to become active participants in their care and partners in decisions that affect them.

Most agencies prepare **clients' rights** handbooks for those receiving services to keep as a reference. A hospital for the mentally ill would include the right to be released from the hospital as soon as care and treatment in that setting are no longer required. Nearly all agencies inform people that they have the right to participate in the development and review of their treatment plan. People generally have the right to participate in major decisions affecting their care and treatment. Most of those who are involuntarily committed to an inpatient setting have the right to refuse treatment to the extent permitted by laws in that state or the right not to be transferred to another facility without clear explanations regarding the need for the transfer. Inpatient units stipulate it is the person's right not to be subjected to harsh or unusual treatment. The hospital may also spell out the fact that the individual may keep and use personal possessions, or the person must be informed about why something is being removed. In most settings, people have the right to handle personal affairs and to practice the religion of their choice.

In outpatient settings, people have the right to a flexible and responsive treatment plan, the right to expect an individualized plan of service, and the right to make suggestions and express concerns. Often there is a procedure individuals can follow if they are dissatisfied with the worker assigned to them or the service plan laid out by the agency.

The following sections discuss some important rights that belong to those receiving services.

The Right to Participate in Planning

When you sit down with someone to begin planning what services would work best or what treatment would be most effective, you each bring unique perspectives to the table. You have a detailed understanding of what is available, which services work better than others, and where people on your caseload can receive treatment or service tailored to them. A person comes to the table with information about themselves that is useful when planning. Putting these two pieces of information together makes for a more effective plan. People who work with you feel that they have had a part in the planning and therefore, some control over the direction of their care. This makes it more likely that the plan will be followed or that individuals will tell you if they see the need for a change.

It is for this reason that we do not just disrespectfully tell people what their service or treatment plan will be without consulting them first.

The Right to Self-Determination

Educating people and informing them about their rights are both done so that clients can exercise the right to self-determination. Paramount to any relationship between professionals and their clientele is the right to self-determination. People have the right to do research about their diagnosis or problem and to question the treatment plan or make suggestions. People have the right to withdraw from treatments and services they find are not helpful. People have the right to decide when and for how long they will use services (unless their involvement with the agency is based on an involuntary court commitment). People have the right to choose their own goals.

Often this presents a problem for a worker who feels compelled to look after the best interests of the individual. One of the hardest lessons you will ever learn is how to let people make mistakes and learn from those mistakes. You can make suggestions and express concerns, but ultimately clients have the right to determine what they will do. You may feel strongly, for example, that one individual is not ready to walk away from the agency; and you may feel certain that the person's doing so prematurely will result in further problems with alcohol. In fact, your client leaves treatment against your advice and eventually does end up with another DUI charge. Although your worst fears and predictions came true, you cannot know for sure that the work with you and the new charge were not important learning opportunities. In other words, people have the right to test the waters, so to speak, and to learn that they are not as ready as they thought they were. In fact, one of the ethical considerations for workers in the field of addictions is to be willing to view crises as opportunities for change and growth. It is useful to consider crises in other fields in this light as well.

Increasingly, however, self-determination means more than this. More and more funding sources and governments, as we saw in Chapter 1, are asking case managers to go beyond simply arranging for services in collaboration with the client. They are asking case managers to encourage people to articulate what their vision of a healthy, productive future would look like. As people do better on medications and remain in their communities, how they function in those communities—how they contribute, feel secure, and pursue their own interests—becomes more important. Self-determination now takes on the future beyond the social and emotional problems that were the original reason for seeking help. Now people are being energized by their case managers to explore and create a better tomorrow of their own making.

Informed Consent

A person receiving services always has the right to consent to these services or withdraw from them. In making this decision, the person must be informed enough to make a wise decision. When the individual is informed and consents to treatment, we call that *informed consent*. Making certain that a person can give informed consent begins with the intake, during which the agency policies are explained and choices of treatment or services are outlined. This level of information should continue throughout the entire relationship between the individual and the agency until termination. This means that people informed about treatment or services can make their own decisions with regard to the services.

The following list contains items that should be addressed when relevant to the person's services. The person has the right to be informed about:

1. Any side effects, adverse effects, or negative consequences that could occur as a result of treatment, medications, or procedures
2. Any risks that might occur if the person elects not to follow through with treatment or services
3. What is being offered to the individual, including what the treatment is, what will be included, and any potential risks and benefits
4. Any alternate procedures that are available

Some of the people with whom we work have a limited capacity to understand all the details of service and treatment. It is our task to find an appropriate balance between too much and too little information and to make our information clear and easy to understand.

Informed consent consists of the following three parts, or criteria. All must be present in order to say that the individual gave informed consent:

1. *Capacity.* The individual has the ability or capacity to make clear, competent decisions in his or her own behalf.
2. *Comprehension of information.* The person clearly understands what is being told to him or her. To make sure that this is so, give your information carefully and always check to be sure the person understands what you have told him or her.
3. *Voluntariness.* The person gives his or her consent freely with no coercion or pressure from the agency or the professional offering the service.

Currently laws and courts are recognizing more and more often the person's right to self-determination. When we fail to tell those we serve the information they need in order to give informed consent, we run the risk of being found negligent, particularly if the treatment or service involved was unusual.

The Right to be Informed of Changes and Decision

There are times, however, when you are working with a person who cannot participate. People who are confused, have severe developmental disabilities, or are psychotic often are not capable of planning or giving input. Letting them know what will happen is still respectful. One case manager's client, Mindy, was hospitalized when she became psychotic as a result of schizophrenia. Mindy barely responded to treatment and the decision was made to transfer her to a long-term care facility. The nurses on the hospital unit made the arrangements for transportation and who would help with the transfer from the unit to the ambulance.

The case manager had agreed to the transfer and came in a few hours before it was to take place to talk to Mindy about it. The nurses were incredulous. Mindy was in a room with only a mattress on the floor because she had taken her room apart several times. She was uncommunicative and had been yelling at voices she heard. "She isn't going to understand a word you tell her," one of the nurses remarked. Nevertheless, the case manager went to Mindy's room, sat down on the bare mattress on the floor and began to describe in some detail what was going to happen. Mindy grew quiet. She never looked at her case manager, but she appeared to be listening intently. When the orderlies came to take her to the ambulance she went without resistance.

Did Mindy really understand what her case manager told her? Is that really the point? The point is that this case manager respected her client's right to know what plans had been made for her. The right to know what treatments and services have been planned and the right to participate to the degree a person is capable are important ways professionals demonstrate respect for the people they serve.

Confidentiality

Confidentiality is both an ethical principle and a legal right. It is the most basic right of any person, either in treatment or receiving services, to know that what the person is sharing in your office will remain confidential. It is important to protect individuals to whom we give service by not disclosing their personal situations without the people having authorized such a disclosure. Today, under new laws discussed in the text that follows, agencies have very specific guidelines for protecting confidentiality.

Release of Information Form

Not many years ago a person seeking services, particularly from a public agency, signed a blanket permission statement allowing information to be shared with others as the agency and the worker saw fit. Today, release of information forms must state specifically to whom the information is being released and must be time limited (good for 3 months, 1 year, and so on). Do not use forms that are not specific in this manner. New regulations now stipulate what is permissible on a release of information form.

Release of Information Regarding HIV/AIDS

In most states, release of information forms for releasing information regarding a client's HIV/AIDS status must specifically state that you may release information regarding the person's HIV status. All references to HIV/AIDS must be deleted from the record unless the client signs a separate form that specifically states that you have permission to release this information. If you are asked to release information about a person who is HIV+ and the person signs a release form, the law in most states specifies that it is not good enough to simply remind that person that his case contains references to his HIV status and get his verbal permission to release the information anyway. You also must have his written permission. If the person has not given you written permission, you must delete all references to HIV/AIDS, including the fact that he may have been tested and the test was negative.

If your state does not have such a law, you are still responsible for protecting your client and must be alert to the possible harm such a release might cause the person. In such a situation, it is wise to involve the person in a discussion about the release of this sort of information or, if the individual is unable to participate in such a discussion, to take steps to protect that individual from undue bias.

In some instances, workers have informally notified their friends and acquaintances in other agencies of a person's HIV+ status, thinking they were doing these people a favor. In fact, this behavior is entirely unethical and can lull other workers into believing they know who is and who is not HIV+. We can never actually know this for certain because of the length of time it takes for the disease to register positive on a blood test. A person can be positive early in the illness and still have negative blood tests. For this reason, workers should use universal precautions with every client when those precautions are called for. Workers who fail to use universal precautions on the false assumption that they know the individual is not HIV+ place themselves at undue risk.

ASSIGNMENT

Find out what the laws are in your state for releasing information about a client that contains references to the client's HIV status.

Collegial Sharing

Out of respect for individuals, you should ask them for permission before sharing information with colleagues from whom you are getting opinions or supervision, unless the case is going to be discussed in the normal course of supervisory meetings with a regular supervisor. Likewise, you cannot share information with student interns without making certain the students have signed agreements to observe strict confidentiality while acting as part of the agency. Suppose you are working in an agency and have been asked to give a student a view of what you do. To illustrate what you have told the student, you show her several case files.

She reads the cases and discovers that one of them is the boyfriend of her cousin. What she reads in the file is alarming to her, and she decides her cousin should not be dating the client. She leaves the agency and begins to share information with the cousin, causing considerable conflict among family members and anguish to the cousin, who knew part of the story but not all of it. This kind of sharing of information is unacceptable, and most agencies do not allow students or volunteers to read anything before they have signed a pledge to honor the confidentiality of the clients and you feel these students thoroughly understand the critical importance of protecting confidentiality.

Guarding Confidentiality on the Phone and in Other Conversations

Other situations also provide opportunities for violating confidentiality. For instance, a person receiving services from your agency may also be receiving services from a local physician. Suppose someone calls, claiming to be the physician's nurse and needing to know at once what medications the client is taking. She may really be the physician's nurse, or she may be a person posing as the nurse in order to determine that the person is using your services and the level of his problem. Even if she is the nurse, the person may wish to keep his physician uninformed about the involvement with your agency. All agencies have procedures for such situations in the event of a real emergency. You, however, must never openly and automatically acknowledge that an individual is being seen in your agency, no matter how important and official the other person seems to be. In the case of a seeming emergency, refer the call to your supervisor unless you know the emergency workers or emergency room personnel well enough to recognize their voices.

When a request for information is presented in a situation that is not an emergency, here is how you might handle the request:

YOU: Hello.

CALLER: Hi. This is Ann Taylor. I'm a counselor at Marlboro Middle School, and I'm calling about Jimmy Smith. Did he and his mother keep their appointment with you today?

YOU: I'm sorry, I can't help you with that. Would you have Mrs. Smith sign a release of information form stating what it is you need to know, and if Jimmy Smith is known to us, we can send you that information.

In this situation you do not give any hint that the client is known to your agency. By saying "if this person is known to us . . . ," you do avoid letting on whether the client is or is not known to you.

Another way to violate confidentiality is to talk about your cases with your friends and relatives, leaving out the names. Others may be able to piece together the identity of the person you are talking about based on other information they possess. In this way, they may discover far more about a person than that person ever intended them to know.

Minors and the Infirm

Take special care to protect the confidentiality of minors and the infirm (individuals who are frail, sick, and are unable to fully participate in decisions about their care). Not all systems respect confidentiality to the degree that we in the helping professions are committed to doing it.

In one children's case management unit, parents were routinely urged to sign blanket release of information forms. When the school requested information on a child being seen, all the information was sent to the school. It was stamped in red letters with the word *confidential*, and it was sent to the school psychologist. Nevertheless, school clerical personnel assisted in typing and filing information for the psychologist and generally read the information sent by the case management unit. Having no training in confidentiality, these clerical people talked among themselves about students, sharing personal information they had learned. Many times they passed on to teachers tidbits of what amounted to gossip. This information shared outside the professional context and without professional understanding jeopardized the progress of the children and the relationship of their parents with the school personnel. As these children moved through the school system, the gossip followed them. Always be very careful about what information you release. Remember that information given about a child can follow that child all through school, prejudicing responses to that child.

In another case, a woman with a developmental disability got a job at the police department as a cleaning woman. She was told that she needed to bring in her "records from mental health" so the police could know why she went there. She arrived at the case management unit, pleased about the job and ready to give all her records away. The case manager talked to her about the wisdom of retaining most of the information as confidential. In the end, a short statement was released, with the client's permission, giving only the most general information about her relationship with the mental health case management unit. It is important to remember that older people or individuals who do not have the capacity to protect themselves can be easily led to sign releases regarding information that might best be kept confidential. The responsibility belongs to you to protect your clients from unnecessary intrusions into their personal information.

Minimum Necessary Rule

Before releasing information, ask yourself whether you are about to release more information than is needed for this other business or organization to accomplish its work with the client. For example, Melissa was a case manager who knew a worker in a remedial education program where one of her clients, Jill, was attending. Although the program needed to know why Jill was referred and what goal the referral was intended to accomplish, they did not need to know that Jill was arrested once for a DUI and that Jill's father was in prison for murdering a neighbor. When dealing with other organizations not engaged in treatment, release only what that organization needs to work effectively with the client. If the client authorizes you in writing to disclose more, only then would you do so.

When You Can Break Confidentiality

The law in all states does make exceptions. The following are circumstances that allow you to break confidentiality:

1. When you must warn and protect others from possible harmful actions by the client. For instance, you or your agency must warn another party if your client is intent on harming that other party. In addition, you should notify the police.
2. When the person needs professional services. For instance, if the person has taken an overdose of medication and is in the emergency room (ER), the ER staff may call, needing to know what prescriptions the client was taking in order to give the proper antidote.
3. When you must protect people from harming themselves. An example might be people who are threatening to take an overdose of their medications with the intention of committing suicide or people who appear so depressed or desperate that they are talking about ending their lives.
4. When you are attempting to obtain payment for services and the payment has not been made. Your agency would refer a person for nonpayment only after reasonable attempts had been made to remind the person of this obligation and only if the individual had made no effort to arrange even minimal payment.
5. When obtaining a professional consultation from your supervisor regarding how best to proceed with a case in the course of normal supervision.

Privacy

Privacy is very much related to confidentiality. Siegel (1993, p. 105) calls it "the freedom of individuals to choose for themselves the time and the circumstances under which and the extent to which their beliefs, behaviors, and opinions are to be shared." Stadler (1990, p. 102) calls it "the right of persons to choose what others may know about them and under what circumstances." Kirst-Ashman and Hull (2009, p. 376) define privacy as "the condition of being free from unauthorized observation or intrusion." In the social services, privacy is invaded or altered under some circumstances, and people need to be informed of those circumstances. The point you should stress with the people you serve is the fact that third-party payers will have access to diagnoses and, in some cases, to actual records or summaries of records. The agency must provide this access in order to be paid for the services it has rendered. Many individuals are unaware of this fact or unaware of the extent of the information being shared. They should have this situation explained to them. This allows people to make an informed decision about whether to pay for services themselves and not involve the insurance company.

Health Insurance Portability and Accountability Act

The federal **Health Insurance Portability and Accountability Act (HIPAA)** was passed in 1996 in part to ensure that people did not lose medical coverage when they changed jobs. Title II of the act contains the security and privacy mandates. These contain stringent rules for protecting a client's health information, and most social service agencies must adhere to these rules. Where state laws are more stringent than this federal act, the state laws take precedence. Failure to follow the guidelines set forth in HIPAA can result in fines from $100 to $250,000 and from 1 to 10 years in prison for those individuals and institutions with the ultimate responsibility for safeguarding patient privacy.

The new rules apply to case management and to care coordination and cover not only formal records but also personal notes and billing information. When you begin work at your agency, they will see that you are informed of their policies and procedures under this act.

Disclosure

Under the new rules, "disclosure" is defined as occurring when health information is released, transferred, or divulged outside the agency. This includes allowing access to patient files to others not working for the agency. The material in question is often referred to as *protected health information*(PHI).

Agency Requirements

In order to comply with HIPAA, every agency must have the following:

1. A statement of the agency's privacy and confidentiality procedures, particularly as it relates to releasing patient information. This statement must be given to every client of the agency. It is considered a notice clarifying how health information will be used and stipulating the client's privacy rights. This is a public document and can be posted in waiting rooms and on websites.
2. A form that people sign and return to the agency indicating that they have received the statement on confidentiality policies.
3. A privacy officer who is familiar with HIPAA requirements and can oversee implementation within the agency and resolve privacy issues as they arise.
4. A set of safeguards to protect client records.

The privacy concerns addressed by HIPAA were raised because of the increasing demand by insurance companies, employers, and others for detailed information on clients and patients, often in excess of what was necessary to process claims.

Security and Privacy

Security in the act refers to procedures to protect health information from inappropriate access by others. These procedures usually include controls on who has physical access to the records, security of work areas and record storage areas, and destruction of duplicate or obsolete files. Electronic security measures are also instituted, such as changing passwords and encryption.

Privacy refers to the person's right to keep specific information private and includes the agency's release of information policies and the rights of the individual in this matter. The law is stringent because personal health information has been taken and used for illegal purposes. Workers with access to PHI have used this information for identity theft, sold it to others who then exploited the patient's situation, used the information to threaten or blackmail others, or took the information to use at another time and location. The penalties can be stiff.

For example, in 2011, a 62-year-old psychiatrist was charged with HIPAA violations when he discussed a woman's psychiatric case with three individuals seeking information on behalf of the patient's employer. While the woman's discharge summary indicated she was not a threat to herself or the public, the psychiatrist exaggerated her condition without the woman's knowledge or permission. In this case the psychiatrist was fined $5000.00 and ordered to complete an 8 hour course on medical ethics.

Sometimes workers share information to be helpful or to feel important. Frieda knew that the other parents in her daughter's day care were concerned about the behavior of another toddler in the class. Frieda knew the toddler's home conditions were chaotic because she was the case manager for the family. The parents were getting a divorce, and the mother seemed to be somewhat unstable emotionally. As a consequence there had been long custody battles and a bitterly acrimonious home life for this toddler. Frieda confided this information to parents who expressed concern about the toddler in her hearing. She felt important that she knew the background to this toddler's situation and had information others did not have. Frieda also felt that people would like her because she had chosen to confide private and "secret" information to them. Confiding personal health information to others in violation of privacy stipulations outlined in HIPAA can create legal issues for you and for your agency. The toddler's family, who could agree on nothing else, did come together to sue Frieda's agency for the violation of their confidentiality and Frieda lost her job and her reputation as a competent professional.

Oral Communications

The law states that agencies are to make "reasonable efforts" to safeguard clients' information. This extends to oral communications. Taking precautions to protect oral communications means:

- Not discussing a person's personal health information where others can hear
- Avoiding situations with clients where there is no privacy, particularly privacy from other clients
- Lowering your voice when discussing clients with others in the agency

Release of Information Form

Under HIPAA, the form signed by the person or a legal representative of the person must adhere to the following:

- The entire form must be in plain, understandable language, and it must be signed and dated.
- There must be a description of the information to be used.
- The form must name the recipients of this information.
- Those who will disclose the information, such as the agency or a therapist, must be named on the form.
- The form must have an expiration date.
- There must be a statement describing the purpose of releasing the information.
- There must be instructions telling the client how to revoke the form.
- A statement must be included to indicate that the information may not be as protected once it is released.
- If the agency will receive money for the information (for example, payment from an insurance company), this must be stated on the form.
- The form must make clear clients' rights to a copy of the authorization they have signed.

Digital Download Download from CengageBrain.com

It is assumed that reasonable steps will be taken to release only the minimum information necessary to support the purpose of the release. When the purpose is continued care of the person or when the person requests that more information be released, it is expected that more information will be released. Any request for the entire client record, however, needs detailed justification.

Individually Identifiable Health Information

Individually identifiable health information includes demographic information (such as age, gender, income, or race) and other information that identifies the individual or could reasonably be thought to identify the individual. Information that relates to an individual's past, present, or future condition is also included in this category.

People have the right to ask that their information be restricted. They may indicate, again in writing, that information is not to be shared with family or friends. These requests are generally honored except in medical emergencies. In addition, clients may ask, in writing, that mail from the agency not be sent to their home address or that calls from the agency not be made to their home telephone, and the agency must honor these requests.

A client may ask, in writing, for a written list of how their PHI was disclosed. The request can extend as far back as 6 years. Note that clients can specifically request how information is to be shared or restricted, but must always do so in writing. A person not able to write such a request may need the help of a case manager.

Accessing the File

Under the HIPAA guidelines, people now have a right to:

1. Read their files
2. Make copies of their records
3. Make corrections or additions to their files, as long as the changes are accurate

As noted earlier, such requests must be presented in writing to the agency and must be accommodated within a specific time period. Individuals who are going to amend their files must state the reason for amending the record in the written request. Client representatives, such as guardians, have the same access and rights as do clients. There may be times when the person will need the help of the case manager to formulate that request.

The rights discussed here are guaranteed under federal law; thus, it is illegal to discourage or threaten people when they make these requests. Currently there is evidence that people who have read their charts and received clear information are less likely to sue for malpractice or create other legal problems. It is not a good idea, however, to just hand someone a chart and provide no explanations for technical information that may be written there. This potentially creates misunderstanding. If at all possible, sit with the individual and carefully review the important points in the chart. Answer questions and explain what has been written so the person understands what is written and does not draw erroneous conclusions or conclusions that could lead the person to believe there is an adversarial relationship described in the chart.

Social Networking

We think of social networking as something we do with friends, entirely unrelated to our work, an activity we engage in on our own time. Contributions to Facebook, Twitter, and other social network sites are assumed to be private and just among friends. In reporting on nurses fired for posting on Facebook, WHTM abc27News (Harrisburg, Pennsylvania) noted, "So you're in your own home, on your own computer, on your own time, typing on Facebook. It could be your undoing."

During the winter of 2010, a group of about 13 emergency room workers were fired from a major hospital in Harrisburg, Pennsylvania, for their social networking activities, activities they assumed to be private. In this case, the emergency room workers had established a Facebook page where they discussed their day's activities with one another. While no patients were actually named, patients were referred to in exasperated and derogatory terms and their illnesses and personal characteristics were described in some detail.

The article quotes one nurse as saying, "The one posting I put was, 'That lady was crazy.' There was no name mentioned, those were the only four words I said." However, the hospital fired this woman who was shocked. "I would never have thought that what I posted in the privacy of my own home would have ever ended up being the big mess that it is," she said.

The workers contended that because the page was unrelated to their work at the hospital and activities on the page took place on their own time, they should not be fired. The hospital argued that the page violated HIPAA laws in that anyone who had access to the page could put enough information together to identify individual patients. Social networking pages are generally not as private as we like to assume. Friends of friends can gain access, sometimes inadvertently. In this case the nurses and others did not exercise good ethical judgment. Anyone coming across this site would not have felt comfortable using emergency room services at this hospital.

Many ethical codes have not caught up with social networking as an ethical consideration. That does not excuse you or others from exercising sound judgment about when, with whom, and how you discuss your clients.

Privileged Communication

Clients and workers alike talk about privileged communication without truly knowing what it is. First of all, it is a *legal concept*. It protects the right of a person to withhold information in a *court proceeding*. It is a right that belongs to the client. It does not belong to the worker or the agency.

All states have a law that stipulates what communication between a client and professional shall be considered privileged in order to protect the client from the disclosure of confidential information during a court proceeding. These laws designate who is to be considered a professional. A number of years ago there was a case in which a man who committed a murder confessed this murder in an Alcoholics Anonymous (AA) group. He tried to invoke the right of privileged communication, but the courts denied it because the state law did not specifically name AA as a group protected by this statute.

Only clients can invoke privileged communication in order to protect themselves. Professionals and agencies cannot use it to protect themselves. If the client waives the right to privileged communication, the professional or agency has no grounds to withhold information. Clients waive this right if they sue your agency or if they use their condition as a defense in a legal proceeding.

When You Can Give Information

At certain times, you can provide information about people in a court proceeding. In some situations, you are required to do so.

Legal Proceedings

In a legal proceeding, you may give information about people under the following conditions:

1. You are acting in a court-appointed capacity, such as that of guardian or payee.
2. You or your agency is sued for malpractice.
3. The court mandates that you turn over certain information.
4. The individual uses a mental condition as a defense or as a claim in a civil action.

Protecting Clients and Others from Harm

Other situations in which you can give information about people relate to your responsibility to protect clients and those connected with them from harm. These situations are:

1. When you believe the person intends to commit suicide
2. When a child under 16 years old is believed to be the victim of a crime such as sexual or physical abuse or sexual exploitation
3. When you determine the person needs to be hospitalized for a mental condition
4. When the person has told you of his intention to commit a crime, harm another person, or harm himself

Intention to Harm Another On October 27, 1969, Prosenjit Poddar killed Tatiana Tarasoff and set in motion court proceedings that brought about changes in the way confidentiality is viewed. That October, Poddar was a patient of Dr. Lawrence Moore at Cowell Memorial Hospital at the University of California, Berkeley. Moore, a psychologist, was told by Poddar of his intention to kill Tatiana Tarasoff. Moore contacted campus police, who briefly detained Poddar but released him when he appeared to the police to be rational. Apparently, Dr. Powelson, Moore's supervisor, directed that no further action be taken to detain Poddar. No one warned Tatiana Tarasoff of the danger she faced, and as a result she lost her life.

The Tarasoffs brought charges against the professionals in this case for failure to warn the victim of the impending danger. When the California Supreme Court eventually heard the case, the court ruled that

> therapists cannot escape liability merely because Tatiana herself was not their patient. When a therapist determines ... that his patient presents a serious danger of violence to another, he incurs an obligation to use reasonable care to protect the intended victim against such danger. (*Tarasoff v. Regents of the University of California*, 1976)

The steps the court included were warning the intended victim, warning others who would apprise the intended victim of the danger, and warning the police. The court went on to state:

> We recognize the public interest in supporting effective treatment of mental illness and in protecting the rights of patients to privacy, and the consequent public importance of safeguarding the confidential character of psychotherapeutic communication. Against this interest, however, we must weigh the public interest in safety from violent assault.

The opinion closed with the following:

> We conclude that the public policy favoring protection of the confidential character of patient-psychotherapist communication must yield to the extent to which disclosure is essential to avert danger to others. The protective privilege ends where the public peril begins.

This case established a "duty to protect" for individuals who treat patients who appear to present an imminent danger to an identifiable person or persons. The ruling appears to apply mainly to therapists, but here the waters are muddy. Human service professionals in all states have taken the position that if such circumstances were to occur in the course of their work, the courts would find them negligent if they had not exercised the precautions laid out in the *Tarasoff* case. Most states now have statutory or binding case law that establishes the duty to warn, but some do not. Regardless, you must assume that the courts would find you or your agency negligent if you failed to take the precautions outlined in the *Tarasoff* ruling. It is unlikely that you would be excused from liability because you are a case manager, and not a therapist.

Rarely would you make the decision to warn alone. If you believe a person poses an imminent danger to another identifiable person or persons, you must take the matter up at once with your supervisor. If your supervisor is not available for consultation and you believe you cannot wait, notify the police.

In a step down unit for the mentally ill, a man living there left one evening. No one knew where he was going, and when he didn't come home that late evening, the staff became alarmed. He had been talking about going back to the farm where he grew up to "evict those people who put us out." In fact, the family had sold the farm, and the people living there were the owners. At this point, the worker determined that the family at the farm should be warned. The supervisor, unfamiliar with the law, resisted, even though she would have ultimately been responsible had something happened. Later the worker ran into the director of the agency and asked her opinion. Immediately the director told the worker to contact the people at the farm and let the police in that jurisdiction know he might come to the farm. In fact, the man did show up and talked about the need for the owners to move out. He did not pose a threat, but the fact that he might have done so was important to consider. In this case, the police returned him to the step down unit where his behavior was discussed with him. This supervisor's lack of understanding about the law could have caused problems for the client, the people living at the farm, and the agency.

ASSIGNMENT

Find out what laws exist in your state regarding your duty to warn or your duty to protect a third party from harm by your client. There are databases available online that outline the laws or lack of laws in each state. Some states have no laws but rather policies that leave decisions up to the practitioner. If there are no laws on the books or if the laws are extremely permissive, what is common legal opinion regarding the duty to warn?

Mandated Reporting

All states have laws that require professionals to report the abuse and neglect of children. These laws make it a criminal offense to refrain from reporting abuse and neglect if you suspect these are going on in a child's life. In other words, you, as a social service professional do not have a choice about whether to report or not to report. The law mandates that you must. In some states, laws require human service workers to report elder abuse as well. The definition of child abuse and elder abuse varies from state to state. Professionals who must report abuse and neglect under the law are called "mandated reporters." The laws in each state stipulate who is a mandated reporter; variations exist among the states in regard to which professionals are considered mandated to report.

Although there are laws on the books that require you to report abuse and neglect, workers are often guilty of not making the report. The reasons given are many. Some workers feel that reporting abusive parents will ruin a good relationship they have established with these parents. Some workers feel they do not have enough "proof" of abuse or enough information to report. However, it is not your responsibility to determine if abuse and neglect are taking place. If you have reasonable cause to suspect there is abuse or neglect you must report it. The agency that receives the report, such as child protective services or the district attorney's office, will make the ultimate determination. If no abuse is found this is not considered in anyway a mark against you. In reporting what you suspected could be going on you have followed the law and the ethics of your profession.

Even in states where there is no mandate to report elder abuse, there may be protective services for the elderly to which you can report suspected abuse of an older person. You have an ethical responsibility not to ignore abuse of this type, regardless of the law. It is your responsibility to protect clients, particularly individuals who cannot protect themselves.

ASSIGNMENT

Find out your state's definition of child abuse. Learn which professionals in your state are considered mandated reporters of child abuse. What are the laws in your state regarding elder abuse? Who must report elder abuse in your state?

Diagnostic Labeling

Agencies that rely on a diagnosis in order to be paid for service by a third-party payer (such as an insurance company, Medicare, or Medicaid) need to inform people of that fact. People rarely understand that labels are used in this way, and most people do not know what the labels are or what they mean. They are rarely clear about the fact that the information will be passed on to their insurance companies. People need to know this, so they can then decide whether to continue to receive services from the agency. Some individuals may elect to leave the agency or to pay for the services themselves, without involving their insurance companies, as a means of ensuring their privacy. Unless they are informed, they will not know they have these choices.

Another point about diagnosing people is that practitioners use the categories of illness to know which treatment to use and how to develop the most effective treatment plan. Much research has been done to link the best treatments with each of the diagnostic categories. People will appreciate the need for a diagnostic label if they understand this. What may appear to people as simple respect, kindness, good communication, or personal support on the part of their therapists may actually be the use of well-developed treatment modes.

Involuntary Commitment

Generally, an involuntary commitment occurs to a facility that specializes in inpatient mental health care. It could be a unit in a general hospital in the community where the clients live, a private psychiatric hospital, or, in some cases, a partial hospitalization program where people receive treatment during that portion of the day they are most at risk.

Patients have a right to expect the least restrictive form of treatment. If they need hospitalization but not a locked ward, they should not be locked up 24 hours a day. If they can get the care they need in a partial hospitalization program, they should not have to go into the hospital. In discussing the movement to deinstitutionalize mental patients, Bednar, Bednar, Lambert, and Waite (1991, p. 100–101) wrote, "treatment should be no more harsh, hazardous, or intrusive than necessary to achieve therapeutic aims and to protect clients and others from physical harm."

The courts take seriously their responsibility to commit individuals in need of psychiatric care who are unable to obtain it because of a current severe impairment. In making the commitment, the courts make it clear that the purpose is treatment, and not punishment for behavior. For that reason, court commitment proceedings are often less formal and more pleasant than criminal proceedings. Students may observe these proceedings if they choose, as the proceedings are public. If you are involved in a commitment procedure, be sure to document all the steps you take in order to protect yourself from liability.

The criteria for committing someone against her will are as follows:

1. The person poses a danger to self or to others, *and possibly one or more of the following:*
2. The person has a severe mental illness or a mental illness that is currently acute.
3. The person is unable to function in occupational, social, or personal areas. The impairment is severe enough that the person cannot provide adequate self-care.
4. The person has refused to sign a voluntary commitment for treatment, so that an involuntary commitment is the last resort; or the person is incapable of signing such a commitment or of choosing appropriate treatment.
5. The person can be treated once committed; that is, known treatments and medications can relieve the acute condition the person is experiencing at present.
6. The commitment adheres to the criteria of the least restrictive treatment setting.

Digital Download Download from CengageBrain.com

ASSIGNMENT

Find out what the commitment laws are in your state. Look at the various types of voluntary and involuntary commitments. Find out under what circumstances the client can leave a facility when voluntarily committed. Find out what constitutes due process in your state for those being committed involuntarily.

Ethical Responsibilities

Responsibility for the individual's welfare while the person is in your program is yours. The person views you as an authority. No matter how inexperienced you feel, when people work with you, they will see you as the person with all the answers. For this reason, you will have considerable influence over what your clients decide to do. It is important to keep their needs at the forefront of your planning and delivery of services.

Burdening Clients with Your Problems

Sometimes people in human services use clients to meet their own needs. You could, for instance, burden the person with your own problems. You might say things such as "Oh, that happened to me too" or "Wait until you hear what happened to me!" You might have had a bad day and want to talk to someone about it, and so you tell your client all about it, as Kent did with his client Lucy.

Meeting Your Needs

Do not ask the individual to do something that meets your needs or is not in the best interest of your client. Because of the influence you have with this person, it is easy to influence a client to do something that is beneficial to you. You might get the client involved with a friend of yours who sells insurance, or you might ask the client to go on television with you or to do an interview about the client's condition for the paper. There might be some payoff for you, but because the person could have considerable difficulty saying no to you, it is imperative that you never place your client in this situation. Often the media wants to interview a person with schizophrenia or recovering from alcoholism. Inform members of the media that they will have to locate their own interviewees.

Insisting on Your Solutions

You may have a need to look efficient, innovative, or particularly therapeutic, and so you might try to give your clients solutions to their problems. You may have had a similar problem at one time and feel there is only one good way to resolve it—the way you used to resolve it, a way your clients can use without experiencing the hard knocks you took figuring it out. You might be tempted to lecture, to discuss your situation and how it compares with theirs, or to warn your clients. None of these actions will help your people to grow by finding their own solutions.

Another way you can make people do what you want them to do is to treat them rudely if they fail to use your solutions or to move quickly enough toward a solution. Being rude is not the same thing as being firm. You can set limits, but it is inappropriate to treat people brusquely for not improving or for not taking what you suggest as healthy measures.

Exploiting Dependency

Clients are naturally vulnerable. They come to you at a time in their lives when they are hurt, upset, and disorganized—a time when it is easy to come to rely on another person. You are in a position to exploit this vulnerability by maintaining the individual in a dependent position long after such dependency is useful for that person. For example, you might enjoy having people call you about the details and decisions of their lives. It might make you feel important or needed. You might encourage them to lean on you for assistance in matters they could manage themselves. Be very careful not to allow this sort of relationship to develop.

In one support group run by a psychiatric nurse, individuals gathered once a week to discuss their problems. Most of the participants were also depressed. A student, Grace, from the local college joined the group and was an active participant for about 2 years. During that time, the nurse who led the group often went out to lunch with Grace and was extremely encouraging. It appeared in retrospect that the nurse had developed dual relationships with a number of group participants, eating with them, inviting them to her house, and going to plays and concerts with them. The nurse explained that this was her way of supporting her clients.

Grace completed her associate's degree and her bachelor's degree before she was accepted at a graduate school in another state. She told the group and the nurse in charge that she was no longer depressed and that she felt she was ready to move on now. She shared her good news about her acceptance to graduate school. Instead of showing pleasure and encouragement, the nurse became angry. She told Grace that she was trying to deny her need for the group and for the nurse. She ridiculed Grace's acceptance to graduate school, telling Grace she was not ready for such a large step and would surely fail. When Grace continued with her plans, the nurse stopped speaking to Grace and encouraged others to stop speaking to Grace as well. This is an example of a group leader, a worker, who could not tolerate the fact that her clients would not always need her. The group was meeting her needs, which she was clearly putting before the needs of her clients.

Protecting a Person's Self-Esteem

We can all agree that denigration of people, whether through verbal or physical abuse, is unethical, and in some cases illegal. You may believe that you are highly unlikely to encounter such behavior except in extremely unusual cases. Nevertheless, it is wise for you to understand that some workers are tempted to treat people this way. There are four reasons this is likely to happen.

Unpleasant People

Just as in any other walk of life, there are people in social services who are not pleasant people. They are unpleasant in many different aspects of their lives, and they are insensitive to the toll it takes on others, particularly clients who are uncertain of their self-worth.

Need for Power

There are workers whose own sense of self-worth seems uncertain to them. They choose fields where they will have a degree of power over others. In this way they seek to elevate themselves at the expense of people they can clearly believe are poor souls. These individuals make life difficult for clients simply because they can, because they have the power to do so.

For example, one student reported that while she was on a fieldwork assignment, she and the staff and clients were all having soft drinks together. A client approached the worker in charge and timidly asked if he could have more ice for his drink. The worker responded with, "No, you don't need any more ice. If your drink is warm, it is because you are so slow drinking it. Go back and join the others and drink up." After the client turned and walked away, the worker leaned over and helped himself to ice for his drink, laughing as if this was a joke.

Lack of Support for Workers

Social service workers who lack support from their supervisors or administration often lean together for support. They tend to develop a we–them attitude with regard to the clients, feeling the need to do whatever they must to support each other and hold clients apart. In one unit, adolescents were housed together after committing offenses. The least experienced staff came on in the evening, and that is when the teens would challenge the authority of the staff. Calls for help and requests for information and training on how to better handle the evening shift were ignored by the administration. Left alone with little support or knowledge, the staff resorted to coercion, often physical coercion, to manage the disruptive situation. In the end, the staff was blamed for using excessive force, but the lack of support and interest in these workers by supervisors and the administration contributed significantly to the way these adolescents were treated.

Isolated with Unpredictable Behaviors

Workers in group homes, partial programs, or evening residential programs often are left alone without support when clients are exhibiting unpredictable behavior. Fear and a need to control the behavior and the situation can lead workers to use verbal, and even physical, abuse. For instance, two workers, one a student in a fieldwork placement and the other out of social work school only a year, were in a group home for the mentally ill. One evening, after one of their clients clearly became manic, the student made a number of calls to supervisors. Supervisors responded irritably. It was their time off, they complained; the workers would have to figure it out for themselves. That's what they were being paid to do. A subsequent request to call in the crisis team was similarly denied. The two students spent the evening and all night with a client who was increasingly out of control with no supervision or support. Agencies that do not provide good support for less skilled workers are open to having workers band together against the clients in self-defense.

One problem new workers can encounter is finding themselves working for the first time in a place where they question the treatment of clients. If this happens to you, those in charge may tell you that this is the "real world" and that what you learned in school is impractical and does not apply. In these situations, the new workers clearly have entered situations that developed among the other workers long ago. Workers who encounter such abusive situations have a choice of either reporting the abuse or looking for work in a place where people are treated ethically, but they do not have a choice about their own behavior.

Ethically, you are charged with the care of the client. That includes the person's feelings of worth. Ethically, your behavior toward people should help enhance their view of themselves as worthy. Behaviors on the part of social service workers that subtract from a person's sense of self-worth are entirely unethical.

Stealing from Clients

It goes without saying that it is illegal and certainly unethical to take money or things belonging to clients. We have looked closely at how people lose their privacy. Gossip, giving information to strangers without a release form signed by the client, and releasing more information than is needed are all ways that people can lose their privacy or lose control of their information when they are being served by social service personnel. As previously noted, HIPAA laws outline the right to privacy and the protection of peoples' health information. However, workers can steal from individuals in other ways without even thinking about it. When you enter other people's lives and those people are not in a position to protect themselves, they are just as vulnerable as you are when you allow service or repair people into your home when you are not there to protect it.

There are two ways workers steal from clients. Both of these are theft and are entirely unethical:

1. Workers can steal a person's privacy.
2. Workers can steal a person's esteem and sense of worth.

Consider how people can be robbed of their self-esteem and self-worth. Vulnerable and unsure of themselves, perhaps feeling awkward and dismayed over needing to ask for help, people come for assistance with precious little self-confidence and self-esteem. What self-esteem they do have is needed for support in their struggle to regain their health or recover from bad habits. Workers have an opportunity at this point to reassure and encourage or to steal that sense of self-esteem and self-worth. It happens when people are denigrated, spoken to rudely or brusquely, called names, or ignored when they are present. It happens when people are made fun of, treated cruelly, shamed, and ridiculed or forced to perform actions they are incapable of performing at the time.

Let's look at some examples. Kimberly had a long-standing battle with schizophrenia. When her mother was diagnosed as terminally ill, she called a crisis hotline to talk about this pending loss. The day her mother died she called again and the worker replied, "Didn't we discuss this before?" When Kimberly said they had "but my mother died today," the worker went on, "well, do you have anything else to talk about because if you don't you are wasting my time."

Peter had been sober for 2 months when he began to drink again. He felt bad about it and fearful that he would go on a binge, so he sought out the worker at the detox unit assigned to him. When the worker finally took him into his office, he said to Peter, "So you couldn't stay off the bottle! What a loser. I guess you know that by now."

In an after school program for teens with behavior problems, Curt was telling his worker that he could not return to school until he had completed the program. "I have no time for you. Grow up and complete it," said the worker, and with that she walked out of the room. For the rest of the afternoon and evening she refused to acknowledge Curt, invite him to eat with the others, or respond when he approached her. She would look past him or turn to another client.

I am sure that as you read about these incidents you felt these were egregious examples of workers mistreating clients or patients, but in many settings rude and often unkind communication is used frequently, either because workers feel harried or because they see this as a way of motivating people or because it makes them feel important. For the truly professional worker, it means that you will decide consciously that you will never knowingly subtract from a person any sense of self-worth or self-esteem. If you can make this promise to yourself, you will be conscious of how even your most casual remarks can either steal something of value or enhance the health of the people you help.

Competence

A significant characteristic of professionals is their ability to clearly know their limitations. Ethical professionals do not try to do work for which they have not been trained. Recognizing the limits of one's training and experience is very important. This means that you will be aware of areas where you could use some help or direction and that you will seek assistance when you need assistance. You will ask those who have more experience or education to assist you rather than attempting to do work for which you are not qualified.

In addition, seek additional training throughout your career. Most certification and licensing programs require that individuals obtain further training on a yearly basis. Even if you are not part of such a program, you have an ethical responsibility to increase your skills, knowledge, and understanding of the field in which you work.

Responsibility to Your Colleagues and the Profession

We all have an ethical responsibility to protect our clients. Sometimes our clients need to be protected from those who are charged with their care. Nearly every professional code of ethics contains statements supporting the ethical responsibility of the professional to take action when a colleague is no longer able to function effectively or is openly violating ethical guidelines.

Impaired Workers

A social service worker is considered impaired when he or she is no longer functioning effectively due to substance abuse, mental illness, or personal problems.

In such cases impaired workers are so consumed with their problems that they are no longer able to focus on the needs of clients. In other words, they are distracted, focused on things other than their professional responsibilities, and often neglectful to the point of endangering clients.

If Someone with Whom You Work Becomes Impaired If you find yourself in a situation where a coworker appears to be impaired and therefore unable to be effective, you have an ethical obligation to take action. Generally, the first action to take is to talk privately with the person who seems to be having problems. Point out your concerns and listen to any explanations you receive. Explore ways to help the person resolve the problems.

Usually agencies have established procedures for handling concerns about colleagues who are thought to be impaired. Sometimes, if the person holds a professional license, the licensing board is notified so it can take appropriate steps to curtail the individual's opportunity to practice until the personal problems are resolved.

In one outpatient unit where clients received medications, it became obvious that one of the RNs was taking some medication for herself. At first the staff was not sure how to handle this. The RN was the supervisor. She did the pill count, but the workers noticed that clients ran out of medications sooner than expected with numerous seemingly reasonable explanations. The staff was torn between wanting to let someone know and fearing that they could be wrong. Finally, in a staff meeting one member remarked that she was concerned that the clients were so often out of medication and she wanted to better understand how that happened so the agency could take steps to correct it. When the RN became defensive and refused to participate in the discussion, the staff went with their concerns to the administration.

If you have concerns about how to proceed, it is useful to discuss your concerns with a senior professional. Here you may be able to clarify whether and to what extent clients are endangered by the behavior you have observed, and how the behavior indicates that your colleague is impaired.

If You Become Impaired

Ethically, you have a responsibility to refrain from activities that may lead to your own impairment. Should you become impaired for whatever reason, you have a further ethical responsibility to resolve your problems if they will interfere with your ability to practice. Practicing with clients if your physical, mental, or emotional problems will interfere to the detriment of the clients is unethical. It is important to have good self-awareness and be alert to the possibility that personal problems are interfering and having a negative impact in your work with clients. If this is the case, you have an ethical responsibility to seek help and to limit or cease your work with clients until your own problems are resolved.

Roy had had a drinking problem off and on all his adult life. He managed to hide it well enough to function in college and in his work as a case manager for many years. When his wife left, however, he began to drink more and missed work more consistently. A coworker noticed the problem and talked to Roy about getting help, but Roy brushed him off. Soon after, Roy began seeing one of the clients who also had a drinking problem and had come to the center for both her depression and her alcoholism. Roy kept this relationship secret, and the couple drank in bars that other case managers would not frequent. One night Roy and his girlfriend got into a fight at the bar where they had gone after dinner. The bartender asked them to leave; the fight moved to the street, Roy beat his girlfriend, and the police were called. Only when Roy ended up in jail, his career and marriage in a shambles, and his addiction out of control did he sober up enough to agree he needed help. Roy served his time, was terminated from his position as a case manager, and began outpatient treatment for his addiction. He is currently working as a night watchman for a furniture store.

It is often difficult to admit that we have problems, particularly when we work in a field where people expect us to have healthy answers to their issues. Nevertheless, problems are part of life and they always present opportunities for growth. Denying our own problems is not healthy and further impairs our ability to be useful social service workers in the future. Address your own problems as they occur as part of a lifelong pursuit of health and wisdom.

Ethical Violations

This chapter has put forward some of the common ethical standards and issues you will encounter, but it is not entirely comprehensive. You may encounter situations in which you have questions about what is ethical and what is not. Consult your code of ethics and talk to senior professionals about your concerns. Not all situations present clear-cut ethical options.

Sometimes you may have a colleague who is seemingly violating an ethical principle. A discussion with your colleague about your concerns is often the first step you might take, describing what you have observed and your concerns about your observations. If no satisfactory resolution results, you must then express your concerns to senior professionals or the administration in order to end the unethical behavior. Your agency will likely have a procedure for reporting unethical behavior; if so, that procedure should be followed.

Professional Responsibility

Finally, remember that you represent an agency and that it is your responsibility to establish a relationship with your clients and your community that is befitting of the agency. This will affect your relationship with clients in two ways.

First, know the parameters of your agency and operate within them. If you work for an agency that gives out food and fuel to the poor, do not attempt to do mental health counseling. If you are a case manager in a drug and alcohol unit, do not attempt to arrange foster care for one of your client's children except through the agency designated to handle that. If you work in a shelter for battered women, do not try to do drug rehabilitation. When a person needs services that fall outside the particular focus of your agency, make a referral to another agency that can best handle the problem.

The second way your relationship with the client is affected is related to dual relationships. Remain professional. Limit your contact to the focus of your agency and to the focus of that particular person's problems. Do not invite people home to dinner, take them home with you for the night, or become socially involved with them because you feel sorry for them.

Perhaps you are in the ER giving assistance to a woman who has been raped in her home. You are working for a rape crisis center. It is late. The woman you are interviewing is terrified of going home. You call crisis intervention to arrange for temporary lodging, but they are currently out of the office on a call and will have to "get back to you." The woman has tried unsuccessfully to reach two family members but has reached only their answering machines. Finally, in desperation, you take the woman home with you. You would rather do this than sit in the ER all night because you have to be at a meeting in the morning. The woman is educated and seems very pleasant and refined. She goes home with you, spends the night at your house, and returns with you to the agency in the morning, where they help her obtain temporary housing and see that she gets safely to work.

Two months later you begin to receive calls from the woman, who seems to want a friendship with you. She has found your number in the phone book. Soon after this, you receive a call in the middle of the night. It is the same client. She has had a fight with her boyfriend, and now she wants to stay with you. You tell her she cannot do that, and she becomes hysterical. In the next several weeks, she appears at your house several times, asking to stay with you. Always there is some reason she cannot stay where she is currently living. She knows your phone number, so she calls frequently. You have to be very firm in order to set limits; sometimes it is hard to do.

This story about a worker taking someone home is not all that unusual; it does happen. Rarely, however, is a person who is hurting and vulnerable able to reestablish a professional relationship with such a worker, complete with boundaries and limits, once the worker has extended this kind of friendship or kindness.

Summary

This chapter is particularly important because it involves your ethical obligation to the client and outlines some legal concepts you must follow to protect your client and yourself. The primary issue is always the welfare of the client. That must come before all other considerations. When we choose this line of work, we deliberately choose to work with vulnerable people who cannot be expected to protect themselves or to know their rights. It becomes our responsibility to see that clients are well protected and are treated or given service under the highest ethical standards.

Common codes of ethics, including the *Code of Ethics* of the National Association of Social Workers (NASW) and the *Ethical Standards of Human Service Professionals*, can be found in *Codes of Ethics for the Helping Professions* (Brooks/Cole, 2004).

Now that you are thoroughly familiar with your ethical and legal responsibilities, it is time to turn to case management as a basic area of practice in which ethical behavior is expected and informs your decisions.

Video Examples

To view the videos that accompany this book, go to CengageBrain.com.

- "An Ethical Issue"—Because ethical decisions are often not clear or vary from one situation to another, it is a good idea to discuss your thoughts and concerns about how to proceed with another person. Another person is often able to sort out things from a different perspective. In "An Ethical Issue," two case managers are discussing whether or not to break confidentiality. They look over the good reasons to do so and reasons breaking confidentiality might not be appropriate.

Exercises

These exercises can also be filled out online at CengageBrain.com.

Exercises I: Ethics

Instructions: The hypothetical practice situations that follow are designed to stimulate thinking and discussion on the issue of confidentiality. Each situation is followed by a multiple-choice list of possible responses you might make. Choose the responses that you consider the best. In some cases you might want to use more than one of the responses listed. Others may choose a different answer. Discuss with your fellow students the different possibilities and what might present the best outcome for the client.

1. Paula is a 17-year-old client in the daytime partial hospitalization program. Her mother phoned and requested to know Paula's psychiatric diagnosis so that she could inform the family's physician who is treating Paula for diabetes. You should:

    a. Advise the mother of the diagnosis and the name of the psychiatrist who made the diagnosis.
    b. Call the family physician directly and advise him of the diagnosis.
    c. Ask Paula to sign a release of information form giving consent for the physician and the mother to be advised of her diagnosis.
    d. Refuse to release the information at all.

2. Kelly requests a copy of his current treatment plan. You should:

    a. Have Kelly put the request in writing and discuss the issue with the treatment team.
    b. Make a copy of the current treatment plan and give it to Kelly.
    c. Discuss the treatment plan with Kelly and then see that he has a copy.
    d. Refer Kelly to the attending psychiatric physician.

3. A 13-year-old boy requests that his school counselor be sent a copy of his initial interview and discharge summary. He signs a release of information form, documenting his written consent for the information to be transmitted. You should:

    a. Forward the material to the school counselor.
    b. Give the information to the boy who can deliver it to the school counselor.

c. Have the medical records department forward the information to the school counselor.
d. Refuse to release the information until a parent cosigns the release of information form.

4. In the case above you would

a. Release the complete chart
b. Release just the discharge summary
c. Release only those portions of the discharge summary that the school needs for their work with the boy and not release the initial intake summary as it contains information that is no longer relevant but taken out of context could be damaging to the boy's family
d. Meet with the parents to go over what you have decided to release as outlined in C before anything is released to the school.

5. Mary Smith is a depressed elderly woman who was admitted to Polyclinic Hospital due to severe back pain. She was advised she might need surgery to correct the problem. You are her case manager at the Office of Aging, and she calls to say she is considering suicide. The constant back pain has made her feel like "just giving up." Mary is currently at home, awaiting a surgery date. You know Mary has a supply of pain pills, and she says she wants to take all the pills. You feel there is a substantial risk that Mary might follow through on her threat. You should:

a. Contact the Polyclinic orthopedic staff who are currently seeing Mary in the outpatient clinic.
b. Maintain frequent contact with Mary, but respect her wishes to keep her suicide plans confidential.
c. After discussing with Mary what you are about to do, contact crisis intervention.
d. Advise the city police department of Mary's suicide plans.

6. Bill Jones is a client who has been in alcohol treatment programs at your facility. He is currently depressed about his pending divorce and present marital separation. He has signed a release of information form for you to share information with his priest, who is counseling Bill about his religious conflicts regarding the divorce. A man calls you claiming to be Bill's priest and requesting information on Bill's current state of mind. You have never actually spoken with Bill's priest, and you think this might actually be Bill's wife's attorney calling. You should:

a. Give no information on the phone until you have verified the identity of the caller.
b. Refer the caller to Bill, but send the information to the person at the address on the release form Bill signed.
c. Insist upon meeting with the priest in person.
d. Share no information with the caller and contact Bill
e. Get the person's number and call him back.

7. Patty is completing a student internship for her associate's degree in the therapeutic activities program. She asks to review the medical records of the people who were just in her projects group. You are supervising Patty. You should:

   a. Advise Patty that the records are confidential and may not be inspected by students.
   b. Make certain that Patty is well trained in the policies and procedures relating to confidential information, and only then allow her access to the medical records.
   c. Permit Patty free access to the records because she is like part of the staff.
   d. Obtain written consent from each client for Patty to review the records.

8. Jerry was a client who improved and was discharged 2 years ago. You receive a call from the National Can Company. The caller explains that Jerry has applied for a job and that the company would like to hire him. Jerry told them he was in treatment 2 years ago and was discharged after showing considerable improvement. The company wishes to confirm the fact that Jerry did indeed complete the program as he claims. You say to the caller:

   a. "I don't know if Jerry was ever a client of our agency. If you send me a release-of-information form I can look into that and get back to you."
   b. "Jerry was a patient here but I am not at liberty to say any more than that without a release-of-information form."
   c. "I can tell you that Jerry was a client here around 2012 and he successfully completed the program with us. I will need a signed release-of-information form from Jerry to put in his file."
   d. "I don't know what you are talking about. Good-bye."

9. Clark is currently enrolled in treatment, and you are his case manager. He asks you if he may read his medical record. You should:

   a. Ask Clark to put the request in writing, and assist Clark in completing the written request if he seems to have limited skills in reading and writing.
   b. Present Clark's request to Clark's treatment team.
   c. If the treatment team concludes that it will not harm Clark to review his record, allow Clark to read it in the presence of a therapist (after deleting information from sources that asked to remain anonymous).
   d. Decide with the treatment team who will assist Clark in reading and understanding his record. Then follow through by allowing Clark to review his record with that person.

Exercises II: Ethically, What Went Wrong?

Instructions: The following hypothetical practice situations are designed to help you apply what you have learned in this chapter. For each situation, decide what was done in the situation that was unethical.

1. Jennifer had a long day and was trying to get out of the office before 5:00 P.M. She had one more person to see. Dr. Adams had asked Jennifer to give Abdul, a young man recently diagnosed with schizophrenia, a prescription for a new medication. Jennifer had her coat on when she handed the prescription to Abdul in the waiting room. Abdul wanted to know what the medication was and why his prescription was being changed. "Will there be any side effects?" he asked Jennifer. She replied hurriedly, "Oh, no. Dr. Adams says just take this until he sees you next time."

2. Carl is uncomfortable around gay men. Bert, his client, is gay and has just broken up with his lover. Bert, who is 42 years old, had been in a long-term relationship and is devastated and in tears in Carl's office. Because Bert has suffered from severe depression in the past, Carl attempts to have him evaluated by the therapist this afternoon. In the meantime, Bert is still weeping and now threatening to take his life. Carl is particularly uncomfortable with this man's tears and believes this is drama. Carl says, "Oh, c'mon now. Let's get a grip. You can't sit in here all afternoon carrying on. Here, take some tissue and go out in the waiting room until Dr. Paul can see you."

3. Elizabeth visited the home of an elderly man and got him to sign a release of information form so she could process an application to the county nursing home. In the man's records were references to the fact that many years ago as a teenager he was convicted of shooting a man in a bar fight, a crime for which he served 2 years in prison. She knows the people at the home will be titillated over this little tidbit of information, especially her friend Rhoda, who does the intakes. Even though she knows this is not part of the home's evaluation, that the client has led an exemplary life since that time, and that the nursing home staff might take it out of context, she releases the information anyway, based on her client's signature on the release form. She and Rhoda have a good laugh about it the next day.

4. Jim is doing an intake with a man who claims he is depressed. He tells Jim that ever since his wife left he has had trouble concentrating and waking up in the morning. He talks about how lonely it is at home, how much he misses his children, how he is tempted to drink in the evenings, and how little he has to look forward to. Jim nods. He understands. "Yes, my wife left last month too," Jim tells the man. "I know just what you mean. I get to feeling like, well, like there isn't as much meaning. I never knew the kids were so important to me, but I guess they were. On Saturdays, I used to do things with my son and I still get him every other weekend, but it's not quite

the same thing, is it?" "No," the man responds, "I was thinking . . ." Jim interrupts the man to say, "Well, I do a lot of thinking too. I think about what I could have done differently and if it was my fault. Don't you think these women would see that it's hard, too hard I think, to raise kids alone?" The conversation continues in this vein until the end of the interview.

5. Carmen is supposed to see her small caseload of persistently mentally ill individuals at least twice a week. Lately, with school and her mother's death, she has not really seen her clients that often. She has checked in with them on the phone, but she also has used time when she was out seeing clients to do errands at the library and to empty her mother's home. Now one of her clients is in court after committing a crime. The client and the lawyer agree that the client might be able to use his mental health status as a reason for committing the crime, and they ask to introduce the case record as evidence in the court proceedings. Fearing that it will be discovered how little supervision and attention she has given her client, and knowing that ultimately she could be blamed for the fact that her client committed the crime while under her somewhat irresponsible care, Carmen invokes the concept of privileged communication to avoid having to give the file to the court.

6. Ted is in a clinic with his elderly client, Gretchen, for a routine blood workup, which they do every other month. He notices Gretchen is bruised on the face and arms. For a while he makes small talk with her, and then he asks her about the bruises. She is somewhat evasive but indicates, "They weren't the result of no fall!" Without explicitly blaming her daughter and son-in-law, with whom she lives, Gretchen makes it quite clear that the bruises are not the result of an accident. After the blood test, during which neither the doctor, who sees her briefly, nor the technician make any mention of the bruises, Ted takes Gretchen home. He toys with the idea of reporting the bruises to protective services at the county Office of Aging but decides not to. He bases his decision on the fact that the law does not specifically require him to do so, that it would be hard and take a lot of time to have to place Gretchen in another living arrangement, and that the daughter seems like a very nice person whom Ted does not feel like stirring up over an uncomfortable situation.

7. Kitty has a whole list of things to do today and doubts she can get it all done. She hates the way there are always things left to do at the end of the day. It just seems that no matter how hard she works, something new comes up that she cannot complete. One of her clients, Isabel, has told her on the phone that

she wants to sign a release of information form for her lawyer. Kitty has the form ready for the time when Isabel will be coming in at the end of the week. Today a man calls and says he is Isabel's lawyer and he needs just two dates to help him file a brief with the court on the Isabel's behalf. Kitty gives him the two dates and hurries to the next thing on her list.

8. While having lunch in the staff room, Jorge is obviously mad. He spent one morning taking a meticulous social history from a new client. The client, a man in his 20s, was pleasant and helpful. He seemed to genuinely want the assistance of the agency and to like Jorge. Two more interviews followed to set up services, and the man signed a release of information form for Jorge to meet with the client's physician. Jorge cannot understand why this man never mentioned the fact that he is HIV+. This Jorge found out in the conference with the man's physician some weeks later. "How do these people think I am going to help them if they don't tell the whole story?" Jorge fumed. "They come in here and want my help and then withhold information from me. They leave me in the dark. I don't know what's going on, and then they think I'm going to be able to help them."

9. A new worker, Jill, is working at a large residential facility for the mentally ill and has been assigned four clients for whom she is to develop goals and objectives to help these clients move forward to greater independence. She meets with the first two clients and then confides to a worker who has been there longer that she had trouble understanding what the clients wanted to work on. The worker tells Jill, "Just make up the plans. These people are a waste of time. They won't ever get any better. Look at that one. This is his fourth trip through here. No one ever made a difference with a plan, and you won't either. Just put something down to satisfy the insurance company, and come in here with us. There is a good movie on TV tonight, and the staff is going to put the residents to bed early and get together in the patient lounge to watch it."

10. Beatrice, who has suffered from schizophrenia for most of her life, has been placed in a long-term residential facility. One night the worker decides to take the residents to a movie. The residents all get in the van to go to the movies, and the worker waits to leave until everyone has a seat and has fastened their seat belts. Beatrice finds a seat but complains that the seat belt does not fit, that she cannot fasten it around herself. The worker replies, "Well, if you didn't eat so much, you wouldn't be so fat. You always pig out at the table, and this is what you get. I guess you're too fat to go to the movies tonight, Beatrice. Guess you'll have to just stay home."

11. Pedro noticed that his colleague, Antoine, was using clients' spending money to make small purchases for himself. Each resident in the group home was given a specific amount of spending money every month, and it was kept in the resident's envelope. When money was spent from the envelope, a receipt was to be left in the envelope showing where the money went. Antoine was taking money for small purchases for himself—lunch, movie tickets, a gold chain. He was placing the receipts for these purchases in the residents' envelopes. It was not possible for administration, when doing an audit of all the residents' accounts at the end of the month, to determine from the receipt who actually benefited from the expenditure. Pedro thought about telling the administration but felt it was likely that Antoine would deny the allegations, and this would ruin their working relationship. Therefore, Pedro did nothing.

12. Marcella began to drink in the evenings after work when her husband left her for another woman. As the months went by, the divorce became increasingly acrimonious. There were accusations, attempts to take Marcella's money, and attempts to deprive her of custody of the children. The children began to exhibit problems, and there were financial problems as well. Drinking in the evenings expanded to a drink with lunch and later to a drink and then several drinks in mid-afternoon. In time, Marcella could not face the day without alcohol when she first got up. She continued to report for work where she was the sole worker on the day shift in a small residential setting with four clients. Marcella began to ignore the residents. It started with naps in the afternoon, which left the people unsupervised. Later, Marcella found it too hard to fix dinner for her residents and began to allow them to eat junk food for dinner. As the situation deteriorated, Marcella became more and more mired in self-pity, anger at her ex-husband, and alcohol abuse. She continued to work at the group home.

13. Arnie has problems with substance abuse. He considers himself an "alcoholic who likes a little cocaine now and then." He is funny, articulate, and clever. When he comes to the case management unit he seems open about his progress and regressions. He always asks how his case manager is doing, what she did for Christmas, how her little boy is doing. Sometimes he brings in the paper and leaves it for her to read, saying he has read it and is finished with it. On Friday evening some of the case managers go out to dinner at a place that serves alcohol and very good food. They are having a good time unwinding after work when Arnie joins them. It appears that he is drinking a soft drink, but no one knows that for sure. Arnie and the case managers laugh and talk about their work until late in the evening. Arnie is funny and has hilarious insights about some of the clients he has encountered in the waiting room.

Exercises III: Decide on the Best Course of Action

Instructions: Sit with a small group of other students and decide how you will handle this situation. There are many areas both ethically and legally that are not clear, so the discussion you have with your colleagues is much like a discussion you might have in a real agency. There are no "correct answers."

You have been working with a client who is HIV+ and is a regular user of heroin. He needs both medical and substance abuse treatment. However, he is inconsistent about coming for regular treatment and medical care. You suspect he is not taking medications prescribed for him. In addition, he is sexually active with several women. He has asked that you not contact him at his home where he lives with one of his girlfriends. He has stipulated that no family members may have any information about him. You think that if you could commit him to a substance abuse treatment facility he will be out of circulation sexually and he will receive the treatments he needs to save his life. You do not know for sure where he is but know his girlfriend with whom he has been living probably does. Can you contact her? What ethical and legal principles are at play here? What do you decide to do or not do?

Exercises IV: What is Wrong Here?

Instructions: Sit with a small group of other students and decide where there was a lapse in good judgment or a lapse in good ethical behavior. Where are the gray areas? What makes these situations clear-cut or unclear about the proper course of action?

1. A children's case manager writes a letter to a parent of a 14-year-old girl, currently a client at the case management unit. In the letter is this: "While it has not been confirmed or established that your daughter was sexually abused, it is my opinion that she has experienced sexual abuse in the past." What did this case manager do wrong? What could be the consequences to the family members if such a letter was received?

2. A human service worker is asked to plan a recreational activity for elderly residents on the floor where she works. She is not sure what to arrange but finally decides to do a program on supplemental Medicare insurance. She invites a friend who sells supplemental Medicare insurance. The friend gives a complete explanation, handing out brochures on her business and her business card and explaining the insurance products her company can offer individuals on Medicare. What is the issue here?

CHAPTER 3

Applying the Ecological Model: A Theoretical Foundation for Human Services

Introduction

In working with other people, human service professionals apply the *ecological model* to develop a broad understanding of the individuals who seek help. This model, sometimes referred to as *person-in-situation* or *person-in-environment*, looks at the individual person in the person's **context**.

Ecology is a term that comes from biology and refers to organisms and their reciprocal relationships within their **environment**. In other words, biologists look at the organism and focus on the interactions of that organism with the surrounding environment. This clarifies and defines for biologists the organism being studied. Biologists see interactions among various organisms in an environment as keeping things in balance in that particular setting.

In seeking to find a model for understanding people within their context or within their environment, social work and other behavioral health professions began to use the lessons of biology to formulate a clearer picture of their clients. The ecological model employed in biology was found to be a useful way to better understand individuals, families, and other groups seeking assistance. Adopting this model in the 1970s, social service professionals have refined and improved the model so that today this has become a practical **assessment** tool for use in the social services. The idea is that each interaction affects in some way the people involved in the interaction and the way these interactions affect people determines how they will respond. One of the best ways to understand this is the way Greif and Lynch (1983, p. 38) describe what happens.

> "(A)s a person enters each new situation he or she usually adapts to its demands and, by his or her presence, changes the situation at least structurally. A person is constantly creating, restructuring and adapting to the environment even as the environment affects the person."

For example, an irritable older person living with her harried daughter who works full-time and is raising two children alone will be treated differently than an irritable older person living with her retired daughter and son-in-law who can afford to hire people to help them with their mother's care. In the first example, the daughter is likely to become short-tempered with her mother. The children are likely to come first. Mother may conclude that her daughter doesn't care about her. Mother's hurt feelings are expressed by withdrawing from the family and expressing some resentment toward the children. The children, for their part, believe their grandmother doesn't care about them because she rarely interacts with them and seems to resent them. Daughter herself is overworked and expects her mother to do things for herself the mother may no longer be able to accomplish. When this doesn't happen daughter becomes irritated with her mother because she can't afford to hire anyone to help her with her mother's needs. In this situation everyone is exasperated with everyone else. We can see that the interactions each person has with every other person in this family seem to contribute to the exasperation and misunderstandings.

There are a number of ways a case manager might intervene to modify the environment to make the family system work better. For instance, you might find in-home assistance paid for with Medicare for the older woman. You might begin to encourage the daughter to take her mother to school functions and you would look for other opportunities for the children to have more positive interactions with their grandmother. You might make a referral to a therapist for one or two sessions just to explore how they might all understand and work together better as a family. However, if you just hear from the harried and resentful mother or you only see the disappointed and somewhat depressed grandmother you would not see the whole picture and you could not entirely address the problem.

In the second example the interactions will be different. The daughter and son-in-law share the responsibility and have help when they need a respite. They are not working and do not feel harried or overburdened.

If we look at the two daughters, there may be further differences in what each of them brings to their situations. Perhaps the first daughter who already feels overburdened never had a good relationship with her mother or blames her mother for some of her current troubles. How differently she will respond to her mother when compared to the second daughter who may have had a good relationship with her mother, sees her mother's irritableness as simply a part of her not feeling well, and tends to humor her while making every attempt to make her mother comfortable.

What we learn from using the ecological model is that people do not just show up with a collection of problems. They live and work in an environment or a context as a unique person and that context can have positive or negative effects on their actions and those actions can affect others within the context positively or negatively.

The Three Levels of the Ecological Model

The Ecological Model, Figure 3.1, looks at the person and the person's context on three specific levels:

1. ***Micro level,*** where the focus is on the client's personality, motivation, affect, and other personal attributes and how these affect the way he or she interacts with others, how others are perceived, and how these characteristics contribute to well-being or instability.
2. ***Meso level,*** where the focus is on the context immediately surrounding the person (family, church group, close friends, and work group) and how this contributes to problems or provides support and solutions. Here we are seeking to know how the immediate environment enhances or impedes successful functioning.
3. ***Macro level,*** where the focus is on the larger society's characteristics and the way the person experiences these or the way these are brought to bear on the person's situation. Here we are looking at institutions and organizations within the broader society, such as the political system, social stratification, racism, the educational system, and the economy.

Human service workers are expected to be aware of all three levels when assessing a person's situation and to be able to intervene on all three levels when such intervention is appropriate.

FIGURE 3.1 The Three Levels of the Ecological Model

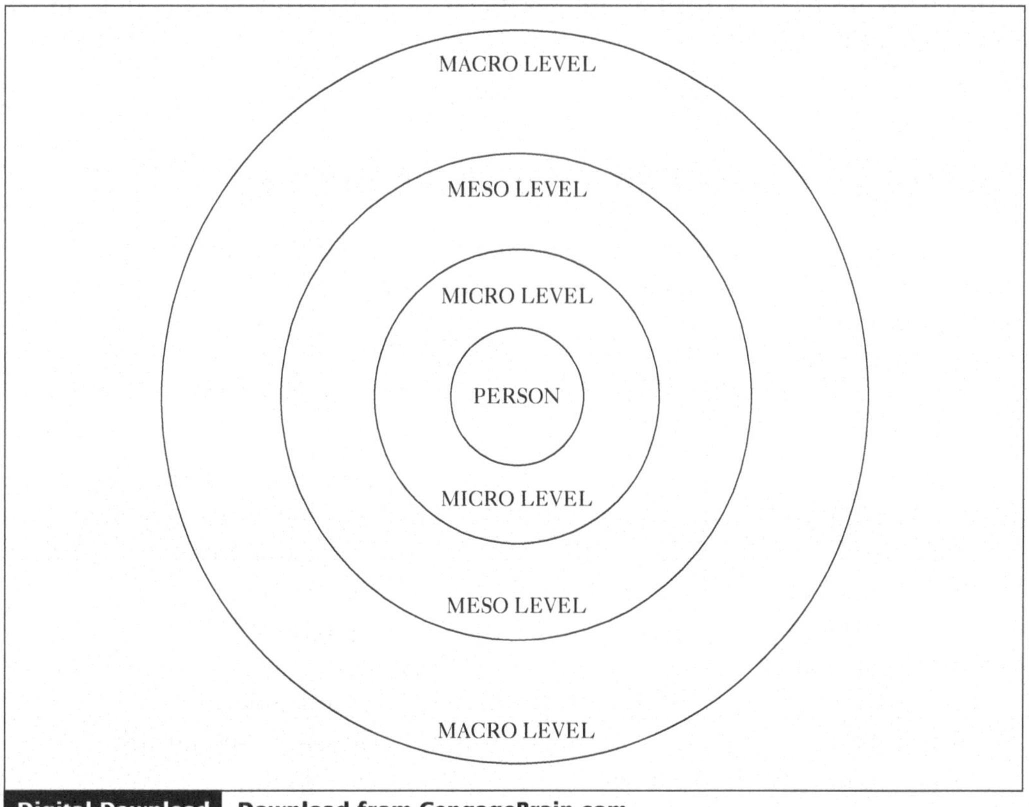

© Cengage Learning®

Digital Download **Download from CengageBrain.com**

The Micro Level: Looking at What the Person Brings

When you do an assessment to open a case or do follow-up planning, a number of individual characteristics will impinge on the problems and the eventual outcome.

These micro-level characteristics can be divided into two broad categories:

Biological Characteristics	Psychological Characteristics
Neurological development	Early shaping experiences
Reflexes	Perception
Genetic makeup	Personality
Degenerative processes	Affect
Illness (chronic, terminal, or temporary)	Cognition
Physical health	Nurturance
Nutrition	Life transitions/current position in the life cycle
	Motivation

This general outline gives you a starting point for understanding what your client has brought to the situation.

People have had different early life experiences, are composed of different genetic configurations, and possess different personalities and perceptions. Each of these differences interacts with the external circumstances of the person's situation to promote self-fulfillment and well-being, to block those goals, or, quite possibly, to have no effect at all.

Looking at What the Context Brings

People function in contexts that are personal to them and they function in a larger social context, which is the larger society. These contexts can be divided into two broad categories:

Mezzo/Meso Level or Personal Context	Macro Level or Social Context
Family	The larger culture of the society
Work group	The larger organization of the church or workplace
Social groups	
Family culture	The larger community
Family values	Government
Family structure	Economy
Religious group	Social stratification
Social class	Prejudice and discrimination
Role status, conflict, and strain	Political system

It is important to obtain information about the contexts in which people grew up and in which they are now functioning. The reason for the individual's problems may lie in the context rather than with the person.

When you learn about the context, you learn more about what motivates your clients, the environmental cues they receive to behave or make decisions the way they do, and what early circumstances shaped their way of responding to their community and their situation. People come from different social contexts. People grew up in different households with different parents and different levels of nutrition and encouragement. People have different ways of looking at things and explaining them. The environment in which they function may place pressure on them to see things in a certain way or to believe certain things about their environments and these may create problems for them in the larger society. The economy may have favored their work or begun to dispense with it. The political system may have awarded your client's subgroup power or disenfranchised that group in some way. Your client may have experienced prejudice, an indifferent medical system, or a poor educational system. On the other hand, this person may have grown up in a wealthy suburb, attended private schools, and received the best medical care money could purchase.

Why Context Is Important

You are well aware of how distorted communication can become when a statement is quoted out of context. It is possible to skew impressions and deliberately create misunderstandings by quoting only a portion of what someone has said and not the entire conversation.

In a class discussion centered around self-exploration and self-awareness, for instance, Matt's teacher questioned him about his homework assignment when he said he was not sure if he could help people with a certain disability. He went on to explain that at one time it was thought that he had that particular disability, and he had worked very hard to prove that he was not disabled and to overcome people's initial impressions of him. Now he found himself feeling uncomfortable with people who suffered from that disability. He also explained that he expected his course of study to help him overcome the problem and that he was very aware that his reactions might be inappropriate.

Later Anne, who was in class that day, confided to Aisha and Alice that she did not feel that Matt should be allowed to continue with his studies. Surprised, Aisha asked Anne why she felt that way. "Oh, because he said in class the other day that he feels uncomfortable around certain disabled people. I mean, if you can't work with disabled people, you need to find something else to do." Aisha and Alice quickly agreed.

Anne's description of what Matt had said was distorted because only a portion of Matt's comments were repeated outside the context in which they were said. She did not include the fact that students were having a frank discussion to better understand themselves. That same kind of distortion takes place when we look at individuals out of context. In your work, every person you will see functions in a context, an environment. You cannot adequately understand that person without also being able to understand the context in which that person functions and interacts.

Urie Bronfenner (1979) described the mesosystem this way: "(A) mesosystem comprises interrelations among two or more settings in which the developing person actively participates" (p. 25). In other words, you are looking at where people function on a daily basis, the groups with which they interact daily and how those affect or do not affect their reasons for seeking help.

It is very tempting to overlook context. Many of us fall into the trap of thinking that *A* causes *B*: "Juan is irresponsible, so he lost his job." If we eliminate *A*, *B* will cease to be a problem. If we make Juan more responsible, he will not lose any more jobs. Or "Jane is too demanding, so her husband left her." If we teach Jane better interpersonal skills, she will have better relationships. Although helping Juan to become more job-ready and helping Jane to communicate better may very well be a positive part of your plan for them, this kind of understanding of their problems and assigning of solutions largely ignores the context in which these problems arose. It also makes it much easier to see the individuals as being entirely responsible or to blame for the problems they have brought to your attention. When we blame others, we nearly always feel less empathy with their difficulties, and we are less inclined to be truly useful in the human service sense.

However, supposing that we send Juan to job readiness classes and send him out newly prepared to get a job and we find that the work he does, basic plumbing, has limited opportunities and those who do hire discriminate against recent immigrants. In fact, Juan did not show up at his last job after being harassed by his fellow employees and given only the most menial tasks. The employer complained that Juan was irresponsible but the employer's refusal to stop the harassment and his occasional participation in it made Juan's seeming irresponsibility somewhat understandable. Juan was responding to the prejudice exhibited toward him. There are other environmental considerations. Suppose Juan comes from a family that does not value work and discourages him from being consistent on the job, finding other things they want him to do for them instead. Maybe the plumbing company really had very little work for Juan and he drifted away when he had time on his hands and became bored. The job complained because he wasn't there when they needed him but they didn't really need him very often. If we fail to look at the context, we fail to see the whole picture. Juan may be more responsible after his training but still facing the same issues in his environment he faced before. In working out his plan, we would want to see if there are ways to address these.

Seeking a Balanced View of the Client

All individuals constantly interact with any number of systems in their environments. All individuals bring to those interactions unique characteristics. Unless the human service worker has a balanced view of both the person and the person's context, important information and constructive opportunities are lost. In the cases of Ralph and Eduardo, we can see how important this is.

Ralph went to prison because of some youthful gang activity. While he was there, he took advantage of every opportunity to change. He went to church regularly, developed a personal relationship with a minister who came to the prison often, and obtained his high school diploma. Ralph was a warm, humorous person who attracted many friends. His outgoing personality attracted people to him who ultimately encouraged him and gave him support. During his time in prison, his mother wrote to him often, pleading with him to change his ways. Ralph felt bad about the trouble he had caused his mother, particularly in view of the fact that she had raised him after his father left home, and he saw her letters as a reason to do better. When he left prison, he enrolled in college courses and attached himself to the church, where he was warmly welcomed.

Eduardo, too, was in the same prison because of youthful gang activities. He was quiet and retiring and did not attract the attention and support that Ralph had secured for himself. Eduardo attempted to get his high school diploma while in prison, but he had trouble asking for academic help when he needed it and eventually abandoned the project in frustration. Preferring not to join groups, he did not go to church or any other group activity that promoted independence and responsibility. Because Eduardo spoke so little and rarely smiled, he was often misunderstood and thought of as being hostile. In fact, he felt shy and awkward around other people. Eduardo's mother wrote to him regularly, and she too pleaded with him to do better and "turn his life around," but Eduardo tended to see these letters as nagging and to blame his mother for the fact that his father left when he was very young. He rarely answered her mail. When Eduardo left prison, he moved back with his old friends and resumed his former criminal activities.

This illustration demonstrates how individual characteristics, that is, the micro level of the ecological model, play a role in the outcome for the client. Part of developing a balanced understanding of the client is being able to see what the person brings to the situation and how that interacts with the larger context of the client's life. Ralph brought a personality that attracted others to assist him. He brought a good relationship with his mother and a motivation to do things more constructively. Eduardo brought a more retiring personality, one that was less attractive to others and often misunderstood. Eduardo's interpersonal skills were not as developed as Ralph's. The individual characteristics of Eduardo and Ralph affected the outcome of their prison time.

Now we will look at Eduardo and Ralph differently. We will look at the meso level of the ecological model, the level that immediately surrounds the person. For our purposes, let us suppose that Ralph and Eduardo are both warm, humorous people. Both make friends easily and enjoy the company of other people. Each of them is sent to prison for youthful gang activities, but the context is different. Eduardo goes to a recently built prison upstate that focuses on rehabilitation. There he is provided with high school and college classes as well as religious and self-improvement activities. He is able to take advantage of many different programs to further his goals. A supportive counselor meets with him on a weekly basis and works with him to create a good set of goals and implement them. The location of the prison has another advantage. Eduardo is now closer to his father, who lives only a few miles from the prison. His father begins to visit, offering his support and a place for Eduardo to live when his sentence is completed. Eduardo leaves the prison on a solid footing and continues his work toward a college degree.

Ralph, on the other hand, is sent to an ordinary prison where the counseling staff is overwhelmed. His counselor sees Ralph's potential but has difficulty enrolling Ralph in high school courses because they are crowded. During the time Ralph is at the prison, the education staff experiences a number of turnovers and layoffs. Ralph never can get into the program and stick with it. He rarely sees his counselor because of the number of inmates with whom the counselor must work. No family member comes to visit Ralph, partly because he has been sent so far from where they live, and partly because they blame him for his incarceration and have lost interest in him. Ralph's mother, sick with severe chronic asthma, rarely writes. Ralph attends church services at the prison regularly, but the prison does not allow inmates to meet with the pastors before or after services because of a strict schedule. The pastors who have formed relationships with some inmates visit irregularly at other times. When Ralph leaves the prison, he has not completed his high school diploma. He moves near some people he knew in prison, and soon he takes up the criminal activities in which he participated before his incarceration.

Here it is the context on the meso level that is different. Eduardo finds himself in a supportive context: a counselor who focuses on his goals and sees that these are implemented, plenty of self-improvement opportunities, a warm relationship with his father, and a prison committed to education. Ralph, however, finds himself confronted with indifference, lack of supportive programs and activities, an overwhelmed counselor, and a family too distant to give encouragement.

The interaction never ceases. The individual makes choices, but the environment prompts those choices. The individual responds to the outcome of those choices, and the environment reacts or adapts to that response. This interaction begins at birth. A fussy baby with calm, patient parents will start life differently from a fussy baby with overworked, anxious parents. An infant with severe disabilities will receive a good start with a large, loving family who has the money to devote their time and energy to getting her the best medical and rehabilitative care. An infant with severe disabilities may arrive in another family where everyone tries their best to give the infant a good start; nevertheless, the disabilities prove overwhelming to the caretakers, there is no cure, and family members find that any semblance of a normal home life or time with other siblings is severely curtailed. The first baby is raised at home; the second one is placed in a good institution.

In human services, the trained eye will look for and see a balanced view of clients and their contexts when assessing individuals' needs. Just as important is the human service worker's understanding of how person and context interact to produce certain outcomes for the client.

This topic is discussed in greater detail in Chapter 16 on assessment; this discussion puts forth the framework you will follow in assessing the individual's contribution to the situation and their current context.

We will turn here to Jane, the person who seemed too demanding so her husband left her. Here is what Jane brought to her difficult relationship with her husband that ultimately led them to separate. Jane was sharp witted but often biting and critical (personality). She expected perfection from both herself and her husband and never failed to point out his mistakes. She had a somewhat flattened affect, never seeming to be happy or engaged (affect). Jill suffered from chronic asthma (chronic illness). She brought memories of having been sexually abused by her two uncles at around age 11 (early shaping experiences). Her way of relating to men was characterized by cynicism and sometimes denigration. They were never good enough. She perceived men as being more interested in sex than in her (perceptions). She was highly motivated to get ahead in her career and her long hours of dedicated work protected her from having to be in close proximity to her husband for long periods. In other words, she used work to retreat and to protect herself (motivations). Here we see some of the personal characteristics Jane introduced into her situations. This is the micro-level understanding of Jane.

Turning once again to Jane, let's look at her personal context or mezzo level. Here we would be looking at the groups or the environment immediately surrounding her. She grew up in a family that valued work and education. They were never entirely aware of the extent of the uncle's sexual abuse of Jane, although several times Jane's mother warned the uncles to stay away from Jane. The family was undemonstrative and seemed to be always busy. They expected that of Jane. She was often chastised for minor imperfections and worked constantly as a child to be perfect and gain the admiration of her parents. Consequently, Jill married a person who was well educated and had a demanding job. Between the two of them it was hard for them to find time to be together. Jane's husband, however, worked to make time, while Jane sought to avoid too much time together. Jane, mirroring the behavior of her family of origin, was extremely critical and demanding of her husband. Here we have looked briefly at Jane's family of origin and the family she created with her husband.

Jane's work group was equally striving and worked long hours with her. In many ways they supported each other and had more contact with each other than they did with their spouses. Further, Jane was rewarded with raises, outstanding evaluations, and public acclaimations for her work. These rewards made work attractive to Jane.

However, Jane experienced role conflict, frequently feeling guilty about not being an attentive wife when she was at work and feeling anxious about her work when she was at home.

Jane and her husband came from a middle-class background that valued education, hard work, and independence. They lived in a comfortable home in a suburb with two cars, friendly neighbors, and wide grassy lawns. Jill did participate in the annual block party each summer.

This is the personal context within which Jane functioned on a daily basis. What about the social context or macro level? Here we are looking at the society at large. Jane and her husband were solidly middleclass and lived that lifestyle. In addition, they were Caucasian and did not experience racial discrimination. Jane was paid less than her male coworkers who were doing the same job, experiencing gender bias, but for many years she was not aware of this disparity. The economy worked for the two of them as they held professional degrees for which there was a demand and found work readily after college.

In this case the larger social context barely impinged on Jane and her husband. That would not have been the case had Jane been a person of color, a person belonging to a large religious group that felt a woman's place is in the home, or a person with a significant disability. The social context would have played a decisive role had Jane lived in a community that confined people of color to specific neighborhoods with fewer community services. The social context would have come into play with regard to the laws related to gender and racial bias and social context would have become important if the economy had no jobs for people with Jane's education and background.

Developmental Transitions

The ecological model is also concerned with normal life changes, often referred to as *transitions*. These changes are called transitions because they are events that move a person from one phase of life to another, requiring the person to make adjustments or to adapt in some way to new circumstances. Many of these events are simply part of the normal development that all people experience from birth to death. These transitions are often expected, and for some transitions there is preparation.

Urie Bronfenbrenner (1979, p. 26) explains it this way, "An ecological transition occurs whenever a person's position in the ecological environment is altered as the result of a change in role, setting, or both." For example, Mrs. Stilmyer goes from being a wife to being a widow. Mr. Hoffman goes from being at home to living in an assisted living facility. The Carters go from being a couple without children to parents of twins. Sally goes from being at home with mother to spending half of every day in preschool. Joanne goes from married to divorced, with a new home and a big mortgage she must pay herself. Each of these is a fairly normal transition.

Some people do not cope as well with the changes brought by transitions as others do. Perhaps they have more going on in their lives than they feel they can handle. Perhaps the events have changed their lives dramatically in ways that are viewed as negative. Many of the people we see in the human service field are going through, or have recently experienced, one or more transitions. Here is a list of some of the transitions that people experience:

Starting kindergarten or first grade	Starting a new job
Going to high school	Getting married
Going out on the first date	Buying a first home
Leaving home for the first time	Experiencing ill health
Losing one's job	Losing a spouse through death
Experiencing a disaster	Divorce
A large mortgage or other debt	Losing some physical capacity
Considerable financial losses	Considerable financial gains
Children leaving home	Children marrying
Birth of grandchild	Death of a child

And there are many more. Although every single one of these events probably will not happen to one person, we can expect that most people, in the course of a long life, will experience a number of the transitions on this list.

It is important to know about the common life stages a person passes through and to recognize where your clients are in their life stages and transitions. Transition problems are common to many people. Often treatment is unnecessary—self-help and support groups can provide the support a person needs to make the transition. Sometimes, however, a person may find the changes overwhelming, completely negative, or intolerable. These people may need professional help to handle these changes and adjustments.

Developing the Interventions

You have looked carefully at your clients' issues and problems. You have come to understand the ways in which your clients have responded to their context and the way the context has contributed to your clients' motivations and decisions. As a case manager, your role is to design a plan with each person that will address the areas of need. Human service workers can and do intervene on several levels.

Many clients of the social welfare system appear to be individuals who have encountered inadequate support in their environment. One or more of the institutions that we believe should support the individual in our society has failed these people or has been unable to supply what was necessary to avoid problems. Institutions such as education, medicine, the economy, politics, and the family may have let this person down in some way.

When this occurs, our society looks to the social welfare system to supply what is needed, to address the unfortunate gaps in a person's life, and to apply interventions that will prevent a worsening of the problems. Your task as case manager is to look at the client and the client's environment, to gather the facts about each of these, and to understand how context and person interact to the detriment of the person or contribute to the well-being of the client.

With this information, a plan is developed with the person that addresses maladaptive interactions between the individual and the environment and that notes those parts of the environment that are positive and usefully supportive. The interventions you design or choose should be two-pronged: personal interventions that strengthen the person to handle the environment, and environmental interventions that alter the context to better accommodate the person. Here are some examples of the types of interventions that can be incorporated into individual plans:

Interventions to Strengthen the Person	Interventions to Strengthen the Environment
1. AA for substance abuse	Family education to support sobriety
2. Parent skills training for the parents of a child they abused Group therapy for abused child	Temporary removal from home to foster care Foster home parents given information on creating supportive foster home environment
3. Job training for the person with developmental disability	Work with the employer to provide a supportive work environment
4. Interpersonal skills training for adolescent in minor trouble	Use the child's interests to develop more constructive in-school activities Bring father into the picture in a positive way Family therapy with the mother
5. Medication for the person with schizophrenia Regular appointments with psychiatrist	Place in a supportive living environment in the community Develop constructive connections with the local church Two family sessions to reinvolve the family in the person's life

These are fairly routine interventions in a person's problems to ameliorate a negative or destructive situation or to enhance the person's self-fulfillment. The point is, however, that a service plan formed without the understanding that interventions take place in two distinct areas could be quite hapless and without focus. When doing a service plan, the human service worker makes certain that both areas have been addressed with appropriate interventions wherever possible and that the interventions are documented in the record.

Working with the Generalist Approach

The ability to recognize and effectively address issues on all three levels is generally referred to as the *generalist approach*. Aware that problems occur between the designated client—be that a family, an individual, or a group—and the client's environment, it is important to see the interaction between the two and to look for ways to intervene on all the relevant levels on behalf of your client. As noted previously, clients are affected by the environment they occupy, and the environment is affected by the clients' response. For example, look at how Ralph affected the people around him, their response to him and his response in return. Contrast that with Eduardo's environment and the response he received as a result of his interactions with his environment. The client affects the environment and the environment affects the client in a never-ending interaction that can have positive and empowering results for the client, or just the opposite.

When case managers look at how people and systems affect the client's problems on each of the three levels, the case manager has correctly made a multilevel assessment. This assessment should lead case managers to develop interventions that will enhance both the identified client and the client's environment. The generalist approach has as its goal the better functioning and increased competence of all parties. By looking at the whole picture instead of just a piece of the picture comprising only the identified client, the case manager has laid the foundation for making solid and long-lasting change possible.

Macro Level Interventions Are Advocacy

It is assumed that the human service worker is not limited to just helping individuals. Your work as a human service professional places you in a unique position to be able to speak to the problems affecting large numbers of people. This happens in two ways.

In the course of your work, you will encounter many who have been damaged by abuse or discrimination. You will see groups of people harmed by poor school systems, a lack of medical care, or scant supervision. Ethically, we have an obligation to speak to the needs of those with less advantage in our society. Having seen the damage firsthand, we are better able to speak to conditions that need to be remedied in our larger society and to keep statistics and information on the extent of the problem for use in persuading lawmakers and others in power to take action.

You will also see areas of service that have been neglected or that require development. Perhaps there is a need for more supported living arrangements for those with mental illness in the community. It may be that mothers returning to work from welfare lack the means to dress appropriately for the job, need day care for their young children, or require transportation to get to job interviews. Who better than the human service professional to bring to the attention of those who develop programs the areas of service that are lacking in your immediate community? You and others can bring to light the unique needs of your community and help to develop much-needed services.

For example, you might have a 12-year-old client who needs an individualized education plan due to his learning disorders. As an advocate for this 12-year-old boy, you would approach the school or support the parents in approaching the school to obtain this specified plan for your client. Your goal would be to make the school system work better for this individual. Your intervention would be on the meso level.

Now suppose that you find that there are numerous others complaining that students with learning disabilities are being ignored within the entire school system. You have six individual clients about whom you have concerns that the school is not meeting their needs. In addition, your colleagues in your children's case management unit are also citing cases where their clients are not getting the academic support they need.

If you all go together to the superintendent of schools and respectfully request a review of the procedures and give him information, you have intervened on the macro level. Now you are advocating for an entire population, many of whom are not clients of your agency. Perhaps the problem involves several school districts or involves schools all over the state and the intervention is with the state Department of Education or, if that is unsuccessful, the state legislature. This is a macro level intervention. You are advocating for a broad population seeking to make a macro system work better for individuals with specific needs.

Unger (2013) considers macro level advocacy and the resulting positive social change an ethical obligation for human service workers. He expresses concern that those in the social services will grow further and further from their obligation to create change that supports a broader diversity of needs. Instead, he contends that these same workers often feel more comfortable confronting individual needs on the micro and meso level, dealing with individual issues and problems without seeking changes that would benefit marginalized populations. Social service work "obligates (social service workers) to put into practice their commitment to broad social change" (Unger, 2013, p. 493).

ASSIGNMENT

Talk to a human service professional in your community about what that person sees as an unmet need in the community. What one service would the professional like to see developed? What specific need would that service meet? What client population would that project address? How would this service or program enhance the environment for clients? How would this enhance the functioning of clients within their environment? How could the clients who need this service be mobilized to work on their own behalf?

Summary

In this chapter, we have looked at a method for assessing your clients in three different dimensions. This method, often referred to as the ecological model, looks at the attributes people bring to their problems, as well as the contributing factors in clients' immediate environments and in the larger societies in which they live and function. Using this model prevents us from blaming only the clients for the problems or seeing the problem as residing in an individual rather than their environment and community. In addition to recognizing all the factors that make up clients' problems, this model also allows us to recognize the broader need for interventions and to develop interventions that address issues on all three levels.

Using the ecological model, you can now produce effective solutions to the problems and the social issues brought to you in the course of your work. In this way, you serve your clients and your society in meaningful and useful ways.

Exercises

These exercises can also be filled out online at CengageBrain.com.

Exercises I: Looking at Florence's Problem on Three Levels

Instructions: Look at Florence's problem as she presented it to the case manager. Decide which parts of her problem are on the micro level, which parts are on the meso level, and which parts are on the macro level.

Florence came in to see a case manager in an agency that addresses child abuse and neglect. Recently her daughter Crystal was removed from the home because of complaints by neighbors that she was abusing the child. An investigation of the situation by child-care workers indicated the abuse was severe. The discipline she was administering was discipline she had experienced and witnessed as a child from her own parents and her aunts and uncles who lived on farms near her family. Florence related that she was the oldest daughter, third in line of nine children, of a farm family of 12 people. Her parents worked hard from sun up until long after dark. Much of the housework was done by Florence and her aunt, who lived with them. Her mother was ill, often in her room in bed. Florence does not know what the illness was, but does not recall her mother ever seeing a doctor. She tells the case manager that she knows her mother and her aunt did not like her.

At 18, Florence ran away with Dave, who did mechanical work on cars. "He was my first and only boyfriend," she explains, weeping. Florence and Dave never married, and they had one child, Crystal. Last April, Dave died in a car accident on the interstate. Florence cries as she describes that night and the way the police came to her trailer and how kind they were to her. She describes how alone she has felt ever since.

Florence receives welfare. She completed eighth grade before her father "yanked me out of school to do housework. Said it was no place for a girl. A girl didn't need no schooling." Florence had enjoyed school, mostly for the companionship of other girls. "I'm shy of people, you know. But at school I had friends." Florence remembers school as hard, and she had trouble with subjects like math and science. "Mostly I sat there and worried about what would happen when I got home from school. It was always something: Mom was worse, I was in trouble, there was some big push to get in a harvest. I was glad when I quit."

Leaving with Dave had alienated Florence from her family. "Dave used to say, 'They're just mad 'cause they can't use you no more.'" For this reason, Florence has not seen her family since Dave's funeral, and they have made no attempt to get in touch with her even though they are only a few miles apart. The welfare agency reports that their workers have rarely seen Florence and have not as yet offered her any services for going to work, although she is on a list of single mothers they would like to make job-ready. Child welfare tells you that they cannot return Crystal until Florence has had intensive parent training and supervised visits with her child. They also tell you that they found her home worn, but immaculate.

Florence confides that she is terrified of going to work, that she feels useless, and that she probably has little to offer on a "real job." She also appears to be depressed, crying at intervals and hanging her head. Socially, she is isolated both because of Dave's death and because her neighbors are fed up with her child-care practices. "The neighbors don't like me either," she says with resignation. The child-care agency is asking for parent training, but it is unclear who will offer that in this rural area.

What part of Florence's problem is a micro-level problem?

What part of Florence's problem is a mezzo-level problem?

What part of Florence's problem is a macro-level problem?

Exercises II: Designing Three Levels of Intervention

Instructions: Look at the four cases below and decide how you would intervene on three levels: the personal (micro), the contextual or social context immediately surrounding the client (meso/mezzo), and the larger environment (macro).

1. Maria is paralyzed from the waist down following an accident three summers ago in a swimming pool. She is hoping to complete her degree in accounting, but she is complaining of depression and an inability to focus on school. When you see her, she looks anxious and tired. Her affect is flat, and she tells you nothing interests her. At the local college, she has had trouble finding appropriate parking and misses many days of class when the weather is bad because of the parking situation. One of the professors she must work with closely has made remarks

about the difficulty of "other people getting around that wheelchair." She believes her boyfriend, who was with her the night the accident happened, has remained with her simply out of pity. When they fight about other things, she throws this up to him, although he vehemently denies it and tells Maria this is a hurtful accusation.

Interventions on the micro level:

Interventions on the mezzo level:

Interventions on the macro level:

2. Mr. Groff is 93 and living alone in his home. He only stopped driving last year. He would like to get out more, perhaps go to the senior citizen center. In addition, he would like to go to the Lions Club and to participate in a foreign policy club he belonged to for years. He tells you sadly that the members of the foreign policy club always seemed amazed at the reading he had done and the sound opinions he expressed, "as though I should be senile!" Since he stopped driving, he has lost contact with them. Right now he sees no reason to go to a nursing home and feels that if he had transportation he could continue to buy his groceries, prepare his meals, and care for himself generally. He tells you, however, that he would like to find a way to be less lonely.

Interventions on the micro level:

Interventions on the mezzo level:

Interventions on the macro level:

3. Margie is in a sheltered workshop for people with developmental disabilities. She does well at work and has many friends. She lives with her mother, and her mother is not happy with the new level of independence Margie is developing. She often goes out with others from work and the supervisors for dinner on Friday night. She has joined a social group for individuals with disabilities much

like hers, and they go bowling and to the movies. Margie has, since she went to the sheltered workshop, learned how to use the phone to make appointments with her doctor and dentist and how to ride the bus to and from both work and the social club, and she has been shopping to buy her own clothes twice with her case manager. Margie's mother complains about all this. She tends to blame Margie for leaving her alone at night and making her unhappy. "Since your father died, you're all I have," she tells Margie. Margie's response to this is to cry and stay home and give up some of her independence. Sometimes she has missed work, hoping to make her absences up to her mother.

Interventions on the micro level:

Interventions on the mezzo level:

Interventions on the macro level:

4. Chris is a single father who is trying to work and raise three small children. His wife was killed 2 years ago in a traffic accident. After the initial shock and outpouring of support from friends and neighbors, Chris found himself alone with all the responsibilities and very unsure of himself. He would like to meet other men who have the same problems but cannot find any groups, even though he has been told about several men who are in the same situation. He tells you he is not sure what the best method is for disciplining his children, whom he describes as "good kids." Sometimes he feels he is too lenient with them, and at other times he is afraid he is unnecessarily strict with them. A local women's health center has groups for bereaved single parents, but Chris believes those would not be open to him. "It would be all women, wouldn't it?" he asks. In addition, he is having a hard time at work balancing the responsibilities there with parenting responsibilities at home. "Of course, I want to do a good job and get the promotions so I can support these kids through college, but I need to be home in the evening, or someone does, and I don't think that is always well received at work."

Interventions on the micro level:

Interventions on the mezzo level:

Interventions on the macro level:

CHAPTER 4

Cultural Competence*

Introduction

Seeing each of our clients as unique individuals is the only way to accurately perceive them and to be constructive in the way we serve them. One key element of individuality is **culture** and subculture.

Most of our attitudes and perceptions are the result of our interactions with others throughout our lives. In time, these interactions come to seem natural. As professionals, we need to become aware of our personal ways of thinking about others and their situations. Is our thinking useful? Will it promote the well-being, self-esteem, and independence of our clients? Because we are a culturally diverse society, it is important for professionals in human services to respect differences and to seek to understand these differences whenever possible.

Culture and Communication

Each of us brings to any situation perceptions and attitudes that are influenced by our own culture. Our own **ethnic group**, family values, outstanding experiences, and cultural traditions all influence both the way we communicate to other people and what we believe other people mean when they communicate with us. Often, we are unaware of the extent to which these factors color our interactions with other people.

In addition, we do not usually take the time to understand that others may come from a culture that differs from our own significantly. We may judge others' actions by the standards prevalent in our own culture. We may expect certain behavior we

---

*The material in this chapter is adapted with permission from *Communicating with Strangers*, 3rd ed., by W. Gudykunst et al. Copyright © 1997 McGraw-Hill Companies.

believe is appropriate and become annoyed when we do not see that behavior. We may misunderstand the communication of others, leading to lost rapport and opportunities. This is dangerous when we have accepted the professional responsibility for giving assistance to other people.

Your Ethical Responsibility

Ethically, you have a responsibility to take the time and make the effort to become familiar with cultures that differ from your own that you have extensive contact with as a social service professional. It is not ethical to simply assume you know all there is to know about a group because you see members of that group on a daily basis or hear about them from the news media. Instead, you need to ask questions, take seminars, and gather information that will enhance your understanding of that group or culture.

When You Are Not Sure

It is not possible, on the other hand, to study and become familiar with all the different cultures you might encounter in the course of your professional lifetime. In your work, it is quite likely that you may encounter someone from another group whose culture is unfamiliar to you and whom you will see only briefly. What you need is a method you can use that will allow you to participate in those encounters and interactions competently.

Where Are the Differences?

Differences among people occur on a number of sociological levels. These differences can be overcome and understood, or they can become obstacles to good communication and understanding.

Cultures

Generally, cultures coincide with national or political boundaries. People living in one country have a culture that differs from the culture of those living just across the border. When we refer to culture, we are really talking about the culture assumed by an entire society.

This means that we in the United States have in common with one another a basic knowledge. We learned this knowledge through the socialization process—from our schools, parents, religions, and even television and magazines. Although each individual may see the culture just a bit differently and no one knows everything there is to know about it, people share enough in common to be able to relate to and cooperate with one another.

By the time we are young adults, the culture we carry with us in our heads is largely unconscious. Our culture influences how we communicate with other people, and it influences the way we determine what the other person means. In other words, what we say is affected by our culture, and, in turn, our interpretation of what another person is saying to us is colored by our culture. Because this process has become automatic for us, we are not aware of the significant influence our culture has on our interactions with others. Furthermore, if most of the time we are communicating with others from our own culture, we will assume that all people mean what we mean and see things as we see them. With this way of thinking firmly in place, there is a tendency to assume that our own culture is the better or correct way to be at any given time.

Subcultures

Within any given society, there are groups of individuals who, for the most part, follow the culture of their society but hold in common with each other somewhat different cultural ideas. This may be a religious group that holds ideas that are somewhat different from mainstream thinking about patriotism and serving in the military. It might be an ethnic group whose subculture is shaped by the discrimination experienced in each generation.

Subcultures usually are not completely out of step with the larger society's culture. There is, however, something about the subgroup culture that sets its members apart. It might be values, traditions, beliefs, lifestyle, or any combination of these.

Race and Ethnic Group

Important in understanding subcultures is understanding the terms *race* and *ethnic group*. According to Gudykunst and Kim (1997, p. 20), *race* refers to a "group of people who are biologically similar," and *ethnic group* refers to "a group of people who share a common cultural heritage usually based on a common national origin or language" (p. 20).

Racial groups often have distinguishing physical characteristics, whereas ethnic groups may be distinguished by their language, religion, or some other aspect of their culture. It is important to keep in mind that race alone is not a factor influencing communication; on the other hand, ethnicity with its culture can have considerable influence on communication. If the racial or ethnic group of your clients indicates a subculture that is unfamiliar to you, the potential for misunderstanding is increased.

How We Develop a We-versus-Them Attitude

During the process of socialization, we learn that some groups are acceptable and others are unacceptable. The acceptable groups are seen by us as in-groups. We are more comfortable with in-groups than with groups we consider unacceptable. We see the members as being similar to us, and we expect the members to hold beliefs and values very much like our own, and to act and think as we would. When we talk about in-groups, we generally do so favorably, holding them in positive regard. We are better at predicting how members of in-groups will respond or behave.

Out-groups are those groups with whom we feel uncomfortable—groups with whom we have less inclination to interact on a regular basis. Generally, we do not hold members of out-groups in a particularly favorable light. We may be suspicious of the motives of an out-group because we do not fully understand its culture. Using our own culture as the standard, we may find the out-group culture inferior. Members of the out-group may appear unpredictable, unreliable, or devious to us.

Strangers

When people do not act or think the way we believe they should, they seem strange to us. Many people you will encounter in the course of your work will seem like strangers to you. For example, Julio is a stranger to nearly everyone he sees on a daily basis. He, his little brother, and his mother live in a small city with others from Puerto Rico. When he is with his small group of friends and relatives, he is not perceived as strange. On the other hand, when he attempts to interact with the larger American culture, many see him as a stranger. His language, his behavior, and, in some cases, his attitudes appear strange to members of the larger culture. He is tolerated, and even given menial work, but he feels set apart. People he must meet and work with every day view him as a stranger because they know little about Julio's culture. They see him every day, yet he is in no way considered part of their group.

Julio went to the case management unit to seek services for his brother, who was diagnosed by the school psychologist as having an intellectual disability. There he encountered some problems. His accent and unfamiliarity with the language made it difficult for him to be understood. While the worker talked about residential placement and education, Julio resisted, indicating the family just needed help with the local school, where he felt his brother had been misunderstood because of a language problem. The local school had suggested Julio and his mother go to the case management unit because school officials did not believe they could provide adequate services. Julio felt their referral indicated insensitivity and an unwillingness to be concerned with keeping the family together and helping his brother function better in English. The vast array of services being offered at the case management unit was bewildering to Julio. He was inclined to simply withdraw from the situation and tell his mother to keep his brother at home.

To the worker at the case management unit, Julio seemed strange. She did not exactly use that word, but she wondered with some exasperation why he did not want to take advantage of the many services available to his brother. Why did he seem so reluctant to keep appointments with both the worker and the school? Why had he withdrawn his brother from school when this was clearly against the law for a child so young? To the case manager, Julio's behavior was inexplicable.

It is always the majority group that defines who is a stranger and who is not. The people who seem strange to us are not strangers to those with whom they hold common cultural traditions. If the situation was reversed and we found ourselves in a place where our cultural ways and values were different from the majority, then we would be the strangers.

Gudykunst and Kim (1997) use the concept of the stranger to define those whom we encounter who seem strange to us, whose ways of thinking and acting are unfamiliar, and who are not members of our in-groups. In other words, they are people who are close enough that we cannot ignore their presence, but they are unfamiliar to us and therefore seem like strangers. (Throughout this chapter, I use the term *stranger* as it is defined here.) If people come from another culture, possibly from another country, it is entirely possible that they do not know enough about your culture to be able to relate easily to you. In addition, it is quite likely you do not have enough information about their culture to be able to make the new situation smoother for them.

As the human service professional, you are the person who can take the initiative in making the adjustment smoother for those who come to us as strangers. When people we might consider strangers have developed a good degree of competence in the majority culture and can communicate well, they will be healthier. Studies, however, indicate that it takes a long time for immigrants to adjust to the new culture and that if this maladjustment is severe or long-term, it can cause serious mental health problems as a consequence.

As the world community becomes more global, we can expect to encounter people from many different cultures who will seem like strangers, people who are different. Gudykunst and Kim (1997) make the point that we have internalized our own culture to such a degree that we believe it is innate in some way. They write, "anyone whose behavior is not predictable or is peculiar in any way is strange, improper, irresponsible, or inferior" (p. 357).

When people deviate from the familiar, we are likely to notice it instantly. We may feel anxious or surprised and uncertain. We may be forced to look more closely at our own cultural assumptions. Perhaps we are forced to conclude that aspects of our culture that we have taken for granted are not particularly useful. Our cultural identity may be challenged. Obviously, it would be easier to avoid all this and stay away from strangers. Many people do just that, preferring not to experience these unsettling emotions. In human services, however, our work is all about encounters with people, and our purpose is to be helpful. Avoiding strangers would be irresponsible. For that reason, we need to know what to do when we encounter strangers.

Anxiety and Uncertainty

It is common for most people to feel uncertain or anxious when they are attempting to interact with people from other cultures. If we are consumed with our uncomfortable feelings, our communication with strangers will be impeded. We need to be able to manage our feelings during these encounters to provide for a constructive exchange.

Many times we attempt in some way to reduce anxiety or stress in these encounters. We project our notions about what the person means, giving us a certainty that might not be justified. We might try to develop theories about the other person that feature similarities. The more we believe a person is like us, the less likely we are to feel anxious. Thus, we might look for similar psychological reactions, similar group affiliations, and similar cultural aspects.

When Misaki came from Japan to study at a local college, she was the only person from Japan on campus. Other girls in her dorm invited her to join them for meals and to walk to classes with them, and Misaki did so. When the other girls laughed and talked about trivial matters, Misaki was silent. She rarely made small talk with the girls. They began to interpret her behavior as "too serious" and worked even harder to draw her into their discussions. Misaki was always pleasant but contributed little to these exchanges.

To the girls in her dorm, she seemed too serious; but to the resident assistant, Misaki appeared depressed. Ann, the resident assistant, based her opinion on the fact that Misaki never looked up when she spoke and never looked directly at Ann. Ann asked Misaki if she would like help with her sadness or depression. Misaki said "yes," so Ann made a referral to the campus clinic. There an intake worker decided that Misaki was indeed depressed and concluded it must be about leaving her homeland. To each question the intake worker asked, Misaki answered "yes." Yes, it was hard being here in America; yes, she missed her parents; yes, she had trouble understanding everything the professors said in class.

A counselor at the clinic recommended to Misaki that she join in more with the girls in her dorm and learn to "loosen up and have fun." In Japan, however, people who talk a lot are not viewed as particularly trustworthy. Those who use silence more frequently are considered discreet and trustworthy.

The girls in the dorm, the resident assistant, the intake worker, and the counselor in the clinic did not understand Japanese culture well enough to refrain from judging her by the standards of their own culture. Furthermore, in Japan people often assess what it is the speaker wishes to hear and answer "yes" as a means of keeping social harmony. They might not mean yes in precisely the way an American might interpret it. In addition, Japanese people often do not look directly into the eyes of someone with whom they are conversing. To look directly at another is a sign of defiance or aggression. Looking away is a sign of respect. The Americans took Misaki's behavior as an indication of shyness or depression because that is what the behavior would most likely mean in American culture. By fitting Misaki's behavior into American cultural meanings, the Americans did not have to feel anxious about how to interpret the behavior of a stranger.

American girls talk to each other frequently and often about trivial matters as a way of cementing their ties to one another. To the girls who befriended Misaki, this seemed normal. Her silence did not. In order not to feel anxious about her, they projected their own theories about her behavior onto her and concluded that she was sad about leaving her home in Japan. This was a normal reaction, understandable from their standpoint. The theory made Misaki seem more like them; thus, their theorizing reduced their anxiety.

We will unwittingly go to great lengths to resolve our anxious or uncertain feelings. Often what we do is inaccurate and not useful in promoting clearer communication.

Thoughtless versus Thoughtful Communication

First, we need to find a way to control our anxious feelings to allow us to really be able to listen and communicate. If we are likely to feel anxiety when we talk to strangers and if this anxiety is going to interfere with a realistic understanding of these strangers, we are going to hear and communicate in a skewed or inaccurate manner. What is worse is that we may be only vaguely aware of this problem. The following sections discuss areas you can evaluate to make your communication more thoughtful and accurate.

Recognizing Our Tendency to Categorize

Think about what we do when we are communicating without thought. We categorize people; we assume there is only one correct or normal way to view things; and we are closed to information that does not fit with our cultural perspective. When we encounter someone who is different, we dump that person into one of our large categories or **stereotypes**. When someone's behavior does not fit or that person's thinking is strange to us, we are thrown off balance. To prevent that, we have categories all ready into which we can place such people. Because we do this habitually, we are not completely aware of our categorizing.

Actually, as you work with various sorts of people, you may find that you need to add many new categories to your conceptualizations of others. These categories, if more specific and definitive than those based on stereotypes, will be better predictors of behavior.

Looking for Exceptions

To become more thoughtful as you communicate, start to look carefully for the exceptions to your categories. If you think all Hispanics are loud, think about times when you have encountered Hispanic people who were not loud. If you believe all Muslims are militant, look for times when individual Muslims have expressed cooperation. If you believe all Jews horde money, seek out the times that Jewish people have been generous. In other words, recognize that the categories you have been using are likely to be much too broad to account for all the specific differences you may encounter in the people you serve.

Another way to look for exceptions to your categories is to seek differences in each specific individual. If you dismiss a stranger as coming from a group that is generally believed to be resistive to your efforts to help, you will have little success compared to recognizing the resistance and then looking for times the stranger was not resistive. If you have categorized someone as too talkative, look for times when that person was listening instead. If you are sure the people in a particular group are stupid, look for times individuals in that group made wise decisions or choices. Seeing others as individuals will make these exceptions important to you. As a competent worker, you will diligently seek these exceptions to gain a more accurate understanding of the people you are serving.

Checking Our Attributions

Most research shows that when we see a stranger's behavior as negative, we are inclined to blame that behavior on the stranger's character or disposition. When we see the stranger's behavior as positive, we are more likely to think this person is an exception and attribute the exceptional behavior to the environment or the circumstances. In other words, it appears that for many of us, giving up our stereotypes is very hard. We would rather see the exceptions to our stereotypes as something external to the stranger, and we are likely to blame behavior that seems to fit the stereotype on the personality of the stranger.

The opposite is often true as we go about attributing causes to our own behavior and the behavior of those we consider to be members of our in-group. If we see negative behavior in these people, we are likely to blame the environment or circumstances, while we often consider positive behavior a reflection of the person's character. The following list summarizes how we often see things:

Our positive behavior	Attributable to our good character
Their positive behavior	Attributable to the environment or the circumstances
Our negative behavior	Attributable to the environment or the circumstances
Their negative behavior	Attributable to their poor character

When we do this systematically, it reveals **prejudice** on our part. In addition, these systematic errors in attribution cannot have come about thoughtfully. They are thoughtless, automatic ways of looking at other people. As you become more thoughtful in your communication, become aware of how you explain the behavior of others.

Evaluating Scripts

We have all learned certain scripts for the activities we engage in frequently. For instance, if you meet someone you see often, but do not really know very well, you might say "Hi" as you pass that person. She might say, "Hi. How are you?" You would probably say something like, "Fine, and you?" She might then respond with "Just fine, thanks." By the time this exchange is completed, you may be several yards apart and walking in opposite directions. This constitutes a script for passing someone you see every day but do not know very well.

There are scripts for a variety of everyday activities. We carry them in our heads to be used when the appropriate situation presents itself. We have learned them first by observation and then from our own participation in these activities. The exchange demonstrated in the previous paragraph did not take much thought. Two people may pass each other every day and go through much the same exchange. They do not stop and consider what to do as each encounter presents itself.

We expect that people from our culture will respond to our "hello" with a "hello" of their own. If one individual does something different, we are thrown off balance. People from other cultures, however, have learned different scripts. For instance, a common area of misunderstanding relates to the fact that different cultures have different nonverbal ways of indicating that they do not want to be approached. Suppose you are indicating through your body language to someone from another culture that you do not want to be approached, and this person approaches you anyway. You may feel pushed and invaded when, in reality, the stranger could not recognize the signals. You could make a similar mistake. You might see a person you want to join you. You might wave to the person and point to your group, indicating that the person should come over and join you. To a person from an Asian country, this would be insulting. Waving people over, and especially using one finger to do so, is considered rude. We are looking at two different scripts.

Behavior or communication that seems strange to you may simply be a different script presenting itself. Stop and think about what the unexpected behavior means to the stranger. Is this offensive behavior, or does the stranger mean something quite different? Is there a possibility that you are misreading the signals or cannot recognize the signals from this stranger? Is it likely that the signals you are sending are not familiar to the stranger?

Checking Perceptions

Gudykunst and Kim (1997), who have written extensively on this subject, recommend that we simply check our perceptions with strangers to see if these perceptions are accurate. Instead of assuming that we know what a stranger means, we need to check. These authors recommend a three-step process:

1. Describe the other person's behavior, being careful to simply describe what was observed without evaluating or labeling the behavior.
2. Tell the stranger how you interpreted the behavior. In doing so, be matter of fact. Refrain from any hint of a negative evaluation of the behavior.
3. Ask the stranger if your perceptions are accurate.

Checking your perceptions is a good way to keep the communication between you and the stranger accurate and meaningful. It is important not to assume you know what the stranger means or what the stranger feels. Check with that person to see if what you perceive is correct.

Allowing Differences

It cannot be stressed enough that thoughtful communication is extremely important in reaching real understanding with strangers. Obviously, the better the understanding between you and a stranger, the more likely it is that you will be effective and competent in your assistance to that person.

Not all strangers will respond the same way to their new environment. Differences between the culture of the stranger and the culture of the host society may account for how a stranger responds. Large differences in verbal and nonverbal behavior, in norms or language, or in political and religious orientation can make adapting to the new surroundings more difficult. Where the differences are small, things may be easier for the newcomer. For example, someone from Canada would have less trouble adjusting to the United States than someone from Botswana. When you take this into account, you are able to look more thoughtfully at the stranger's attempts to adapt and be more helpful in that process.

In addition, recognize that there is a lot you do not know and be open to finding out more. When you are communicating with someone who is a stranger to you, put aside your goal for that conversation and begin to listen carefully for new information the person might be providing to you.

Finally, accept that there is more than one way to view something or to understand something. People have different perspectives, but that does not mean that some are superior to others or more correct than others. We may have been taught that this is so, but look at other ways to explain behavior besides your own perspective. Try to understand what perspective the stranger may have. This can be done only if you communicate thoughtfully.

Dimensions of Culture

Researchers in the field of communication have looked for ways to help us understand cultural differences even when we do not know the details of every culture. They have proposed that cultures have an underlying foundation of individualism or a foundation of collectivism. Cultures fall along a continuum, with no single culture being all one or the other; but many researchers believe that communication can be facilitated between people of different cultures if we know whether the stranger with whom we are communicating is from a culture that is primarily an individualistic culture or a collectivistic one. This tool is particularly helpful when we do not know all the particulars of a specific culture.

Individualistic and Collectivistic Cultures

Using information from Gudykunst and Kim (1997), we will look at some of the characteristics of cultures that are predominantly individualistic or collectivistic. Figure 4.1 summarizes some of the general differences between these two types of cultures. Figure 4.2 lists some examples of countries that tend to be individualistic and some that tend to be collectivistic.

FIGURE 4.1 Characteristics of individualistic and collectivistic cultures

Individualistic Cultures	Collectivistic Cultures
Individual More Important	*Group More Important*
• Individuals should look out for themselves and their families	• Members of in-groups look out for each other in exchange for loyalty
• Promote self-fulfillment	• Require that people fit into the group
• Emphasize individual initiative and achievement	• Emphasize belonging to groups
• The in-group influence is very specific to times and place	• The in-group influence is very general over all situations
• Individual goals are emphasized	• Group goals are emphasized
• Tend to apply their value standards to everybody (universalistic)	• Tend to apply different value standards to members of their in-groups and to members of out-groups
• Emphasize needs and goals of the individual over the group	• Emphasize the needs and goals of the group over the individual
• Support unique individual beliefs	• Shared in-group beliefs
Have a More Vertical Culture	*Have a More Horizontal Culture*
• People are expected to stand out from others	• People are not expected to stand out from others
• Value is placed on freedom	• Value is placed on equality
• Maximizing of individual outcomes	• Cooperation with in-group members

Source: Adapted from Communicating with Strangers, 3rd edition, by W. Gudykunst et al, McGraw-Hill Companies.

Digital Download Download from CengageBrain.com

FIGURE 4.2 Individualistic and collectivistic cultures

Countries That Tend to Be Individualistic Cultures	Countries That Tend to Be Collectivistic Cultures
(based on predominant tendencies in the culture)	*(based on predominant tendencies in the culture)*
Australia	Brazil
Belgium	China
Canada	Columbia
Denmark	Egypt
Finland	Greece
France	India
Germany	Japan
Great Britain	Kenya
Ireland	Korea
Italy	Mexico
Netherlands	Nigeria
New Zealand	Pakistan
Norway	Panama
South Africa	Peru
Sweden	Saudi Arabia
Switzerland	Thailand
United States	Venezuela
	Vietnam

How Individualistic and Collectivistic Cultures Differ

First, individualistic cultures tend to place a higher value on the individual than on the group. Collectivistic cultures, on the other hand, tend to place more value on the group. Another difference lies in the way in which society is viewed. In individualistic cultures, there is ranking and hierarchy; collectivistic societies tend to be more egalitarian.

There is also a difference in the way the two cultural types use the surrounding context in communicating. In individualistic societies, the communication tends to be so direct that a person rarely needs to check the context to fully understand the meaning. In more collectivistic societies, context is extremely important. See Figure 4.3 for a summary of the communication differences between the two types of cultures.

Figure 4.4 highlights some of the specific elements that make communication different between individualistic and collectivistic cultures. As the comparisons in the figure indicate, there is plenty of room for misunderstanding. A person from a culture that values clear, explicit information might suspect someone from a high-context culture of being manipulative or confused. Someone from a horizontal culture might find someone from a vertical culture rude and boorish or incredibly selfish. If a client from a collectivistic culture waited to engage in services until he had group consensus, the worker from an individualistic culture might mistakenly think the client was resisting treatment or uninterested in help. If a worker from an individualistic culture encouraged a woman from a more collectivistic culture to look out for herself and leave an

FIGURE 4.3 Communication differences between individualistic and collectivistic cultures

Individualistic Cultures: Low-Context Communication	Collectivistic Cultures: High-Context Communication
• Low-context communication is more precise • Direct and explicit	• High-context communication uses understatements, pauses, silences, or a shortage of information • Indirect and implicit
Context Is Only Minimally Important	*Context Is Important in Determining Meaning*
• Tend to use very direct communication • Listener does not have to use context to obtain meaning • Communication is less ambiguous • More concerned with clarity as necessary for effective communication • Communication is about the same for in-groups and out-groups • Value saying what you think • Value truthfulness • Verbally direct, precise, and absolute • "Yes" means agreement	• Tend to use more indirect communication • Listener must use context to obtain meaning • Communication is more ambiguous • More concerned with avoiding hurting others or imposing on others • Communication is very different for in-groups and out-groups • Value avoiding confrontations • Value courtesy • Verbally indirect, imprecise, and probabilistic • "Yes" does not necessarily mean agreement

Digital Download Download from CengageBrain.com

abusive marriage, the client might feel helpless and unsupported. Leaving the group might not be an option for her.

One common error is for people from individualistic cultures to assume the person with whom they are speaking from a collectivistic culture is speaking as directly and explicitly as they are. Individuals from collectivistic cultures can make the reverse mistake, assuming the person from an individualistic culture is only implying or speaking indirectly.

Privacy and Self-Disclosure At different times, we feel open to interaction with other people or we feel closed and seek privacy. Different cultures regulate privacy needs in different ways. While individualistic cultures do so with physical boundaries, collectivistic cultures do so by psychological means. For instance, in collectivistic societies, people who might be encountered in general public situations are often seen and treated as nonpersons and simply ignored. In this way, the individual is protected from unwanted involvement. In individualistic societies, this would be seen as rude.

Time Time is conceptualized differently in different cultures. How people conceive of time determines how they are likely to use it as well.

FIGURE 4.4 Specific communication differences

Individualistic Cultures	Collectivistic Cultures
Privacy Regulation	*Privacy Regulation*
• Use of physical barriers such as doors, walls, private rooms and offices, fences, hedges	• Use of psychological barriers such as speaking softly, treating one another with decorum, treating people in public as nonpersons, sending nonverbal cues that approach is not desired
Self-Disclosure	*Self-Disclosure*
• More likely to self-disclose because privacy is protected through physical barriers	• Less likely to self-disclose to protect accessibility of others
Monochronic Time	*Polychronic Time*
• Time is seen in discrete compartments • Compartments are used to schedule events one after the other • Actual time on the clock is more important • Emphasize adherence to schedule • Punctuality important	• There are no compartments • Can do more than one activity at a time • Activities are more important than the time • Emphasize completion of tasks • Punctuality not so important
Face (public self-image)	*Face (public self-image)*
• Less emphasis on respect for elders and superiors • Concern with saving one's own face	• Emphasis on respecting or giving face to one's elders or superiors • Concern with saving the other's face
Persuasion	*Persuasion*
• Focus on the person they are trying to persuade • Direct requests for what is desired • Likely to threaten the person's security • Likely to state negative consequences to a person if the person does not . . . • May ingratiate themselves to the other (I really value you, therefore . . .)	• Focus on the context in which the persuasion is taking place • May use altruistic strategies (for the sake of the group, company, etc.) • Likely to appeal to duty, concern for the whole group • More likely to promise positive consequences if a person does . . . • May imply a "good" person would do this
Expression of Emotion	*Expression of Emotion*
• Less concerned with using emotions to further group cohesion or harmony	• Display those emotions more likely to support group cooperation • Display more of a variety of emotions • More likely to express positive emotions with members of out-groups

(*continued*)

FIGURE 4.4 (*continued*)

Information Seeking

- Try to get to know the person, such as characteristics, beliefs, past experiences, and attitudes
- Look for personal similarities with members of an out-group
- Tend to self-disclose to strangers

Conflict

- Prefer to deal with conflict directly
- Not too concerned that all parties save face
- May look for ways to integrate conflicting views or compromise
- Not as tolerant of free expression of a variety of emotions
- Negative emotions toward others are expressed privately so as not to reflect badly on the in-group
- Negative reactions to members of the in-group are withheld so as not to disturb the harmony of the group
- Negative reactions to members of the out-group are more often expressed to increase the cohesion of the in-group

Information Seeking

- Try to get to know the person's group affiliations, age, and status groups
- Look for group similarities with the out-group member
- Tend not to self-disclose to strangers

Conflict

- Prefer to deal with conflict indirectly
- Concerned that all parties save face
- May try to avoid the conflict or give in to the other

Digital Download Download from CengageBrain.com

Face It is common for people in individualistic cultures to talk about saving face. In sociological terms, *face* refers to a public self-image. In collectivistic societies, there is an emphasis on protecting the face of others, a concept that is less emphasized in individualistic cultures.

Persuasion Different cultures use different methods for persuading people to undertake certain activities or to comply with specific requests.

Expression of Emotion Different cultures express emotions differently and use the display of emotions to further the cultural values.

Information Seeking In all cultures, people attempt to gather information that will clarify situations and reduce anxiety. Members of individualistic and collectivistic cultures go about this task differently. Interestingly, research indicates that in America, European Americans tend to self-disclose more than African Americans do. When close friendships are formed, the opposite is true.

In close relationships, African Americans will self-disclose more than will European Americans.

Conflict Look at the differences among cultures in dealing with conflict. Research suggests that "Chinese prefer bargaining and mediation more than North Americans, . . . Mexicans tend to avoid or deny that conflict exists, . . . Canadians prefer negotiation, . . . Nigerians prefer threats more than do Canadians" (Gudykunst and Kim, 1997, p. 282). In the United States, people are more likely to deal directly with conflict, looking openly for ways to resolve it. With all of these cultural differences in the preferences for handling conflict, it is important to approach conflict thoughtfully.

Obstacles to Understanding

We use a number of different mental mechanisms that can block clear communication. These mental mechanisms are used primarily to reduce anxiety when we encounter strangers. Often we are not aware of the extent to which we resort to these mechanisms. They are particularly present when we are engaging in thoughtless communication. The following sections discuss some of the obstacles that prevent real understanding.

Stereotypes

Some stereotypes are positive, and some stereotypes are held only loosely. Communication is most likely to be obstructed by rigidly held, negative stereotypes that can lead a person to make inaccurate assumptions and predictions about another person. Sometimes a person fits our stereotype of a particular group, but many other persons in the same group may not fit the stereotype at all. Becoming aware of our assumptions and questioning them is important.

Ethnocentrism

Ethnocentrism means that we use the standards common in our own culture to judge the behavior and culture of other people. It is important to understand that we are all ethnocentric to some extent. It is common to look at others through the lens of our own culture. The way our culture is arranged seems "normal" or "correct." Deviations from our culture, therefore, seem abnormal and incorrect. It is not that we consciously decide to employ ethnocentric tactics, but rather we are socialized into viewing the world in a particular way.

The antidote to ethnocentrism is cultural relativism. Using cultural relativism, we try to understand the meaning of others' behavior and communication within the context of their culture, not our own. When we use ethnocentrism to judge people from other cultures, we create barriers and distance. When we use cultural relativism to understand others, we diminish barriers and distance.

Prejudice

Gudykunst and Kim (1997) define prejudice as "judgment based on previous decisions and experiences" (p. 124). People hold prejudices against whole groups of people and against individual members of those groups when they are encountered.

Prejudice involves an attitude that generally stems from a negative stereotype. If you think all the people in a particular group are pushy and devious, you will probably decide that you do not like those people. Not liking those people is a prejudiced attitude that is based on the stereotype you hold in your head about members of that group. This easily leads to discrimination, in which you take pains to avoid being around these people whom you do not like. You might deny a member of this group a job for which the person is qualified or deny a family of this group housing in your neighborhood.

Conflict

When we are already suspicious or uncomfortable with a group of people, misunderstandings can turn into hostility and conflict very easily, particularly since we are likely to attribute the negative behavior of strangers to their personal characteristics, while attributing the negative behavior of in-group members to the situation. All the mental mechanisms described previously serve to make other groups seem less worthy of being understood and enhance the possibility that conflict will occur with individual members of a particular out-group.

Be aware of two points. First, misunderstandings may stem from mental mechanisms that are inaccurate or have obstructed real understanding on everyone's part. Second, once a conflict has occurred, the approach to resolution may be quite different from one culture to another.

Changing Attitudes

It appears from recent research that we can change our attitudes toward strangers or members of out-groups through a number of opportunities to interact with them positively. Stephan (1985) talks about ways to increase understanding and to create more favorable relationships among groups. For instance, an emphasis on cooperation, rather than on competition, is helpful. It is useful if those coming together have about the same status within their own groups and have some similarities in common with each other. Supporting the individuality of each member helps smooth things out. Voluntary contact and contact that is focused on substantive issues as opposed to superficial issues are more useful. Everyone involved should work toward a positive outcome.

The goal of intergroup cooperation and contact is to learn to see members of other groups as individuals rather than as representatives of our own biases and stereotypes. Gudykunst and Kim (1997) address this in their dichotomy between uncertainty-oriented people and certainty-oriented individuals. They write about an uncertainty orientation:

> Uncertainty oriented people integrate new and old ideas and change their belief system accordingly. They evaluate new ideas and thought on their own merit and do not necessarily compare them with others. Uncertainty oriented people want to understand themselves and their environment. (p. 185)

A certainty orientation is quite the opposite. The authors write:

> Certainty oriented people, in contrast, like to hold on to traditional beliefs and have a tendency to reject ideas that are different. Certainty oriented people maintain a sense of self by not examining themselves or their behavior. (p. 185)

When we are communicating thoughtfully, we can make a conscious choice to acquire more of an uncertainty orientation toward new situations and strangers. By remaining open, we can learn more.

We need to go one step further, however, by offering confirmation to others with whom we communicate. When you are working with people who are strangers to you, confirm for those people that they are valuable to you as individuals, that their experiences and concerns are important, and that you are willing to become involved in helping them to resolve their problems. When we deny that another person's concerns and experiences are valid and therefore imply that they are insignificant, we demean that person as an individual, and the opportunity for meaningful resolution and rapport is lost.

Competence

Workers who are adaptable to situations and flexible in choosing how to respond to situations do better in cross-cultural communication. These workers are intuitive and sensitive to what others might mean or need and are open to considering the interaction from a number of different points of view. Figure 4.5 highlights some of the points to remember about individualists and collectivists that might make this process easier.

Competence in cross-cultural interactions depends very much on the individual worker's commitment to give high-quality service to every person who comes for assistance. As you begin to practice, you will encounter more people from specific minority groups or cultural groups with which you are unfamiliar. Ethically, you are responsible for developing an understanding of their cultures or subcultures, at least along the various dimensions outlined here.

Until such study has been completed, it is important to consciously and thoughtfully monitor your interactions with strangers. Make certain that you hear the significance of their concerns and experiences, that you respond in a way that lets them know that they have been heard, and that you provide a respectful environment where problems can be resolved.

FIGURE 4.5 Points to remember in cross-cultural interactions

Individualist Characteristics

- There is a degree of emotional attachment from their in-groups.
- Behavior cannot be accurately predicted based on group membership.
- Out-groups are not seen as extremely different from in-groups.
- Equal relationships where status is equal are preferred.
- People take pride in their own accomplishments.
- Arguments that emphasize harmony and cooperation are not very persuasive.
- Saying negative things about others is more likely.
- It is easier to separate criticism from the person being criticized.
- Long-term relationships happen less often.
- Relationships that contain more rewards than costs are more likely to be maintained.
- Initial friendliness is not considered a sign of an intimate relationship.
- Respect based on age, position, or sex is generally not as likely.

Collectivist Characteristics

- Emphasis is on group membership.
- Group membership can be used to predict the collectivists' behavior.
- When group membership changes, the collectivist behavior changes.
- Unique relationships, where one has more status than another, are comfortable.
- Competition is seen as threatening.
- Harmony and cooperation are emphasized.
- Keeping a positive public self-image (face) is a concern, and individualists can help them to maintain that.
- Criticism and the person being criticized are not seen as separate; thus, avoiding open confrontation is best.
- Deliberately cultivated long-term relationships work better.
- More formal initial interactions are preferred.
- Forced self-disclosure will cause a negative response.
- Respect for age and position is important.

© Cengage Learning®

Digital Download Download from CengageBrain.com

Summary

We know that it is not possible to know the particulars of every culture and respond appropriately to individuals from these cultures. While it is your ethical obligation to know more about the culture of people you see regularly in the course of your work, you will encounter people from different cultures from time to time whose culture you know nothing about.

To avoid problems, it is helpful to know whether a person comes from an individualistic culture or from a collectivistic culture. Once this is established, you are able to respond in keeping with the characteristics of those cultures. In addition, it is always a good idea to check your understanding when conversing with a person from another culture. What seems reasonable to you from the perspective of your own culture may not be reasonable at all to someone from another culture.

Students do better working among many cultures if they keep an open mind and are willing to discard stereotypes and listen for new information. This chapter is designed to give you a starting point for interactions with people from other cultures, but the hope is that you will learn more about other people and their values and norms as you take the opportunity to interact with people from cultures other than your own.

Exercises

These exercises can also be filled out online at CengageBrain.com.

Exercises I: Testing Your Cultural Competence

Instructions: Look at the culture of each client described in the following scenarios, and decide what might be the underlying issue. What are you thinking about the client as you read each description? What are your first ideas about what constitutes the client's problem? Do you have a problem personally with the behavior of the client, and if so, in what way? The following brief explanations about cultural behaviors may help you answer the questions when you read the scenarios that follow:

- In many Asian cultures, members do not talk about family problems, feeling that these are private. They may also pretend that no problems exist.
- In most Asian cultures, crossing the legs and pointing the toe at another person is considered extremely rude. Because members of Asian cultures like to maintain harmony, they would not be likely to tell you directly that they were offended. In addition, in most Asian cultures, waving at another person or indicating that a person should join you by calling the person over with your hand or a finger is also considered extremely rude.
- Asians are not likely to make changes in the family or to engage in discussions about the family unless the male head of the household is present or is consulted. Furthermore, they are likely to tell you things are all right in order to maintain harmony. Things may not be all right. They may also tell you that you were helpful to them because they assume that is what you want to hear, not because it is true. Telling you what they believe you want to hear will maintain harmony.
- In most Asian cultures, group needs and considerations are more important than individual needs and considerations.
- In most Hispanic families, the man makes the major decisions and expects to be consulted about anything affecting the family. He would not be likely to take his wife's ideas or concerns into consideration. She would be expected to defer to the husband.
- Many Hispanic families allow mental health problems to persist for a very long time rather than admit that there is a psychiatric problem.
- In Hispanic culture, it is often believed that depression is due to a lack of religious faith.
- In Hispanic culture, mental health problems are often attributed to sin.

1. A man from Vietnam is in your office because his 11-year-old daughter has been having trouble in school. The school suggested the daughter be tested by your agency. You are doing the intake, but only the father has come into the office. He is very reluctant to tell you any specifics, but talks instead in extreme generalities. What might be the reason for his reluctance to talk to you in detail about his daughter's problems?

2. A Japanese family in the emergency room is seeing you because of a serious accident in which their teenage son was severely injured. During the course of the conversation, you cross your legs so as to be more comfortable. The family continues to talk to you in a polite but superficial manner, and gradually each member drifts away—to get a soda, to use the restroom, and so on. They are obviously resistant to sitting down with you again.

3. A Chinese woman is hospitalized for a serious infection, and her doctors think she seems depressed over possible home problems. You talk to her, and she appears to reassure you that everything at home is fine. When you come in the next time, she wants you to talk to her husband instead. He, too, is reassuring and pleasant. Later you ask the woman if your talking to her before was helpful, and she smiles and tells you it was. You are not sure.

4. You work in the school counseling office, and you have been asked to help a gifted young Vietnamese student fill out applications to several prestigious colleges. She is in line for a number of scholarships. She works on the applications, but with obvious reluctance, and indicates she cannot consider going away to school until the family has decided what she will do.

5. After the birth of her fourth child, a Puerto Rican woman is referred for depression by her family doctor. At the intake interview, her husband comes and answers all questions. He indicates that he will decide what she needs and what is to be done. The wife says very little and speaks only when she appears to feel her husband wants her to do so.

6. A Mexican American family brings an elderly aunt to the emergency room. The older woman is severely depressed and emaciated. She also appears to be nearly catatonic. She is admitted immediately to the hospital on an emergency commitment. You learn that this woman has been in severe depression for years and wonder why the family waited until things became so serious. Later, when you talk to the aunt, she is somewhat improved. She indicates that her pastor visited her and told her that her depression is due to her lack of religious faith. She tells you she agrees with this assessment. She tells you she believes that if her faith was stronger and she was less sinful, she would not feel like this.

7. A young Hispanic woman is admitted to the hospital, and the doctor believes she is showing obvious signs of schizophrenia. She is hallucinating and has not been eating. Her family tells you this problem is the direct result of her sinful behavior. According to them, she is too friendly with the boys in her class. The students often call each other to compare homework, and sometimes a group of boys and girls will go to a party or several boys and girls will walk home from school together. The family has tried to get her to cut off these friendships and to stay home and help her mother more. Because she did not listen, she is being punished.

CHAPTER 5

Attitudes and Boundaries

Introduction

The way we see other people and the way we relate to them as a result will affect how things turn out in a relationship. The **boundaries** we erect and the boundaries we fail to observe can similarly have an effect on outcomes. In this chapter, we look at **attitudes** that facilitate relationships and boundaries that are useful and we consider those attitudes and boundaries that are adverse. The purpose is to give you an opportunity to increase your awareness of attitudes and boundaries and to become more observant of your own ways of viewing other people, particularly other people you intend to help.

Understanding Attitudes

Attitudes are extremely important. The feelings you have about other people are bound to be communicated to those people one way or another. If your attitudes are positive and supportive, you will be more likely to establish rapport. If you feel superior or disdainful, no matter how well you try to hide those feelings, they will eventually be communicated to your client, and you will lose a working relationship.

Good human service workers have learned about themselves—their fears, sensitivities, and errors in judgment. In facing those things about themselves, these workers have come to understand themselves in a way that enables them to feel understanding and **warmth** toward themselves and toward others. If you are able to forgive yourself for the mistakes you make and the struggles you have had and see them as an important part of growing, you will recognize that you are basically all right. It is then much easier for you to understand others who are making mistakes and who are struggling with issues and problems. You know that the problems you have faced have provided valuable lessons. These personal struggles have helped you to grow into a more sensitive and insightful person. Problems and unfortunate decisions happen to everyone. With this personal understanding, you will be more inclined to see others' personal struggles as productive of growth, and not necessarily as a reflection of personal inadequacy.

Begin, therefore, with yourself. Be tolerant of your mistakes. Look at yourself objectively. Forgive yourself for errors in judgment, particularly errors that have taught you important lessons or helped you to grow. Recognize that part of being human is to struggle with issues and transitional problems, and that through this process we are often strengthened and given new insight.

Basic Helping Attitudes

You need to bring three basic helping attitudes to your work: warmth, **genuineness**, and **empathy**. Studies show that even in the absence of much formal training, workers who genuinely care for their clients and are committed to them will be able to help their clients make important changes and move toward better circumstances and increased emotional health.

Warmth

A worker needs to be friendly, nonjudgmental, and receptive. These three attitudes create a warm atmosphere, one that serves to put the client at ease.

In your presence, clients feel valued by you as a person. You communicate a belief to them that they are worthy of being understood. You do this through your actions, body language, and the way you listen to each person. In addition, you refrain from judging what clients say and the actions that they take.

For instance, a warm person would smile at a new client, offer the person a chair and would show genuine interest in why the person came in today. Your capacity to smile at another warms a room and eases tension immeasurably. The atmosphere is cold when there is no smile or the offer of a chair and the worker begins mechanically with a series of questions found on a form.

In addition, you show clearly that you are receptive to what people have to say. You listen to what they tell you they have done and felt, what brought them to seek your assistance in the first place. When you say, "Tell me more about that," you are inviting the person to open up and talk safely with you. Lacking warmth, a worker might be inclined to judge what the client has said: "It sounds like your decision wasn't a good one." Such a response immediately puts the other person at a distance.

Part of being warm is establishing that the client has something to offer in the relationship as well. You come across as respecting your clients' right to make their own decisions. You may facilitate better decisions than the ones they might have made alone, but ultimately the decisions about their lives are their own, and you will respect that. Sometimes we are painfully aware that a decision is not the right one for the person to pursue. Later in this book, we discuss how you can give clients some of your thoughts in ways that people can hear and use your ideas more easily. Nevertheless, people may choose to ignore you and make their own decisions, and you will respect that and support the person if need be through the consequences. We have all made unfortunate decisions and concluded, as part of our growing process, that they were not very useful.

Some people think being a warm person means that they must stand by passively, never confronting or giving the person another way of looking at things. Warmth means you are able to facilitate change with your support, and change is often painful. As you learn the skills in this course, you will pick up methods to use that help people see things from more than one perspective, to change their direction and possibly see characteristics about themselves that are difficult to face. Warmth is what makes it safe to do that in your presence.

Genuineness

You have heard the expression "be yourself" many times. This is a must for those who help others. Nobody relates well to a phony. People sense when you are not being authentic.

Perhaps you use slang that sounds forced, phrases you do not usually use. Maybe you are using them now to make clients think you are familiar with their lifestyle or culture. Maybe you put on a phony dialect or use profanity to seem more down to earth. You might pretend to be a physician, wearing a white coat and allowing your clients to call you "doctor." A client calls you "doctor," and you do not correct the impression. You might pretend to have degrees you do not have. You could use big words you know the client does not understand. None of this is authentic. You will be seen as a person who is pretentious or rather foolish at best, and untrustworthy at worst.

Tom found himself in a hospital setting as a case manager working among doctors and therapists, all of whom wore white coats. Tom began to wear a white coat as well and allowed the patients on the psychiatric inpatient unit to call him Dr. Rolland. In doing so, he styled himself as superior to the patients using words they could not understand and giving advice rather than listening. Tom was a likeable person and could have been a real asset to the treatment team. Under these circumstances however, he was avoided by staff who were resentful of his pompous demeanor. Patients too seemed uncertain about what he meant when he spoke so his work with them was not particularly beneficial to them.

Be open and truthful. Strive to be yourself, to present your authentic self to the other person. If you do not know something, say so. If you lost a client's forms, tell the person that and apologize. If someone asks about your credentials, matter-of-factly tell the person what they are.

Empathy

Empathy in its most basic form means being able to put yourself in the other person's shoes, but clearly empathy is much more that this. For example, Susan Gair (2012, p. 134) writes that empathy is a "quest . . . to be able to hear, feel, understand and value the stories of others and to convey that felt empathy and understanding back to the client . . . ." It is, according to Gair, "feeling with the client rather than for the client (sympathy)" (p. 135). Other researchers describe empathy in much the same way.

Empathy, therefore, is not sympathy. It does not mean that you are so sad for people that you take their situations home and fret about them when you are away from work. It does not mean that you feel sorry for people and communicate the belief that they are poor souls or that their situations are without hope. Sympathy is what we often have for our friends and relations. Empathy involves assessing where the client is at any given moment and being able to express that and support it. Sympathy is the common feeling we have for others in pain. Empathy is a basic clinical strategy for supporting people through difficult times.

Elliot et al. (2011, p. 44) writing in the journal *Psychotherapist*, describes empathy as "an active on-going effort to stay attuned on a moment-to-moment basis with the client's communications and unfolding experience." They go on to write that empathy means that the "primary task is to understand experiences rather than words" (p. 47). Other writers refer to empathy as "engagement" with the client on the part of the worker. Still others talk about "making a connection." Some writers suggest that empathy comes from our common humanity with others and it is this common humanity and our personal awareness of that which allows us to practice empathy.

In summary, empathy involves much more than listening to what another has said and understanding the spoken words. Instead, empathy is the capacity to feel what the client feels and to grasp the significance and impact the client is experiencing. To do this well, you need to be able to comprehend the other person's point of view and see clearly what that person's needs and feelings are. In human services, we often say that we are able to listen with the third ear. We hear more than what the client is telling us. We hear the underlying emotions, desires, and worries. We see the body language and tune into the tone of voice. With practice we become proficient at this, and we develop a special sensitivity. As Elliot et al. (2011, p. 48) point out, empathy means attending "to what is not said, or what is at the periphery of awareness as well as that which is said."

What do you do with your empathic understanding of the client's experience? The next part of empathy is being able to accurately communicate to the person an understanding of their emotions and experience. If you can put into words the feeling the person is experiencing, you are practicing empathy, but it is much more than words. For example, Lila was talking to her new case manager about the loss of her mother. Lila, the only child of a single mother, was 15 years old and home alone the night her mother died in the final stages of leukemia. Visiting nurses and hospice had all been in the home earlier and hospice came again when Lila called them. However, at the moment her mother died, Lila was alone with her and feeling loss and helplessness as her mother slipped away from her.

Her case manager listened intently. She leaned forward toward Lila, her facial expression registering real concern for this young girl. Her words were, "Lila, that must have been so hard for you" and Lila felt heard and understood. She nodded and began to cry. "That must have been so hard for you" can be said in many different ways but said empathically it is heartfelt and accompanied by body language and tone of voice that say clearly that this case manager understood the pain Lila felt that evening.

Further along in this textbook you will learn the skills to respond to people who tell you their problems and worries, but mere words, while they can convey understanding, are not enough for the empathic listener. Empathy is expressed with more than words. Body language, tone of voice, and an underlying sense of the pain or the desperation or the sadness of the other person are all part of empathy. Elliot et al. (2011, p. 44) write that empathy is, "an active, ongoing effort to stay attuned on a moment-to-moment basis with the client's communication and unfolding experience." And they talk about empathy being "grounded in authentic caring" (p. 48). In her study of the effects of empathy, Warmington (2011, p. 16) writes, "the display of emotions such as kindness or compassion revealed in tone of voice or facial expression is likely to form a stronger bond than the recitation of lines learnt in communications classes alone."

Empathy and Safety

Most of the research into the effectiveness of empathic listening shows that empathy creates a safe space in which the client can express genuine emotions and tacitly gives permission for them to do so. When one is understood, he feels safe enough to explore his problems and issues (Elliot et al., 2011). Empathic engagement with the client creates the space and the permission to fully explore and discuss the issues with which the person is grappling. "(A) sustained effort to understand the kinds of experiences the client has had, both historically and presently" (p. 44).

Empathy and Compassion

Many writers studying this topic link empathy to compassion. When we are empathically engaged with another person who is distressed, we come to feel compassion for that person and a desire to help in some way to alleviate their suffering. Gair (2012) notes that several researchers have "concluded that people feel empathy and want to help" (p. 135). Warmington (2011) talks about a "desire to act to alleviate (the other's) suffering" (p. 16) and she goes on to say that compassion comes about naturally when we are empathically engaged with the other person. She writes, "(C)ompassion could be said to be complementary and intimately related with empathy" (p. 16). In other words, when we truly see things the way our client sees things, understand her point of view, feel her emotions, a desire to help that person naturally follows.

Empathy and Boundaries

Carl Rogers (1980) wrote at length about empathy. Here are some of the definitions he used to help us understand what empathy is:

- "(Empathy is) the therapist's sensitive ability to understand the client's thoughts, feelings, and struggles from the client's point of view" (p. 85).
- "It means entering the private perceptual world of the other (and) being sensitive, moment to moment to the changing felt meanings that flow in this other person" (p. 142).
- Empathy is "an active on-going effort to stay attuned on a moment-to-moment basis with the client's communications and unfolding experience" (p. 142).

Nevertheless, Rogers raised the concern that given this degree of engagement with another person, workers could lose the boundary between themselves and the client. He talks about not losing the "as if" quality of the relationship. In other words, the case manager is fully attuned to the feelings and perspective of the other person "as if" these were his own. However, he does not make them his own.

Numerous writers echo this concern writing that good empathic listeners, such as case managers, could lose themselves in the other's problems. This can happen when a case manager identifies too closely with the client. Perhaps the case manager is overcome by the client's feelings of helplessness and therefore cannot construct a solution with the client. Maybe the client's issues and pain are too close to feelings the case manager has had and therefore the case manager cannot separate her own problems and feelings from those of the client (Warmington, 2011).

The answer to this appears to lie in knowing our boundaries. Gerdes and Segal (2011) tell us that, "true empathy cannot exist without a strong sense of self as separate from the other" (p. 145). They go on to write, "Self awareness allows us to disentangle our own feelings from the feelings of others . . . " (p. 145). The concern is that the feelings of the client can become the feelings of the worker. Gerdes and Segal (2011) point this out. "Without perceptual boundaries (case managers) risk experiencing a client's feelings of anger, depression, anxiety, or joy as their own feelings" (p. 145).

When we do not have a strong sense of self, we may experience the emotions and issues of the other as if they were our own and we might begin to project our ideas, our motivations and prejudices onto the other person as if these belong to him (Gerdes and Segal, 2011).

In summary, to practice empathy well, you communicate in a way the other person can understand and accept, not in a way that is threatening or judgmental. You communicate tentatively, aware that you might be wrong or a bit off the mark. You readily accept correction by the client. Empathy is expressed "in the context of positive regard and genuineness . . . . (G)rounded in authentic caring" (Elliot et al., 2011, p. 148).

On Being Judgmental

Harriet blamed the mother of a child she was seeing for the child's problems. Lily, aged 8, had pica, a condition in which the person eats substances that have no nutritional value, substances that are not food such as dirt, paint, or paper. Harriet was cool and dismissive of the mother when the parent attempted to explain that she felt something physically was wrong with Lily. What Harriet saw was a parent who lived in poverty, who could not possibly, therefore, know what was going on with her child. Harriet thought of the mother as an inadequate parent based on her poverty and lack of schooling. It wasn't until a psychological evaluation recommended that Lily be tested for anemia or other medical causes for Lily's behavior that Harriet was willing to consider a medical basis for the problem. Based on the psychologist's recommendation and not on the mother's concerns, Harriet set up a medical evaluation for Lily. It turned out later that Lily was extremely anemic and medical interventions were successful in stopping the behavior.

There are human service workers who sit in judgment of clients, applying their own standards to people who are less educated, poorer, or sick. There are clients who have been through trauma, or who simply don't know about alternative ways to approach their lives. There are people who say they are in human services to be helpful, but who are actually wary and distrustful of their clients. This is an attitude that has no place in the helping relationship.

For example, Rose, a new worker, was part of a planning team for a new shelter that was about to open for children who needed a place to stay before foster care could be found. These were children removed from their homes for physical or sexual abuse or extreme neglect. During the planning, it was decided that the children would earn points for good behavior and lose points when they violated policies of the residence. Children with 10 points or more would get special privileges, and those with no points would stay in their rooms temporarily and miss special activities. The director asked the group, "Where shall we start a child when she comes in new? Shall we start her with 10 points or none?" Rose was convinced that each new child should start with zero points and earn them. "How about starting each child with 10 points and letting her work to keep them?" the director suggested.

When the planning group ultimately adopted the director's suggestion, Rose was upset, thinking that the children would "just come in here and take advantage of us." She asserted that the children "need to know we mean business right from the beginning." The assumption that the children would automatically misbehave and take advantage of the staff was judgmental—adding that judgment to the problems the children were already experiencing would have been cruel. Rose's need to curtail and punish before any misbehavior had taken place would have put her negative attitudes into action in a destructive way.

A judgmental worker is one who measures people by rigid or irrelevant standards. Often, these standards have to do with what the worker assumes about certain people or how the worker feels about others. In addition, judgments are often based on ideas about what people should do or how people should act or live and not on the actual context in which the person lives and functions.

Reality Check

Some human service workers say they have gone into this work to help other people, but the minute a client behaves in a difficult manner, they complain that they should not have to deal with this sort of person. Clients who don't cooperate, who do not show gratitude but are demanding instead, or who are rude throw these workers off balance.

Rita refused to sit down with the mother of a child who was in treatment and work with her on a solution to the child's behavior problems. Rita complained that the mother was a difficult person to work with. "She's always complaining about what we're doing. She thinks she has a better way. I just want to say to her, 'Look, if your way was so hot, you wouldn't have the problems you've got with Benny.' All she ever does is make stupid suggestions. I just wasn't going to get into that with her."

In human services, we are trained to deal with people who seem dissatisfied and upset. We expect to meet people who do not see things our way, who are challenging, or who question our decisions. Many of the people we work for will have an inaccurate perception of reality or difficulty expressing what concerns them, seem unduly sensitive, or be confused about following a plan. This is largely the reason our clients sought our help in the first place. If we only want to work with people whom we like and who agree with us, people who give us no trouble, we will be barely effective and more likely harmful. Just because people do not behave exactly the way we would like is not an excuse to provide poor service.

Rita and others like her have the false impression that clients should be grateful and cooperate with suggestions. For workers like Rita, the reasons for rude behavior or a lack of cooperation are reasons to deny good service and complain about feeling unappreciated. Their attitude is not realistic for human services work.

How Clients Are Discouraged

There are many ways to discourage another person. You could set up a competition comparing the client to others or to yourself. You could push, force, or shame the client into moving toward some goal. You could spend an inordinate amount of time focusing on the person's mistakes or demand that the person do more or try harder. You could insist that clients do things your way, dominate clients by taking over or demanding perfection or unrealistic outcomes, or intimidate the person with threats. You could treat clients like "poor souls"—incompetent, bumbling people who need you to do everything for them.

You could be discouraging with your insensitivity by failing to notice positive changes, ignoring the good things people accomplish, failing to mention the positive changes or your clients' strengths. You could refrain from ever giving feedback except the most negative type. All of these responses are discouraging and introduce a negative element into what should be a helping relationship. **Discouragement** is never helpful.

Today, many individuals, some of whom hold degrees in fields unrelated to human services, are being asked to give direct care to people who seem odd and unusual to them. For this reason, the clients may be frightening to these workers. In order to counteract feelings of uncertainty, such a worker may become extremely dominating and coercive. This gives a false sense of control. The domination and need to order clients' lives in inappropriate ways are discouraging to people who are learning to take charge of their lives and make decisions.

An example of this occurred in a program for individuals recently released from an institution for those with intellectual disabilities. A case manager visited her clients every other day in their apartments, where she offered support to help them remain in the community. She enjoyed working with her clients, and she had received good human service training. When the clients began to make jokes about the pounds they had gained over the years in the institution, she suggested they might want to exercise. Together they decided that walking around the apartment complex might be fun and a good way to meet their neighbors. For weeks, the clients walked almost every day, but at different times of the day and not on days when the weather was too cold or snowy.

Later, the case manager's supervisor, a person with little human service experience, was upset that the case manager had not "scheduled the exercise." The supervisor felt that the clients should not be allowed to just say they would walk every day "because they won't do it. We need to put that in their daily schedule, let's see, at 11:00 to 11:30 every morning. That way we can check on them and be sure they are doing it." The case manager asked, "But what if they skip one day, or feel like going at 3:00 in the afternoon?" The supervisor replied, "My point exactly. This way they have no choice, and they'll meet their goals. We'll look good for seeing that the clients' goals are implemented consistently. Set up a daily schedule for them, and see that exercise goes into it at 11:00." A person beginning to live a normal life in a community, who has decided to take up an exercise program, and who has demonstrated a commitment to that program, does not need a professional to oversee the timing of it. This is an example of inappropriate control. In this case, the walking program fell apart as clients had other things to do or felt coerced by the program director.

A very grave example of discouragement occurred in a partial hospitalization program. Kitty, a client in the program, suffered from severe schizophrenia and depression. Often she was immobilized with sieges of despair and delusions, with voices of many others talking to her in what she called a "confused conversation." Kitty described herself as afraid and appeared to the staff as dependent. The staff surmised that because Kitty had a master's degree, obtained before her first episode of depression, she was really capable of more independence. They developed a series of goals for her to follow, such as riding the bus, shopping at the mall, and handling arrangements for her insurance and transportation. The final step on the list of goals was for Kitty to prepare her tax returns because her degree had been in business.

From the start, Kitty had problems managing the goals. Feeling extremely depressed and occasionally hearing voices, Kitty found it alarming to be on her own in the city. In group sessions, Wayne, the group leader, held her up to ridicule. He encouraged the other clients to scold Kitty and accused her of refusing to help herself more. Kitty asked to reexamine her list of goals, but this was greeted with a refusal on the part of the staff and further insistence that she "get out there and try harder." Finally, Kitty decided to withdraw from the program. When she told the staff of her decision, they told her that unless she cooperated with the program set forth for her, she would not be allowed to come to the clinic for her prescriptions. These prescriptions, partially underwritten with public funds, were important in sustaining Kitty's connection to reality. Kitty finally called a friend of hers, a psychologist who worked in the state mental health system. When the behavior of the staff came to light, staff members were reprimanded and Kitty obtained her prescriptions and counseling services elsewhere.

We need to examine what went on in Kitty's case. First, a client has the right to ask that goals be reexamined. It may indeed be that the goals are not truly in line with the capabilities of the client, and setting goals should always be a collaborative effort. Second, a client can always withdraw from service if the client determines the service is no longer useful or, as in Kitty's case, is actually harmful. The client has a clear right to determine what is good for and helpful to her, and what is not. Finally, the use of medication to coerce the client into doing what the staff has determined she will do is highly unethical.

All these factors combined in Kitty's case to create a discouraging atmosphere. On top of that was the denigration by the staff, particularly Wayne, who pointed out Kitty's deficiencies and ascribed manipulative motives to her failures without ever really working with her to plan goals at which she could succeed. It would have been hard for a person like Kitty to attempt any goals in such a negative situation. Kitty was deeply discouraged by the actions of the program and made no progress there.

Another example of discouragement happened in a transitional living arrangement for the mentally ill. This transitional living situation was one of many living arrangements with varying degrees of independence that helped clients move from hospitalization to independent living. Mario, who had persistent schizophrenia, had been placed in the last step toward independence, a small house with four other clients. At the time, Mario was on new medication, begun while he was in the hospital, which increased his ability to function considerably. While in the program, he adjusted well to the medication, obtained a job, and began to look for a new apartment.

This was all part of the plan. Finding a new apartment was difficult, however; while extremely ill many years before, he had presented problems for one landlord after another. Now, in spite of the obvious improvement, Mario had a reputation for being a problem when he became ill, and no one was willing to risk his becoming a tenant. In the meantime, Mario collected furniture for a new apartment and continued to look at ads in the paper. Staff in the program assisted.

The time for Mario to remain in transitional housing expired. At that point, the case manager called the assistant director of the agency. The case manager was belligerent. He said the client was obviously "high functioning" and was therefore "stalling" in finding a place and moving on. He stated he knew nothing about the client's past history with landlords and was not interested. He reminded the program director that case management was responsible for the bills for the transitional housing, and they were going to stop paying for this service for this "manipulative" client.

"He obviously has no intention of moving and thinks we're all too dumb to see it," the case manager claimed. "Let him know he has 7 days to be out." Without asking for a meeting, without sitting down with Mario or with the staff in the program, without looking at what might be going on and how the agencies could work together to facilitate a positive transfer to conventional housing, the assistant agency director wrote a peremptory and hostile eviction letter, making it clear that Mario had 7 days to find housing or he would be "evicted," and stating that Mario's case manager supported this decision and thus Mario should make no appeal to his case manager for help. Coming home from work, Mario found the letter on his pillow and immediately deteriorated. Staff at his transitional living program had not been informed of the letter and discovered the client in a frenzy in his room, throwing things into garbage bags and crying. Sometime later he left the house.

When he did not return that evening, the staff became alarmed. They found and read the letter left in his room, but were unsure what to do. They notified the evening supervisor of another of the agency's programs, and together they were able to track Mario down in another state where he had gone to be with his sister. Much later, case managers and others began to put notes in the record that indicated work had been done to find housing with this client. The notes were made to appear as if they had been written long before he left the program and that the case manager had actively tried to assist Mario. The notes were back-dated, a highly unethical practice. These notes contained indications that the client had been uncooperative during this time, something the transitional housing staff firmly denied.

In time, the staff in transitional housing was able to let the executive director of the agency know of the true nature of the incident. The executive director had been told this was a smooth leave-taking by the client. Reprimands followed, but these in no way made up for the damage done to the client's sense of self-esteem and confidence or the ground lost in this client's move toward independence. Mario was discouraged in the sense that the circumstances related to the eviction caused him to lose ground and confidence in himself. What had been gained had been lost.

A Further Understanding of Boundaries

There is always a danger that we will see ourselves in the clients we serve. Many of us will work in agencies that serve people who have been through something we went through ourselves. We may meet a person whose situation is different but that person reminds us in some way of ourselves. Perhaps the person is our age or has the same interests we do. When clients remind you of yourself and you are unable to separate your circumstances from theirs, you will become an obstacle to the person you are supposed to be helping. Sometimes we have not entirely resolved the issues in our own life. We may seek employment in an agency that deals with similar problems just to continue to heal. This is not useful. It means that the clients will be treated in the light of your own issues rather than with a completely objective focus on their own personal issues.

Sometimes to protect ourselves, to feel superior, to exercise power we erect unnecessary boundaries that are really barriers to good service. As the case manager, you are the person responsible for maintaining useful boundaries and refusing to erect those that are not helpful.

Seeing Yourself and the Client as Completely Separate Individuals

The Client Reminds You of You

A young woman volunteer who was in training to answer the hotline at a rape crisis center had been raped some years before. It was still an overwhelming event to her, the defining event in her life. She was not ready to begin to work with others who had been raped while her own emotions were so raw and intense. During the training, she would fret and stew over the information being given. She would constantly remind the trainer that this would not have worked for her in her situation because she was too upset or too badly injured. She pointed out her own circumstances and told the trainer to focus more on situations similar to hers. She gave details of her rape at every opportunity. She used her own situation to illustrate the trainer's points.

It was apparent that she was not ready to serve as a volunteer who would have direct contact with the clients. Why? She had not yet recovered from the trauma of her own rape, and it seemed likely that she would impose the circumstances and emotions of her own rape onto those of the caller. In that case, the caller's real problems might get lost as the worker went off on her own concerns and feelings. She was asked to take other work in the agency, and ultimately she left because she never was able to separate her rape from those of the clients the agency served. Her rape and trauma were the same as all other rapes and traumas and she could not distinguish the difference among them.

Sometimes the client is trying to do something we accomplished ourselves long ago. The client reminds us of a time when we were vulnerable and uncertain. For example, in a welfare agency, a young woman, recently off welfare and now working as an income maintenance clerk, was helping a welfare client make plans to get off welfare. The worker was somewhat irritated by the client's concerns about day care and transportation. "Look," she finally blurted out, "If I can work every day, you certainly can!" Here the worker was viewing her client's life through her own life and circumstances, blurring the boundaries between them.

In another situation, the similarities were too much for Yolanda, a human service worker, to tolerate. She went to great lengths to find differences between herself and the client. Working at a program for individuals recovering from alcohol abuse, she had trouble understanding how one of the clients could have "gotten into a mess like this." The client had regressed over the weekend and was drinking again. She came in seeking Yolanda's help, but Yolanda had been through recovery and this woman's "slip" reminded her of the hard fight she had made to stay sober and clean. What made it harder for Yolanda was that this woman was Yolanda's age, had two children the same ages as Yolanda's children, and lived two blocks from Yolanda. Instead of listening to where the client was at the moment, Yolanda became angry, asserting how hard she had fought to get where she was and telling this person to do the same. This woman's relapse threatened Yolanda's view of her own success. It became important, therefore, to demand that the client straighten up and never, ever think of doing this again. Her anger and refusal to start where the client was at the moment alienated the person. In subsequent meetings, Yolanda pestered the woman to tell whether she had relapsed since her last visit, and she admonished the woman frequently about her "tendency to drink." Gradually the client drifted out of the program.

When the client reminds us of ourselves, we may push people to do things that we did to successfully resolve our own similar problem or push the client to take a particular course of action. Our life and circumstances are superimposed on the client's life and circumstances, blurring the distinctions between the two.

The Client Reflects on You

There are other reasons that a human service worker might not separate from the client. As a case manager, you might become extremely involved in clients' problems and solutions because it makes you feel more important or competent if the clients solve their problems successfully. It may impress others if all your clients do well, and so you may cross the boundary and force people to use your solutions. Sometimes, in spite of what we know about how good it is for individuals to learn from their errors and struggles, and in spite of our belief that people are responsible for their own lives, we also mistakenly believe that all our clients must be happy and successful if we are to look good. Again, we are focusing inappropriately on ourselves, and not on the clients' needs. Pushing people so that we ourselves look competent is unprofessional.

Erecting Detrimental Boundaries

False Attributions

Some boundaries are artificial and inappropriate. There is a very human tendency to make these two extremely unfortunate assumptions about people:

1. People who look like me will think and act like me.
2. People who do not look like me are not like me at all, but very different.

Neither of these assumptions is necessarily true. A person who comes from your race or culture may not share your values or circumstances. We need to watch for signs that these assumptions are not creeping into our thinking because these ideas erect negative boundaries that can make you ineffective.

Be very careful what you assume to be true of another person. A person may be of another race or culture and yet share many of your values and circumstances. A person who looks like you may be quite different in tastes, opinions, and way of life. If you assume a person of another race or religion has certain stereotypical characteristics, you will be dealing with a stereotype, and not with a real person.

When you view people in light of these assumptions, you no longer see individual differences. All Catholics, African Americans, Jews, Quakers, and Indians are not alike. If you view people based on your stereotypes of the groups from which they come, you will fail to discern individual differences. Failure to perceive individual differences is a failure to accurately perceive reality. Aside from this being a fundamental characteristic of mental illness, it is impossible to give excellent service to a person you see only as a stereotype.

False Power

Another unsuitable boundary that acts as a barrier or more like a barricade is our own need to display authority, competence, and power. In this way we may intimidate our clients and threaten them as Wayne did to Kitty. We may stop collaborating and give orders instead, as Tom did on the psychiatric inpatient unit. This way of working with people may make us feel important and all-knowing, but it will not help the client.

Transference and Countertransference

Transference

Sometimes, when working with other people, you will find that you remind one of your clients of someone in their past. Sometimes clients are only dimly aware of that. They simply know that there is something about you that they really like or really dislike. This is transference, a collection of feelings and attitudes the client holds about you. Positive transference occurs when the client likes you, and negative transference occurs when the client does not. Understanding this is one more reason why it is important that you not take the feelings of clients personally.

In addition, clients can start out with positive transference toward you but then find those feelings changing. Perhaps a client hopes to make the relationship with you into a friendship. When you maintain professional boundaries, the client may reverse the positive feelings and the transference may become negative.

As a competent case manager, you will want to accept transference when it exists. It is not a good or a bad thing. It is something that commonly takes place in helping relationships. In other words, practice acceptance rather than becoming threatened and defensive. There will be times when clients' behaviors seem irrational, hostile, or even overly seductive. Acceptance of such behaviors while maintaining professional boundaries prevents negative barriers from developing and allows people to continue to feel safe. In that environment, the clients may be able to reevaluate their feelings about you.

It is helpful, when clients respond in unexpected and seemingly somewhat inappropriate ways to the situation, to use reflective listening. This allows people to know that you are not judging their behavior or attitudes and that you have heard them and want to understand. This, in turn, creates the safe environment you are seeking to maintain.

Sometimes this is how dual romantic relationships get started. The client sees in you a romantic partner based on experiences he had outside the agency. His seductiveness may be just what a particular worker needs at the moment and a romantic relationship begins. For this reason, remind yourself of transference when a client behaves in ways that seem inappropriate to the situation.

Countertransference

In countertransference, the case manager projects onto the client certain emotions and attitudes because the person reminds the case manager of someone from the case manager's past or because the client's issues and situation cause the case manager to identify with that person. This can arouse both positive and negative emotional responses from the case manager toward the client. A case manager may give special favors to someone who is reminiscent of an elderly aunt the worker once loved and who is now deceased. On the other hand, the worker may be inappropriately demanding when the client reminds him of a younger brother who was always bullying the worker when they were kids.

Acceptance of yourself and your feelings is extremely important here. We will meet people throughout our lives who remind us of other people. When it affects the way we work with our clients, then we need to be very aware of our feelings. Countertransference feelings are often a good warning signal that we have old issues we need to resolve. It is never acceptable to allow these feelings to interfere with service to your client.

Summary

Many, many good therapeutic approaches have been developed after years of study and trial and error. In addition to all of these, researchers have found that warmth, genuineness, and empathy have a profound effect on clients' ability to move forward and heal. Many students begin this course of study intending someday to become therapists and counselors. Learning how to convey these three essential elements of a therapeutic relationship is where a true therapeutic relationship starts.

Using the ideas in this chapter, you will recognize discouragement and know how to encourage and motivate your clients. Putting aside what you want for your clients and starting where your clients are gives them the footing they need to move to something better. The three basic elements of warmth, genuineness, and empathy form the underlying foundation for everything else you will do with people. Even when you are uncertain about how to proceed, if you are using these elements, your actions will nearly always be viewed by others as supportive.

In addition, it is important, for all the reasons stated in this chapter, to view each client as an individual person and to work very hard to understand what makes this person a unique human being. Through careful listening and observation you come to understand what it is that makes your client a unique person. Professionals work against relying on stereotypes, assumptions about groups of people, and personal feelings about certain problems and their solutions that will color the work they do with others. Take the time to see and hear the characteristics, circumstances, and interests of this person who is trusting you to assist in some way. Focus on what makes your clients separate individuals. Give your service based on what you know about the person as unique.

When workers begin to identify too strongly with their clients, or feel that they must assume some of their clients' responsibilities, they have blurred the boundary that makes both the worker and the client separate individuals. They cross that boundary in unproductive ways every time they assume that a person will be like them or will react as they would. These workers breach the boundary each time they handle a client's problem without collaborating with the client, whom they fail to see as distinct.

Exercises

These exercises can also be filled out online at CengageBrain.com.

Exercises I: Demonstrating Warmth, Genuineness, and Empathy

Instructions: There are five grade levels of responses you might give to people who have come to you needing assistance in sorting out their problems and feelings.

Grade A. These responses are the most useful in establishing rapport and encouraging a continued dialogue:

- Centers on the client entirely
- Stays on the topic (responds to the client's feelings or the content of what the client has said)
- Addresses what is most important at that moment to the client
- Is respectful (indicates the client is an equal; indicates the client is a person worthy of being understood)
- Invites collaboration
- Shows confidence in the client

Grade B. These responses are helpful but could be better:

- Is somewhat confident of the client's abilities
- Minimal invitation to collaborate
- May briefly stray off the topic
- Is just a little superior

Grade C. These responses are usually made by someone who means well, but they are not especially helpful responses:

- Is pleasant, but superior
- Overly helpful without collaborating
- Introduces new topics that seem to the worker to be more relevant
- Misses the feelings
- Does not address the content
- Confuses the worker's situation with the client's situation

Grade D. These responses are not useful in establishing rapport and do not encourage a further exchange:

- Takes over with solutions
- Spends little time listening; is abrupt
- Moralizes and preaches
- Ignores the client's assets and strengths
- Shows minimal interest
- Does not indicate respect for the client

Grade F. These responses are mean-spirited and damage the relationship irreparably:

- Uses denigrating labels and descriptions of the client or the client's actions
- Shows no interest in the client
- Denigrates feelings of the client
- Denigrates the content of what the client has said
- Intimidates, humiliates, or threatens the client (berates and scolds the client)
- Leaves the topic for one entirely unrelated

Following are some vignettes that demonstrate the various grade levels of responses. Look at each response, and assign a grade to each one. Next, using the preceding material on grade levels, tell specifically why you think the response should receive the grade you assigned to it.

Vignette 1

A man has come to your agency for help after he lost everything in a fire. The worker asks the man to tell her what happened. He describes the night the fire took place, but as he approaches the actual incident, he finds it more and more difficult to talk.

FIRST WORKER'S RESPONSE: "This is really difficult for you. Would you like to wait a minute?"

Grade: _____   Reason: _____

SECOND WORKER'S RESPONSE: "Now, this is all over, Mr. Brown. It happened days ago. You need to be thinking about moving on and getting on with your life."

Grade: _____ Reason: _____

THIRD WORKER'S RESPONSE: "This must have been awful for you! Excuse me a minute." Turning to the secretary in another room, "I heard the phone ring, Sue. Was that the attorney calling? Tell him that we need that file before we can do anything for his client." Turning back to the client, "Now, where were we, Mr. Brown?"

Grade: _____ Reason: _____

FOURTH WORKER'S RESPONSE: "I'm really wondering if you can handle this! I'm going to call mental health and set up an appointment for you. Why, you're a wreck!"

Grade: _____ Reason: _____

FIFTH WORKER'S RESPONSE: "I can see it's difficult for you to talk about this. I'd like to work with you to see if we can find some ways to help you. I have some thoughts that I think might help, and I'm sure you do too."

Grade: _____ Reason: _____

Vignette 2

A young woman enters a shelter after she and her boyfriend, with whom she is living, have had a fight. She has been badly beaten. She seems to want to talk and remains in the office even after the worker has completed the admitting forms. She is rather quiet, however, and does not volunteer much information.

FIRST WORKER'S RESPONSE: "Yeah, another case of the violent boyfriend. Here we go again. You'd think you women would stop seeing these guys before it gets to this."

Grade: _____ Reason: _____

SECOND WORKER'S RESPONSE: "Did he ever beat you before? I was just wondering because I'd think that if I'd been through this before, I would have left before now."

Grade: _____ Reason: _____

THIRD WORKER'S RESPONSE: "It sounds like you've had a rough evening. Do you own your own home? No, I was just wondering if you own your own home. It says here you live in a house, and not an apartment."

Grade: _____ Reason: _____

FOURTH WORKER'S RESPONSE: "I was in your shoes once. Believe me, it was a long and difficult battle to get out of that mess. But I did. I just decided that it wasn't worth living like that—life's too short, and I got out!"

Grade: _____ Reason: _____

FIFTH WORKER'S RESPONSE: "What you need is a good lawyer. You tell the staff that comes on in the morning that you want to talk to a lawyer. You can't just sit around and take this stuff!"

Grade: _____ Reason: _____

Vignette 3

A rape victim sees the volunteer in the emergency room. As the volunteer talks to her, she learns that the victim is afraid to go home. The man who raped her is an acquaintance in the neighborhood, and she is afraid that now that he knows she called the police, he will come after her.

FIRST VOLUNTEER'S RESPONSE: "It sounds like you could use a place to stay tonight. Would you like me to try to set something up for you?"

Grade: _____ Reason: _____

SECOND VOLUNTEER'S RESPONSE: "You're really afraid of this guy! Well, you're not going home tonight."

Grade: _____ Reason: _____

THIRD VOLUNTEER'S RESPONSE: "I can see that you can't handle this! I'll make all the arrangements. Don't worry about a thing. I'll get you a place to stay, and I'll set you up with an appointment at mental health."

Grade: _____ Reason: _____

FOURTH VOLUNTEER'S RESPONSE: "How would you like to handle this tonight?"

Grade: _____ Reason: _____

FIFTH VOLUNTEER'S RESPONSE: "Let's get off this gruesome topic. I mean, I know it's important to you right now, but it'll do you good to talk about something else. Tell me about your job."

Grade: _____ Reason: _____

SIXTH VOLUNTEER'S RESPONSE: "You must be feeling so afraid of him."

Grade: _____ Reason: _____

Vignette 4

A young woman comes in because she has been using "crack more than I thought I would." She describes having trouble staying away from crack as her roommate uses it. Recently her roommate has begun prostituting herself to get the money for drugs and many of this woman's possessions have disappeared. She says, "I just think I am headed in a bad direction and decided this morning that I need help to get out of this."

FIRST WORKER'S RESPONSE: "Can you tell a little more about what is going on with your roommate?"

Grade: _____ Reason: _____

SECOND WORKER'S RESPONSE: "How did you get mixed up with this roommate to begin with?"

Grade: _____   Reason: _____

THIRD WORKER'S RESPONSE: "I know this other woman who is in a similar situation and she is having the worst time getting away from her roommate!"

Grade: _____   Reason: _____

FOURTH WORKER'S RESPONSE: "Okay. So what we are going to do here is get you to sign papers to go into a rehab unit. That will get you out of the house and away from the roommate. Then we are going to get your things out of that house. Do you have a place you can store everything while you are in rehab?"

Grade: _____   Reason: _____

FIFTH WORKER'S RESPONSE: "You must feel betrayed by her. Talk to me a minute about what you would like to see happen for you right now."

Grade: _____   Reason: _____

Vignette 5

A man talks to the intake worker in a case management unit about his recent separation from his wife. He comments that he could have been a better husband, that he was stingy and went for days not speaking to his wife if he was annoyed. He thought he was making her see how he felt about things, but she left him, saying he was "uncommunicative." He seems depressed and bewildered by his wife's departure.

FIRST WORKER'S RESPONSE: "Sounds to me like you could use some communication workshops."

Grade: _____   Reason: _____

SECOND WORKER'S RESPONSE: "Did you have to go that long without speaking? I mean, I can't think of what she could have done that made you that mad."

Grade: _____   Reason: _____

THIRD WORKER'S RESPONSE: "You really thought you were getting through to her, so it must be hard to see her leave this way."

Grade: _____   Reason: _____

FOURTH WORKER'S RESPONSE: "Only a fool would think what you were doing was 'communication'! Of course she didn't hear you, fellow!"

Grade: _____   Reason: _____

FIFTH WORKER'S RESPONSE: "See, I think you should have tried marriage counseling before things got this bad. Not now, after she's already gone."

Grade: _____   Reason: _____

SIXTH WORKER'S RESPONSE: "Well, if you had a better understanding of the way women think, you could have avoided this whole thing."

Grade: _____   Reason: _____

Exercises II: Recognizing the Difference—Encouragement or Discouragement

Instructions: Following are two vignettes. Decide what you would *do* or *say* to encourage the person, and then decide what you might *do* or *say* that would discourage the person. Actually picture yourself as an encouraging person, and then as a discouraging person. Remember your actions may be as important as your words. Write in your answers to share with the group.

1. A woman, the mother of two children, has been without a home for a number of months. She tells you she really wants a permanent place to stay. You know there are very few places she can go. You also know she has some talents and interests—strengths that might be to her benefit.

   You encourage her by:

   You discourage her by:

2. A man calls and says he was sexually abused as a child. It has come to haunt him recently, but he is not sure where he should turn for help.

   You encourage him by:

   You discourage him by:

Exercises III: Blurred Boundaries

Instructions: Following are some situations in which the boundary between the worker and the client has become blurred. Identify what went wrong and what needs to happen to correct the situation. Use the space provided to make notes.

1. Alice is very upset because she gave the client some names of people she thought might be helpful in solving the client's problem. Today, she met one of those people at the local deli where she eats lunch and learned that the client

has never been in touch with that person. She has been trying to call the client all afternoon to find out what happened.

2. Bill is feeling very proud of himself. He talked to a man who had very complicated problems this afternoon. He put everything down on paper, while the man sat by his desk and drank a soft drink. Then Bill decided on the best way to handle the situation. The man finished his drink, did just what Bill suggested, and reported that everything is fine now.

3. Mary Lou was once in a very abusive relationship. She was able, through much counseling and grit, to become assertive enough in her own behalf to get out of the situation. Today, she is happily married and the mother of two lovely children. The client she is talking to is in just such an abusive situation and seems hesitant about leaving. Mary Lou wants her to leave and, using examples from her own life, assures the client she is certain the client will have just as happy a life as Mary Lou now has, if she will leave. Mary Lou remembers vividly how she felt when she was in the woman's shoes and tries to make the woman see how much better she will be if she leaves now.

4. Gloria was raped by her stepbrother when she was 16 and he was 23. It was a very difficult situation; law enforcement officials were called, and eventually the situation broke up her family. She is currently working at a rape crisis center and is talking to a client whose situation reminds her of her own. She says things like: "Oh, telling my teacher wouldn't have done a thing for me!" or "I was too far gone to be able to handle it the way you did."

5. Carlos is working with a woman whose father is the president of a large bank in another city. She has become depressed and needs a referral to a therapist and perhaps a psychiatric assessment. Carlos is reluctant to talk to her about needing "psychiatric help" because he assumes this would be offensive to someone as "upper class" as she is. Instead, he suggests she "talk to someone for a little bit." Later, with the woman's permission, her father contacts Carlos and asks if there is anything he can do. Carlos is careful not to suggest short-term psychiatric hospitalization even though the father is offering to pay "for whatever she needs." Carlos is sure such a hospitalization would alienate the father.

6. Candy is working in a shelter for homeless men. She is new at her job and enjoys what she is doing. The fact that many of the men are in poor physical health and are unwashed is of great concern to her, and she works hard to meet the basic needs of the residents, such as food and clothing and a warm place to sleep. When she is working, she never asks the men to help with chores around the shelter even though that is a condition for their staying there. She thinks of her work as very loving and giving, and she sees the clients as hapless and uneducated. Therefore, when Paul comes to her and tells her he graduated from high school and finished a year of college before he got hooked on drugs, she is not certain she believes him. When he asks for help returning to college, she resists giving him information and support for attending the local community college.

7. Marissa is interviewing a woman who has come in for help with her bills and job-readiness training. The woman is the mother of a 2-year-old boy. Marissa is the mother of a 2-year-old boy as well. She considers herself an excellent mother. She prides herself on how she adapts her life around the needs of her child and how his needs come first. She is uncomfortable with the arrangements the woman must make to accommodate her impoverished circumstances, and so Marissa lectures the woman. "Children need stability, not instability and disorganization. They need to know you are there for them and not have continual disruptions. Does your son even know ahead of time when he will be staying with your mother and when he will be staying with you? You need to put this child first in your planning." Marissa is surprised when the woman remarks quietly that seeking help so that she and her son can eat on a regular basis *is* putting her child first.

8. Harold is seeing a man who is coming out of rehab after years of drinking. Remembering how hard it had been for him to find a job and a place to live away from his drinking buddies, Harold begins with a lecture on how to find a job and how to stay away from "the guys you used to drink with." The man looks puzzled and says that he is returning home to his wife and family and his job at a large corporation and came to see Harold for help with services to maintain his sobriety.

CHAPTER 6

Clarifying Who Owns the Problem

Introduction

Before you ever open your mouth, before you ever say a word to your client, you want to be able to discern accurately who owns the problem. Who owns the problem?

*It is the person whose **needs** are not being met.*

It is not the person who is being rude and uncooperative. It is not the person who is ruining a party. It is not the person on our caseload who has begun to drink again. Nor is it the person who is singing off-key and ruining a songfest. It is the person whose needs are not being met. You should know who owns the problem for four very good reasons:

1. *You will know who is responsible for solving the problem.* If you know who owns the problem, then you know who is ultimately responsible for solving it. If you know who is responsible for finding a solution, you will not assume the entire responsibility is yours. In other words, you will not accept responsibility for problems that are not yours. When you take over and try to solve other people's problems for them, or tell them how they should resolve their problems, you may be seen as meddling in their affairs or being pushy.
2. *Meddling is disrespectful.* This sort of meddling is disrespectful, even when you intend it to be helpful. It says clearly that you have doubts about the client's ability to figure the problem out and handle it on her own. You indicate that you are not sure other people have the sense and insight to know what is best for them.
3. *Boundaries are clarified.* Understanding whose problem this is clarifies boundaries. We do not cross boundaries into other people's affairs if we are clear about what issues and problems belong to them. We help where we can and we let people help themselves where they can, growing and learning as a result.

4. *The client loses opportunities to grow*. Furthermore, when you take over with solutions, you interfere with what might be a very meaningful experience for the client. This person may grow from wrestling with this issue. It may be the opportunity needed to gain insight, learn a new skill, or try something that until now has been too frightening. If you attempt to take over with your own solutions and ideas, your client will miss this valuable opportunity. Clients can never say, "I did this myself!" Instead, they will have to say, "My case manager did this for me."

Just because a person tells you about a problem does not mean you must solve it.

Kentaro, working in a sheltered workshop, was learning a new job on the assembly line. He seemed to quickly pick up his responsibilities, but he was having trouble keeping up. Over and over, the worker monitored his progress and gave him tips for improving his speed. The worker stood behind him and grabbed the pieces Kentaro missed. Finally, the worker was called away to the phone. When he came back, he discovered that Kentaro was sitting so that he faced the assembly line from a different position. Now, with better visibility, the client was catching each piece that came toward him and making the necessary adjustments. The worker later said he felt foolish for standing over Kentaro all morning when it turned out the client knew all along how to solve the problem.

Agnes wanted to have a better relationship with her mother. She confided this to her worker one day, and the worker set about helping her solve the problem. While the worker spoke to her about poor communication, mother and daughter relationships, and family therapy, Agnes decided to buy a pretty card and send it to her mother. In the card she told her mother how much their relationship meant to her and how much she wanted them to be friends. She enclosed a little lace handkerchief, and sent the package off to her mother. Soon her mother called Agnes, and they began to talk. Agnes, who knew herself and who had lived a good portion of her life with her mother, understood how best to solve her problem. Listening and helping Agnes talk about the relationship with her mother might have been a better course of action for the worker.

Keep people in a position of authority over their lives to the greatest extent possible. Be mindful of the boundaries between you and let people resolve their issues as much as they can on their own. Remind people who have resolved an issue of how much of the resolution was their own doing. Point it out to them. You might say, "Let's look at all the things you've done to make this happen." Make sure your clients have the opportunity to feel pride in their part in solving the problem. Let people see they can help themselves more (even if it is only a little bit more) the next time, rather than turning to their case manager to solve the problem for them.

Boundaries and Power

Who owns the problem and who should solve this problem is a matter of boundaries. People seek help when they are most vulnerable and, as we have seen, being vulnerable makes people susceptible to suggestions and interference by people perceived to have more wisdom and better answers. It is easy when you are feeling afraid and uncertain to let someone else solve our problems for us. For our part, on the other hand, we sometimes find ourselves thinking it would be easier to just solve the clients' problem ourselves rather than have to watch them struggle with it. This is a danger case managers seek to avoid.

Because we know the system and relate with empathy, our clients are likely to allow us to violate their boundaries and take over. People seek help usually where they feel an agency can be most helpful to them. They seek assistance from a place where the workers are perceived to have good answers. This is known in our work as a power differential. Case managers are perceived to be wiser and therefore can have considerable influence over how people decide to resolve their problems.

Determining who owns the problem and the extent to which people seeking our assistance can solve their own problems or at least aspects of their problems allows for more realistic boundaries where we do not interfere where a person is capable.

If the Client Owns the Problem

Let us suppose your client comes in and tells you that she cannot stand living with her mother anymore. Her mother is verbally abusive and rejecting. Your client is unhappy. Obviously, this client's need for a pleasant home environment and her need to be appreciated by her mother are not being met. The client owns the problem. Does the mother own the problem? She does not appear to. This method of communicating with her daughter seems to work for her. She shows no discomfort or guilt about any pain she might be causing her daughter. It seems to meet some need of hers to communicate in this way. The mother does not have a problem in this situation, as her needs appear to be met.

There are several important ways you can respond to your client's situation. First, listen. Then, rather than providing a solution, be a resource to your client. Give her options. Tell her about services with which you are familiar that might be helpful to her. Ask her for her ideas. What is she looking for? What does she want to happen? Leave the final decisions up to her. In this way, you make sure that the client retains a position of power in her own life, and you act collaboratively. You are the expert on available services, but she is the expert on her own life.

Now let us change the story a little. Your client brings in the same problem, but she has an intellectual disability. The problem still belongs to her, but now you make a conscious decision to get a bit more involved and a conscious decision about the extent to which you will get involved. As a case manager or worker, there will be times when it is important to give more help than others. A wise worker will know how much to help and when to stand back. These are **strategic decisions**.

Deciding how much to become involved is important when the person has a problem with you. For example, a person may come into your office and tell you that he cannot stand the way you sit in front of a cluttered desk and talk to him. He claims he feels disorganized by your clutter and wants you to have your desk cleaned off when he comes in. In this case, you may make a conscious decision to let him own this problem because there is no way to clean the desk off in the middle of a busy day when you know he is coming in. You would thank him for his comments, give a word or two about why that might not be feasible, and tell him you will continue to keep your office as it is for now.

On the other hand, another individual may be upset with you because you are always late. His need for punctuality is obvious; your being on time means to him that you are expecting him and that you value his time. His need is not being met. In this situation, you might decide to help own the problem. You recognize that you have been somewhat disrespectful. You can justify it with your busy schedule, but you also can do better. So you acknowledge the problem, thank him for his comments, and offer to be more punctual. You have made a conscious decision to become involved in the solution to his problem.

It Is Not Uncaring

Sometimes we feel guilty about not doing more. Sometimes others tell us that we should be doing more. After all, we are the person's case manager. Why are we not extending ourselves further? Sometimes the clients themselves are the ones to accuse us of not caring or of being indifferent. Knowing who owns the problem and allowing that person to resolve it is not an uncaring action. In fact, you would never refuse to help a client simply because you determined the problem belonged to the client.

When you allow people to work on their own issues and problems, you respect their right to privacy and self-determination. In addition, you give them an important opportunity to grow and work on their own behalf. Solving one's problems effectively is part of emotional health and maturity. To the extent to which clients are able, we want to encourage them to do as much for themselves as they can.

It Is a Strategic Decision

The extent to which you become involved in helping people solve problems that belong to them is a strategic decision. This is another difference between the professional approach to relationships and a friendship you might have away from work. In the professional relationship with your client, you want to decide strategically how much help to give and the extent to which you will step in. The decision is based on your knowledge about the person's abilities and about how this opportunity can be used to help your client grow.

A woman who is blind might need more help negotiating the transportation system than one who is depressed. A person who is illiterate and from a rural village might need more help working with the Social Security office in the city than would an urban lawyer. A child might require more support than an adult to carry out a personal decision.

The strategy lies in knowing your client's strengths and limitations and tailoring your involvement to those factors specifically. In this way, you avoid taking over simply because that is the easiest thing to do or because you see all clients as helpless. Your involvement is just to the point the individual needs help or ideas and no further.

In certain cases, even though the client owns the problem, you may find yourself taking over and resolving it almost entirely alone. Suppose you are working with a single, 17-year-old girl, disowned by her family because of her pregnancy. She has just delivered her first child for whom she has made adoption arrangements. The child, however, is severely disabled with developmental issues. The doctors feel the care required by the child can never be undertaken by a 17-year-old girl living alone, and the prospective parents have now withdrawn their bid to adopt the infant. In this case, you work out arrangements for the care of the infant, solving the infant's need for a safe, medically appropriate environment and solving the mother's problem of what to do with a handicapped child she believed would be going to the home of another couple. In this case, you would consult with the mother throughout the process, possibly even taking her to see the facility where her baby will receive care, but you would handle most of the actual arrangements. If you did not understand the concept of who owns the problem, you might be tempted to ignore the mother in the process of solving this problem. If the mother were older, married, having her second child with this handicap, and supported by her family (or in any number of different circumstances), your response and the extent of your involvement would change as well.

When the client owns the problem, *carefully* decide the extent to which you will be involved. Test your hypothesis about how much the person can handle alone. Be ready to take on more responsibility or give more responsibility to the individual as you move toward a solution. Watch your involvement to be sure you are not obstructing the client's opportunities to grow or to exercise self-determination and independence.

Be a Resource and a Collaborator

You will have at your fingertips information that can help the person solve a problem. You may have the names of agencies, phone numbers, contact people, and addresses for services. You will also be familiar with policies in various agencies and within large social service systems, such as child welfare and mental health. You will often be more familiar with the law as it pertains to the person's situation. This makes you a valuable resource to a person attempting to arrive at a solution.

Bring the information and facts to the client, and then **collaborate** with the client on the solution. It is the clients who are most aware of which solution will work and which ones are impractical for their circumstances. Together, with your knowledge of the system and your clients' knowledge of their personal lives and circumstances, you will be able to construct a useful approach to clients' problems.

If You Own the Problem

If you are having a problem, that is, your needs are not being met, you will understand that the **resolution of the problem** is ultimately your responsibility. This applies to personal problems, and it also applies to problems you might have in the course of your relationship with your clients. What if it is the client who is always late? Whenever the client is late, you find you are behind for the rest of the day. This is not the client's problem. The client may find it perfectly acceptable to get to your office at approximately the time he is scheduled to see you. He may have scheduling problems or punctuality problems; but in this case, he does not own this problem. You do. Your need to stay on schedule and see everyone you are scheduled to see before 5:00 P.M. is not being met. Therefore, you are the one who is responsible for bringing it up. Do not expect that others will guess there is a problem.

In bringing up a problem we are having with another person, we are asking for that person's assistance in resolving it. Just as you make decisions regarding how much you will become involved in resolving someone else's issue, your clients have the right to determine the extent to which they want to help you. It is conceivable that the client will see your point and make some changes. It is also possible that the client will decide that it will have to be your problem because it is preferable to be late for whatever reason or because being punctual is an inconvenience. There are ways to solve problems like this one; but for now, as the first step, you need to be clear about who owns this problem.

If You Both Own the Problem

Sometimes you both have a problem. Suppose a person needs evening appointments, and you work only during the day. Or perhaps a client wants to shout and yell about her situation, and you find that too unnerving to do a good interview. These are opportunities to negotiate. You, as the worker, have to be able to sort out in your own mind who owns what problem, and you must be able to initiate some negotiation around these issues.

When you both own the problem, you should not view it as a win–lose situation. If the client sees it that way, you need to point out other ways of looking at the situation. Perhaps he can see another worker who does work at night, or perhaps he can come in during the day sometimes and you can stay late sometimes. Maybe she can yell with less intensity, and you can overlook the rest of it. It might work to transfer the client to another worker, one whose schedule is better suited or one who can better tolerate the yelling. There are many ways to negotiate a solution. When you work on a solution collaboratively with the client, you provide that person with an important experience in problem solving. As the worker, you invite the client to join you in this effort.

Margaret had been ill with schizophrenia for a very long time. Rejected by most of the community and most of her strictly religious family, she found solace and support among the workers in the mental health system. In the course of her illness, she had been hospitalized and knew the staff at the hospital well. She had encountered the various members of the crisis team and knew them too. She had a case manager whom she found supportive. Margaret found countless reasons to call workers within the system. Night and day she called with tiny questions, not so much because she could not resolve the problems herself, but because she found contact with these supportive people comforting. Sometimes she would call to ask what time she should go to bed. She might call to ask if she should eat one frozen dinner rather than another. Should she go out for a walk tonight or not? Should she buy a new pair of shoes or not?

Margaret's incessant calling began to create a problem for already busy workers. They grew exasperated. Margaret had a need to feel their support, and the workers had a need to get things done with other clients. Finally, a solution was worked out with Margaret and all the workers in the system who regularly received calls from her. It was decided that a man on the crisis team who shared her religion and genuinely liked her would be the person she would call. When he was off duty, a backup person was designated. Margaret was then allowed only one call a day. She was to save all her questions for that one call. Everyone agreed to this plan.

Although Margaret tested the plan many times initially, everyone stuck to the agreement. Eventually, Margaret began to make the calls more meaningful, asking for help with real problems. Undoubtedly, this one call a day helped to sustain her and helped her to live more comfortably in the community rather than in an institution. It also allowed the workers to focus on other clients.

In this situation, both the workers and the client had a problem, and their needs conflicted. By collaborating on a solution, rapport was not lost, and both the workers and the client gained valuable experience.

Summary

Knowing who owns the problem is an important first step in working with clients. This allows us to understand who is ultimately responsible for resolving the problem. Once we recognize that many of the difficulties our clients bring to us are theirs, we need to determine the extent to which we will assist in problem resolution. Both the determination of who owns the problem and the decision to get involved are the first strategic decisions you make in your work with clients. Your involvement must be tailored to the clients' strengths and capacities so that you do not take over where a person is competent or take from people the opportunity to grow and learn from their experiences.

For many, deciding that the clients own the problem can be seen as uncaring. In reality, you are not abandoning people with their problems. You are, instead, making decisions about how much individuals can do for themselves and where you will be most helpful. In the long run, we want people to be able to take some pride in the fact that they participated in solving their own problems and learn from that experience.

Exercises

These exercises can also be filled out online at CengageBrain.com.

Exercises I: Who Owns the Problem?

Instructions: In the following situations, identify who owns the problem. As you study each case, decide whether it is you, as the worker, who owns the problem; whether the client and perhaps the client's family owns the problem; or whether both you and the client own the problem at the same time.

1. A woman you have placed in temporary housing is angered by the loud music of her neighbors. She appeals to you to do something about it. Who owns the problem?

2. You work at a victim/witness resource center where you assist the victims of crime to handle the emotional and technical ramifications of the crime before they go to court. The husband of a victim, a woman who was carjacked by a teenager one night, takes you aside and asks you to persuade his wife to drop the charges. He tells you confidentially that it would be better for his wife if "she didn't have to go through this." Who owns the problem?

3. The mother of a rape victim, with whom you have been working, calls and says that ever since the rape, her daughter has been crying and unable to eat or sleep. She tells you it is urgent that she know exactly what happened to her daughter, but that her daughter refuses to talk about it. She asks if you can tell her what happened. Who owns the problem?

4. You are talking to the victim of a violent crime in the emergency room. Her boyfriend barges in and demands to know "what's going on." Who owns the problem?

5. You have placed a woman in temporary housing after she left her home following severe abuse by her husband. The husband calls demanding to know where she is and tells you he will get his lawyer and sue you if you do not tell him. Who owns the problem?

6. You are working with a support group. One of the participants tells you on the side that another participant is monopolizing the group's time with frivolous details and asks if you will do something. Who owns the problem?

7. Your client is going to court on his third DUI charge. The family of the woman whose car he hit calls your office before the proceedings because the article in the paper stated your client was receiving help from your agency in preparation for the trial. The family wants you to withdraw your services and advocate with the judge that your client be sent to prison and not to a rehabilitation center. Who owns the problem?

8. You are arranging for housing for a woman who is in a homeless shelter. Her parents come to see you and ask you to see that she also goes to therapy. They tell you she has never "seemed right," and they ask you to give them your opinion of her mental status. Who owns the problem?

9. You have developed a goal plan for a child. The parents agree with the plan, which involves summer camp and other recreational activities over the summer, all with a therapeutic program. The teacher calls to tell you that this child can hardly benefit from school and that sending him to camp is a waste of the taxpayer's money. What he needs, she tells you, is therapy. Who owns the problem?

Exercises II: Making the Strategic Decision

Instructions: Following is a basic situation, with a list of scenarios in which the circumstances surrounding the situation are different. Decide what you would do in each case.

Situation: Hannah recently went blind due to an accident with chemicals at the company where she worked. She is asking for a service plan that will help her regain some independence.

1. Hannah is a PhD chemist with the corporation where the accident occurred. She has received a huge settlement from the corporation's insurance company. The corporation has said she can come back to work if she can be retrained in some way, possibly with computers. Hannah has a supportive husband and many close friends. How do you help?

2. Hannah was a custodian at the small chemical engineering company where the accident occurred. The company had little insurance, and it has no interest in hiring her back for any reason. The company did give her $5,000 at the time of the accident, and the hospitalization plan and workers' compensation helped pay the initial medical bills. Hannah lives alone and has few friends. How do you help?

3. Hannah has an intellectual disability and worked as a custodian at the company where the accident occurred. The company gave her $5,000 at the time of the accident, and the hospitalization plan and workers' compensation helped pay the initial medical bills. Hannah lives with her parents, who are very supportive, and she has two older siblings who also give support. The family has been working with Hannah to help her decide what to do next, and they have found a place where she can answer the phone and give standard information. This company is delighted to have a real person to do this, as the answering machine option seemed too impersonal. Hannah will need some training. How do you help?

4. Hannah is a student working on a chemical engineering degree. She worked part time to pay her school expenses at a large chemical corporation. She wants to remain in school. Her family is supportive of this, but they live in another state. Hannah's roommates seem hesitant about her returning to live with them in their downtown apartment now that she is blind. How do you help?

5. Hannah is a student working in a small chemical lab while completing a chemical engineering degree. She wants to remain in school. Her family is supportive of this, but they live in another state. Hannah wants a seeing-eye dog, has a landlady who is afraid of dogs but who might accept one, and needs to learn how to negotiate the town and the campus as a person who is blind. She has numerous supportive friends. How do you help?

6. Hannah is a student working on a chemical engineering degree and doing a chemical engineering internship at a chemical plant near the college. She wants to remain in school. Her family is supportive of this, but they live in another state. Hannah was using this job to pay for her education. Now Vocational Rehabilitation will help, but Hannah must fill out countless forms. Hannah is depressed and frightened by her blindness and spends days at home alone. How do you help?

7. Hannah had a fairly ordinary chemical technician's job at the company when the accident occurred. She and four other people were blinded by the accident. Hannah has told you she wanted a lawyer while she was still in the hospital, and she also feels the group should meet regularly to talk about the accident and their anger. The others have agreed. Hannah tells you of the state office of services to the blind and wants help connecting to that office. How do you help?

8. Hannah had a fairly ordinary chemical technician's job at the company when the accident occurred. The company offered to pay all her medical benefits and a small stipend to support her while she trained for another kind of work, not to exceed 5 years. Hannah lives alone and has few friends. She makes it clear to you that she is not interested in receiving any help from you. She rejects services that you know could help her to begin training and asks you to leave her alone. How do you help?

CHAPTER 7

Identifying Good Responses and Poor Responses

Introduction

This chapter provides some concrete examples of good and poor **responses**. Those called "poor" are poor in the sense that they tend to block **communication** and prevent the worker from hearing what is really important. The examples of poor responses are followed by examples of good or constructive responses that encourage another person to talk and feel comfortable doing so. Like learning to drive a car, we learn step by step the ways to structure responses that will facilitate good communication.

It is not enough, however, for you to look at these responses and do the exercises. You will soon build a skill in identifying which responses are inadequate and which actually enhance the communication, but this will not teach you to automatically use good responses rather than poor ones. What you are seeking to do is create a conversation between you and your client that flows so that the client feels encouraged to talk, safe about divulging their issues. This chapter is a start, but there is no substitute for practice. In later chapters, you will learn to use constructive responses with more understanding of how they promote rapport and clarity. Even this understanding, however, will not help you to make an automatic response that is therapeutic if you have not practiced.

This chapter clarifies which responses are most likely to promote rapport and which are most likely to promote withdrawal and defensiveness.

Communication Is a Process

We are inclined to think of communication as messages we simply send to another person. In an exchange that person may respond with a message of her own. We know what we mean when we communicate and usually assume that everyone else does as well. If we are misunderstood or our message is not clear when it is received we tend to blame the other person. What got into him today? Why is she overreacting that way? What did he mean by that response? It doesn't make sense?

Likewise, we assume when someone speaks to us that we have pretty much received his message as it was intended. When something people have said doesn't make sense to us we assume they did not frame their message well or their way of communicating this message is garbled.

This is a simple way of looking at communication. A speaks to B and B responds and everyone understands the verbal messages being exchanged. Communication is more than that and for our purposes, where we are engaged in evaluations and empathic interactions, understanding communication on another level is important.

Communication Is an Ongoing Exchange

When we are talking to another person we are sending and receiving messages, often at the same time, in a rapid, ongoing process. For example, you might give your client information about where she can get help with childcare at the same time she may be giving you a look that lets you know she tried those people and did not like them. You might talk about the need to be on time for appointments while your late client shrugs his shoulders. In talking to one another messages are often more than words and the exchange involves both parties in receiving and sending communications during the encounter. You might be explaining a policy to a person who is interrupting you to complain about a different policy. Communication is rarely a neat exchange where A talks, B listens and fully understands the message, then B responds and A listens and fully understands the response.

Communication Involves Meanings

When the client made a face about the agency you suggested for childcare how did we know what she meant? In this case we assumed she was making a face to indicate she did not like this particular agency. What if, however, she meant her raised eyebrows and skeptical look to mean she was surprised *you* would go to all this trouble to help *her?* It can happen when we speak as well. You might say to someone, "I am sorry you had to go through all that traffic to get here" and the other person hears a different message that you are really sorry the person is late and holding things up.

This is especially true when two people involved in a verbal exchange, what we call a conversation, are from different backgrounds, have different cultures, different levels of education, or different life experiences. We can misunderstand each other when one of us is having a bad day, does not feel well, or has different expectations for the encounter.

For example, you might think you are helping your client by telling her when and where she can catch the bus home and after the day she has had she may decide you are trying to get rid of her. You might explain your policy about canceling appointments and your client may think you see him as the kind of person who would not cancel but would just not show up when he has an appointment. A person from an Asian country may approach you to drop out of group and when you try to talk about how you could make the group work better for him, he feels you are hurt by his desire to drop out and quickly agrees to stay, nodding agreeably when he really intended to leave the group. You may leave this encounter believing you have solved his issues with group and he may leave upset that he still is committed to coming.

Communication Is Transactional

Communication is transactional, according to Adler, Rosenfled, and Proctor, 2013. They describe it this way: "(C)ommunication is a dynamic process that the participants create through their interaction with one another" (p. 11). They go on to describe communication as "Something we do *with* others" (p. 11) or "like dancing" (p. 11) where partners can communicate together beautifully or may instead miss cues and misunderstand messages. This process comes in two parts: First, if we communicate what we mean to communicate and we have a good partner who accurately hears the meaning we intended and second, if we hear other person's response and accurately grasp the meaning they intended we have great communication.

If, on the other hand, we know what we meant and the other person hears another meaning and responds to this different meaning and we are confused by this and do not understand their message in response we do not have good communication. When we talk to others we are together creating an ongoing dialogue. Like a dance, this communication can flow beautifully or someone can figuratively step on someone's toes. The dialogue we create with our communication partner is unknowable before we start. The result of this communication can be positive, where your communication partner leaves feeling positively about you and your agency. However, in spite of your best intentions your communication partner may go away angry and hurt, feeling misunderstood.

What we know is that unfortunate communication can happen when we do not think carefully about the message we wish to send and check with our partner to be sure we understand the meaning he or she wants to convey to us. You will be the partner in countless encounters with clients and other professionals. The focus in the next few chapters is to give you the skills you need to minimize the possibility that communication will go awry.

Twelve Roadblocks to Communication*

Dr. Thomas Gordon's (1970) *Parent Effectiveness Training* provided a useful framework for looking at types of communication with the purpose of finding those responses that encourage a person to talk and feel valued and those that block the

---

*Adapted with permission from Parent Effectiveness Training, by Thomas Gordon. Copyright © 1970 Random House, Inc.

communication and discourage disclosure. In his work he outlined 12 specific ways we often block good communication, setting up barriers to real understanding and dialogue. I have adapted these roadblocks to fit the kind of poor communication that sometimes happens in the human service setting. These responses are not helpful in talking with other people. They serve to obstruct rapport and block any constructive resolution of the problem, and they often serve to make our clients feel inferior or demeaned.

In the examples of poor responses provided in this chapter, notice the implied superiority of the worker. Individuals who come seeking assistance already feel unsure of themselves and uncertain about what to do. Workers who come across as all-knowing or judgmental are not helpful. An attitude of superiority—giving people the sense that the worker is "talking down" to them—is harmful to the relationship and makes real communication and rapport difficult.

Notice how hard it would be for a person to trust and be open after hearing the poor responses. What does one say to someone who appears to be patronizing? Most people stop talking or resort to pleasantries after that, and real communication is blocked. Why continue being open if the responses are not empathic?

In addition, you will find that many of the responses are well meaning. They may sound to you like the kind of thing one acquaintance might say to another, but they are not helpful in the case management setting where we want to encourage a person to feel comfortable talking freely, often about difficult subjects. Let's look at some examples of such negative responses.

We look at these poor responses in order to begin to distinguish what we might say that would inadvertently block a constructive relationship between you and your client. It is helpful for you to see these responses, imagine how they might sound to someone seeking help, and recognize how they could stifle further discussion.

Ordering, Directing, Commanding

The first roadblock involves giving the person an order or command. The assumption made by workers who do this is that they have all the correct answers or all the best solutions and ideas. There is no dialogue or **collaboration**. In the following examples, the workers take charge without including their clients:

- "I don't care what anyone tells you! You have to go see that lawyer!"
- "Go right back over to the courthouse and get those forms!"
- "Leave your house and come right down to the office."
- "Look, just go over and apply for the job."

Warning, Admonishing, Threatening

Warning of consequences if the person does something is the second roadblock. Workers who do this do not want their clients to follow certain lines of action, often out of concern for the client. Rather than discuss their clients' inclination to act in a particular way, they warn the client instead. In this next series of responses, the workers here sound as if they know better what is good for the client than the client does:

- "If you take that suggestion, I think you will be sorry!"
- "You'd better not do that."
- "I can tell you from experience that something like that won't work!"
- "You won't get your medications if you don't attend group regularly."

Exhorting, Moralizing, Preaching

The third roadblock is telling clients what they should or ought to do. Again, such workers display the belief that they have all the answers. In several of the responses that follow, you can hear the workers imposing their own moral values on their clients. Listen to the "shoulds" and "oughts." Workers using these two words speak as though what they are saying is a universal given rather than a personal choice or value.

- "You should know that doing that is wrong!"
- "You shouldn't think like that."
- "You ought to see a counselor."
- "You ought to be more concerned."

Advising and Giving Solutions or Suggestions

Telling clients how to solve their problem is a fourth roadblock. When workers believe their clients have nothing useful to contribute to the resolution of a problem, they will make unilateral decisions without their clients' input. Some of the following responses indicate the worker's exasperation with the person. Workers who tell people how to solve their problems believe that their way of seeing the clients' problems is the only way to see them. The clients are treated as though they are hapless or inadequate for not seeing their problems that way as well. Here are some worker responses that illustrate the point:

- "With your husband? We'll set up marriage counseling and you can tell him what you think in those sessions?"
- "I think you just stop seeing this person today."
- "Here is what you need to do. Get your diploma and go on to college."
- "If you listen to yourself, clearly the answer is for you to stop hanging around with them?"

Lecturing, Teaching, Giving Logical Arguments

The fifth roadblock involves trying to influence with facts, arguments, and logic. In this next series of responses, the workers sound as if they believe their clients are incompetent. Only the workers know the whole picture. These responses by workers do not encourage real discussion of clients' feelings and problems:

- "I am going to give you the facts about domestic violence. If this doesn't change your mind, I don't know what will."

- "Look at it this way, the longer you put up with this, the more she gets away with it. People always try to get something for nothing. It is human nature. Your task is to stop it now."
- "Now what you need to do is call the police. That will absolutely circumvent any action on their part and free you to move to another location."
- "Now look, you have two choices. You can either stay or leave. That's what you have to decide. My vote is with leaving because that ends the problem permanently."

Judging, Criticizing, Disagreeing, Blaming

Making a negative judgment or evaluation of clients is the sixth roadblock. In the responses that follow, the workers see themselves as judges of their clients' behavior. Instead of being supportive, these workers are grading their clients' behavior. Their responses can only serve to demean a person who is grappling with problems and feels unsure. Here are some examples of such demeaning worker responses:

- "You aren't thinking clearly."
- "You're very wrong about that."
- "I couldn't disagree with you more."
- "Your plan is faulty because you never did your research."
- "I think you could have handled that better."

Praising, Agreeing

The seventh roadblock is offering a positive judgment or evaluation, or agreeing. Sometimes workers really cannot tolerate the pain clients express about certain situations. In the sample responses that follow, the workers are certainly well-meaning, but their responses cut off any meaningful discussion or further exploration and relieve the workers from having to deal with the real pain their clients might want to talk about.

- "Well, I happen to think you did just fine."
- "You'll figure this out. Don't worry."
- "It will all work out for the best. You'll see."
- "You're smart enough to know what you need to do."
- "You're bright. You'll figure it out."

Name-Calling, Ridiculing, Shaming

Making clients feel foolish is the eighth roadblock. Perhaps workers who give responses like those that follow are just fed up. They may tell themselves they have a right to express such degrading sentiments because they put up with so much from their clients. Only very untrained and unprofessional workers would ever resort to name-calling, but when it happens, the workers generally try to excuse what they have said by claiming they have endured long-term disgust or exasperation. Here are samples of responses from exasperated workers:

- "You're just an idiot if you do that."
- "Okay, you had to go and do it your way. You'll see."
- "What you're doing is totally ridiculous!"
- "You act like you never finished first grade!"
- "So be a fool. It won't get you anywhere."

Interpreting, Analyzing, Diagnosing

The ninth roadblock consists of telling clients what their motives are, or analyzing their actions. Sometimes, in order to feel vastly superior to their clients, workers will engage in surprise revelations. They often do this to show their clients that they know more about their clients' inner conflicts than the clients do themselves. In the responses that follow, the workers are informing the clients of their motives and underlying intentions, as though their clients lack self-awareness.

- "You're just upset because you haven't heard from the lawyer."
- "I think you really wanted to press charges but you just can't admit that."
- "You don't really believe that about her. I know that you are just saying it because you wish it were true."
- "What you really mean is that you don't want to see him anymore."

Reassuring, Sympathizing, Consoling, Supporting

Trying to make people feel better or trying to talk them out of their feelings is the tenth roadblock. The following responses are offered as a way of comforting clients, but they serve to cut off real discussion of painful feelings. Telling clients you know how they feel or you understand what they are feeling is not convincing, even if you have had similar experiences. Listening to the feelings is better than cutting them off. Following are responses by workers that tend to cut off clients' expression of their feelings:

- "You'll feel better in the morning."
- "All new mothers go through this at one time or another."
- "Don't worry. Things will work out."
- "You're not alone. Everyone feels the way you do from time to time."
- "I understand how you feel."

We will examine this tenth roadblock further below, looking at why this kind of comforting reply is not always helpful.

Probing, Questioning, Interrogating

The eleventh roadblock concerns trying to find motives, reasons, and causes. Clients do not always tell their concerns in logical sequence, and there are good ways for workers to go back and fill in the gaps without sounding as though they are prying. In the following examples, however, the workers are actually prying into their clients' motivations and intentions—areas the clients may not be fully aware of or ready to discuss:

- "Just when did you start to feel this way?"
- "Why do you suppose you went there that night?"
- "Do men ever tell you that they feel violent toward you?"
- "What were you really trying to do when you saw her?"
- "So why were you so intent on going there with her?"

Withdrawing, Distracting, Humoring, Diverting

Trying to get clients to focus on something other than the problem is the twelfth roadblock. Sometimes workers are overwhelmed by what their clients have told them. They may feel helpless to make a real difference or offer substantive help. Perhaps they realize that all they can do is listen, but listening is painful and difficult. To save themselves from these uncomfortable feelings of inadequacy and helplessness, workers may resort to responses like the ones that follow:

- "Just forget about it!"
- "C'mon! How are things at church?"
- "Let's see, you could always run over him in your car." (chuckle)
- "Let's turn to other things in your life."
- "You're upset. Why don't we talk about your cat?"

Using the Phrase "I Understand" It is tempting to want to comfort people who are upset with the phrase "I understand." Sometimes workers say to clients, "I understand how you feel." Worse yet, a worker might say, "I understand how you feel. That happened to me once." Often, however, this sounds really trite. Most of us can never fully understand exactly how a client feels.

Another phrase some workers use is "I understand, but . . . " This phrase is even worse. In addition to the fact that we cannot fully understand what a client is experiencing, the *but* in the phrase tends to negate the person's very real feelings and push the worker's perspective instead. Make it a point to refrain from soothing clients with phrases such as "I understand" or "I understand how you feel," and certainly refrain from saying, "I understand, but . . ." Instead, you will learn to feed back to the other person the feelings you believe that person is experiencing, giving the other person a much better sense of being understood.

Useful Responses

In this third section of the book, you will be looking at and practicing responses that enhance communication. Following are several categories of responses that you may find useful as you construct answers to the exercises in this section of the book.

Learning to structure good responses is a little like learning to drive a stick-shift car, rather than an automatic one. For years your communication responses to other people have been automatic. They probably worked because you were in relationships other than professional ones. In our friendships and family relations, people communicate in shorthand. These relationships are generally positive and familiar, so others do not need to guess what it is we are saying.

Now, as a professional, you are responsible for creating an environment that makes people feel comfortable and safe enough to be open. Clients will not know you very well, if at all, when they come to see you. Therefore, you are in charge of communicating in a way that builds rapport and collaboration in regard to the clients' problems. This means that at first you will have to think carefully about how you are going to respond before you do so.

The good response examples that follow contain openers that you can lift right off the page to get you started on the exercises you will find in later chapters on communication. Use these initial phrases to structure constructive responses. Gradually, as you practice, you will begin to sound more like yourself, and your responses will not seem as rehearsed. At first, however, you need to practice effective ways to answer what another person has said. Using the responses provided here will help you to get started in this process.

Ways to Start Responding to Feelings

When people are talking to you about something that involves an emotion or feeling, it helps them to feel comfortable and understood if you can identify their feeling and say that back to them. This is part of practicing empathy. When you do so, it is best to structure a single sentence and say nothing more. Anything that you might add to this could take the conversation away from where they are and over to something you have introduced.

In addition, do not confuse the person by adding more than one feeling. You might say *You must feel so hurt* and that would be fine. If you said *You must feel so hurt and betrayed* your message is less precise and somewhat confusing. Be very careful about this. Here are some useful openers to use when you respond to feelings:

- "That must have made you feel . . ."
- "You must feel . . ."
- "You must have felt . . ."
- "That must have been . . ."
- "That must be . . ."
- "It sounds like you're really feeling . . ."
- "How [sad, upsetting, wonderful] . . ."
- "Sounds like you really feel . . ."
- "You must be . . ."
- "You must feel so . . ."
- "It sounds like you felt . . ."

Ways to Start Responding to Content

There are times when you might indicate you heard accurately by responding to the content of what someone has said. In this way, you confirm that you are hearing what the person has told you and confirm for the person how important the details are to you.

- "So, it's important to you that..."
- "You're really concerned about..."
- "So you were [he/she was]..."
- "Right now you want..."
- "So, in other words,..."
- "So what happened was..."
- "So you decided to..."
- "You really need..."
- "You're hoping that..."
- "It sounds like they [he/she]..."
- "So they [he/she] just..."
- "So you just..."

Digital Download Download from CengageBrain.com

Ways to Start a Closed Question

There are times when you need facts or specific information. Questions that require only a single answer are often referred to as closed questions. Here are some ways to start a closed question.

- "What is your...?"
- "Where did you...?"
- "Who is...?"
- "When were you...?"
- "Where do you...?"

Ways to Start an Open Question

When we are listening to clients give background about their concerns and problems, our questions need to be more open to solicit the information the individuals believe is significant. Here are some ways to start an open question.

- "Can you describe...?"
- "Can you tell me a little bit about...?"
- "Could you talk about...?"
- "Could you describe more about...?"
- "Can you tell me a bit more about...?"
- "Can you fill me in on...?"
- "Could you clarify that a little bit more for me?"
- "Can you tell me something about...?"
- "Could you say something about...?"

Digital Download Download from CengageBrain.com

Ways to Start an I-Message

There will be times when you are concerned about something the client has done or said. You may be worried about something the client intends to do or something the client has not done. In the roadblocks to good communication earlier in this chapter, many of the ways workers brought up their concerns were confrontational and superior. A better way to bring up our own concerns is to indicate that these concerns belong to us. We do this by using the word "I" first. These responses can consist of several sentences and should sound tentative rather than judgmental or decisive.

- "I feel..."
- "I'm just concerned that..."
- "I'm wondering if..."
- "It appears to me that..."
- "I need to understand..."
- "I'm not clear about..."
- "I need to kick something around with you."
- "I'm having a problem with..."
- "I'm uncomfortable that [with]..."
- "I guess what worries me is..."
- "I think what I'm most concerned about is..."

Notice in these responses how tentative they are. The worker is "wondering' about something. The worker uses terms like "I guess" or "I think." This leaves the discussion open for the other person to respond.

Useful Ways to Begin a Firmer I-Message

There are times when you need to act on behalf of clients. Another person may be unintentionally interfering in some way. This may call for an invitation to help, and that invitation must be worded in a way that is more authoritative, but not offensive. Here are ways to be clear about what you want or need:

- "I need you to..."
- "It would be very helpful if..."
- "I wonder if you could help us by..."
- "Could you..."
- "Would you please..."
- "We need your help to..."

Always use please and thank the other person when you make these requests.

Digital Download Download from CengageBrain.com

Ways to Show Appreciation for What Has Been Said

When clients bring something to your attention that is of concern to them, it is a good idea to let them know you appreciate what they have to say. Sometimes they are telling us about something we or our agency has done that bothers them. Nevertheless, it is always a good idea to show appreciation. Here are some ways to start appreciative responses.

- "Thank you for bringing this up."
- "It was good of you to tell me about this."
- "I appreciate your thoughts about this."
- "Thanks for telling me."
- "It's helpful to me to know this."
- "Thank you for letting us know."

Specific Questions Useful in Beginning to Disarm Anger

When clients express anger with us or our organization, it is not constructive to argue with them. Arguing undermines your work with the person and inflames the situation. Instead, showing a genuine interest in what they are telling us is better for maintaining a good relationship. Here are some questions to ask that indicate a genuine interest on your part. By using these responses, you indicate that you really want to understand the problem the client is experiencing with you or your organization.

Do not ask the person all of these questions; one or two of them will indicate a real willingness on your part to grasp the issues and make it safe to discuss issues the client may have with you or the agency. Too many questions can make the client feel he or she is being interrogated.

- "How did I [we] offend you?"
- "What did I [we] do?"
- "When did I [we] do this?"
- "How often did I [we] do this?"
- "What else about me [us] upsets you?"
- "What might I [we] do to clear this up?"
- "Can you tell me more about what happened?"

Examples of Ways to Agree When Practicing Disarming

When we are not acting in a professional capacity, it is common to feel very defensive when someone criticizes us. Usually, however, there is a kernel of truth in what the other person is expressing even though it might seem exaggerated or trifling to you.

Here are some responses you can use to let clients know that you can see and accept the truth in what they have told you.

- "I'm sure I could do better at times."
- "We probably could do things a bit differently."
- "There are people who have had more experience than I have."
- "It may be that we (I) could do things differently."
- "Probably we're not always aware of these problems."
- "It is very possible that (I) we overlooked this."
- "I certainly can be forgetful at times."

Digital Download Download from CengageBrain.com

Sample Response When You Cannot Change

After a heated exchange, an angry client may be expecting that you will change the way you do things. Sometimes you are not able to change things. Maybe you are blocked by the law or because of how a change would affect other clients or staff. When that happens, you need a pleasant way to let the person know you cannot make the requested changes. Here is an example of what you might say:

- "I understand your point. We're going to have to continue this way for now, but it was helpful to hear your concerns."

Sample Response When You Find You Can Compromise

At other times, clients make useful suggestions, and the requested changes can be made. Here is an example of how you might respond in such a situation:

- "I think you have good ideas and there are some ways we can solve this problem."

Ways to Start Collaboration

Nothing really useful can happen for clients if there is no collaboration. Even when clients will be doing most of the work, the word *we* can soften this fact and create a team approach to the problem. In this way, you let clients know they can trust your intention to be supportive without taking over and forcing a solution. Collaboration has another useful purpose. When done well, it prevents your giving the impression that you feel superior and that you see clients as being helpless and inadequate. Here are some ways to begin collaboration:

- "Perhaps we can . . ."
- "Maybe we can [could] . . ."
- "Let's [look at this together, look at your options, see what we can find out about this]."
- "We can [could] . . ."
- "Why don't we . . ."
- "We might . . ."
- "You and I together can [could] . . ."

Ways to Involve the Client in Collaboration

Sometimes clients do not participate. You make suggestions, and the clients simply go along with them. If this happens often, the clients are not collaborating or participating in solutions to problems they own. There are ways to help people become more involved. Here are some examples:

Start with an I-Message

- "I'm wondering if we . . ."
- "It occurs to me that we . . ."
- "I am just thinking that perhaps we . . ."
- "I guess what concerns me is . . ."
- "I think I am worried about . . ."

Finish with a Question or Comment That Invites Collaboration

- "What do you think?"
- "I'm wondering what you think."
- "What thoughts do you have?"
- "Maybe you see it differently."
- "You probably have some ideas too."
- "But it's important to me to know how you see it."
- "But I think what would really be helpful is to hear your ideas about it."
- "How do you see it?"
- "What are your suggestions?"

False Praise versus Positive Feedback

Earlier in the chapter in the discussion of roadblocks to communication, some examples of ways workers might praise clients were presented. These statements do not contain real information clients can use. It is all right to give people positive feedback, but not to say something trivial that contains no information clients can use in the future when they encounter difficulties. When you give positive feedback to a client, structure what you have to say to include information the person can use.

For instance, suppose a client worked on and solved a tax problem. Rather than just praising the fact that she solved the problem, it might be better to point out information about the client she can use in the future. Here is an example of how to do that.

> "I thought you handled the people at the tax office very well. You asked the right questions and didn't get flustered, and I think that really helped to resolve the problem."

Now the client has been given feedback that identifies traits that might be helpful to her in the future: She seemed capable of talking to others and asking the right questions, and she was able to communicate without becoming flustered. This kind of response is much better than one like, "See how smart you are?"

Following is another example of good feedback that contains useful information for the client.

"I think it has taken real determination for you to stay sober this long. I'm just impressed with how you have managed an entire week like this."

Again, this is a statement that gives the client useful information: The worker believes the client has "determination." This is something he can rely on in the future when times are tough. Thus, such a response is better than something like, "Good for you! You're still sober!"

Minor Problems

The following examples demonstrate some of the minor problems you might encounter during your communication with your clients.

Minor Problem One. You assess your client's feelings incorrectly. For instance, you might misinterpret a client's underlying feelings and thoughts. You might say, "You must have felt sick when you saw what the accident did to your car." The client might respond by telling you she really did not feel sick. Instead, she was angry—furious, in fact—at the other driver who stood there and screamed at her. It is always possible that clients may correct you in this way. This is positive. It allows people to make you aware of what they are really feeling and thinking, bringing you much greater clarity.

Minor Problem Two. Your mind wanders. At times you may not be listening attentively. Something has happened at home; you just finished handling a personal problem; or you hoped to go home before 4:00 P.M. and now it is 5:00 P.M. and this new issue may take some time. Your mind leaves the immediate situation and wanders to your personal concerns. Of course you do not want this to happen often, but it will happen. If you are practicing good body language, sitting in a way that communicates interest and attention to the client, these momentary shifts in your focus will not be damaging to your relationship.

Major Problems

The following are examples of major problems that can occur during your communication with your clients that are not at all useful.

Major Problem One. You cannot wait to pass judgment. Sometimes workers listen to clients, but their minds are full of judgments they want to make about what the people have said. Rather than truly listening, these workers are judging the clients in regard to how well the clients handled their situations, whether the clients were smart or stupid, and whether the clients were on top of things or lax in taking care of the situations.

Such workers cannot wait for clients to be quiet so they can give an authoritative judgment. Rather than listening, these workers are preoccupied with what they plan to say in response. This is unprofessional listening. We might do this with our friends and relations, but it has no place in a professional relationship.

Major Problem Two. You ignore the client's feelings. Another way to miss the important issues is to focus entirely on content and never hear the meaning this situation has for the client. In other words, you fail to practice empathy. Instead of commenting, "How difficult it must have been," the worker goes on about the actual details, "So actually this happened in town." There is a place for this kind of listening, listening to content, but when you do this exclusively and never talk about feelings, important opportunities are missed for healing and building rapport.

Major Problem Three. You cannot wait to offer the solution, to give advice. Opportunities are lost for establishing rapport and understanding. If the worker does not acknowledge the feelings a client is expressing and rushes to a solution, the person may not feel they were really heard. Some workers rush right past the feelings to the solutions. Rather than saying, "You must have found that so painful," such a worker would say something like, "Well, you'll need to see a lawyer about this. There is a good one around the corner that we use a lot, and I can probably get you in to see her in the next day or so." Anyone can tell a client he needs a good lawyer. Only a very good listener is able to respond with empathy to the underlying feelings present in the client's story.

Major Problem Four. You feel an overwhelming need to get to the bottom of the problem and solve it. You feel the client is expecting you to do just that and you are trying to figure out the solution and so you begin to ask a lot of closed questions, one right after the other. "Did you ask your aunt about it? Did she tell where your cousin is? Did you try to reach your cousin? Is there a reason your cousin did not call you directly? Where do you think your cousin is?" You ask and the client answers and in your head you are trying desperately to construct a solution for the client. In this case it would have been better to start with one open question, "Tell me more about your cousin's leaving," and follow up with other open questions that seek what the client is thinking would be the best solution here.

Summary

This chapter has provided you with a sense of how good and bad responses sound. Good responses are constructive responses that promote rapport, facilitate collaboration, and build trust; poor responses block rapport, understanding, and further exploration. Knowing how these two different kinds of responses sound, however, and actually using them are two different things. To become proficient, you must practice.

In each chapter in this third section of the book, different types of responses are discussed along with what they are intended to accomplish in a therapeutic sense. Each chapter contains a series of exercises. To do these exercises, at least initially, turn back to this chapter and refer to the examples of useful responses provided in it. Use these sample openers as a springboard to developing constructive replies of your own.

Video Example

To view the videos that accompany this book, go to CengageBrain.com.

- "Case Management with an At-Risk Individual"—To see these responses in an actual interview you can view the video segments online. Those segments (1, 6, 10, 12, 14, 16) where case managers are talking with their clients demonstrate how these techniques are applied when talking to other people.

Exercises

These exercises can also be filled out online at CengageBrain.com.

Exercises: Identifying Roadblocks

Instructions: Following are various scenarios illustrating workers' responses in a variety of situations. Examine the responses in each case and decide whether the worker is blocking communication or enhancing it. Do the worker's responses cut off further communication from the other person or seem to encourage the person to continue? Circle your answer after each vignette.

1. Carlos is afraid his mother is dying. He is talking to the worker in the hospital emergency room about an "attack" his mother seemed to have when she could not breathe and turned blue. She was brought to the hospital in an ambulance, and Carlos is waiting to see whether she will be all right. He is distraught. The worker says, "You certainly did the right thing to call the ambulance. Don't worry she'll be all right. We have very good doctors here." Can Carlos continue to express his anxiety freely?

    Enhanced   Blocked

2. Anita and her family moved, and her parents feel Anita is not adjusting well to the move. Anita wants to return to her old school to be with her friends. Her parents ask the worker to talk to Anita about her desire to return to her former school. Anita talks about how strange the new school is and how much she misses her old friends. The worker replies, "Tell me something about your friends where you used to live." Can Anita continue to tell the worker what she misses about her old school?

    Enhanced   Blocked

3. Elvita has decided to leave an abusive relationship, but she feels guilty leaving her abuser's children behind. She talks about how she knows that once she leaves, she cannot have any more contact with the children; and she is worried about how that will affect these children whom she has come to love and wants to protect. The worker asks, "Just when did you start to think of these children as if they were your own?" Can Elvita continue to discuss her concerns about her abuser's children?

   Enhanced   Blocked

4. Ed suffers from chronic mental illness and needs medications to maintain his mental health. Recently he went to several workshops on the use of supplements and vitamins to maintain mental health. He wants to discuss these ideas with his worker. He makes it clear that he is not really thinking of going off his prescriptions, but he would like to consider trying these supplements in addition to his medication. The worker says, "It sounds like you really got a lot out of that workshop." Can Ed continue to explore the things he learned at the workshop with the worker?

   Enhanced   Blocked

5. Shawna wants to go to college. She attended a poor rural school where most of the students do not go on to college. Her scores for the entrance exams were very poor in math, and she feels unsure that the developmental course being offered to her at the college will really help her catch up. She seems anxious and uncertain. The worker says, "You don't seem to quite understand what developmental courses are. Look at the number of people who take them. Look at how many of those people finish school. You need to think about this a little less emotionally." Will Shawna feel comfortable in the future talking about her concerns about going to college with this worker?

   Enhanced   Blocked

6. Ada is very upset over the divorce settlement. She got the house and the children, but very meager support and only a small amount to go to school to upgrade her skills. As she speaks with the worker, she is crying and expresses the belief that she cannot make it. The worker replies, "Look, get another lawyer. You're going to have to just face the fact that you had a lawyer who wasn't serving you. Go back to court! Reopen the case! Make a stink!" Will Ada be able to talk about her divorce with this worker?

   Enhanced   Blocked

7. Lindsey is an alcoholic and tells her worker she wants to stop drinking but doesn't know how. She tells the worker that for a while she stopped drinking while she was going to AA meetings, but she stopped going and began to drink again when she reunited with old friends. The worker says, "Here's

the bottom line. You either do what you need to do or you don't. Don't come in here crying about how you stopped AA. That's your responsibility to go there—so go!" Can Lindsey continue to talk about options for not drinking with this worker?

Enhanced   Blocked

8. Reynaldo just lost his job and is frantic about how he will pay the rent. He talks to the worker about how he will pay his rent and whether he will be eligible for unemployment. In the course of the conversation, he mentions his fear of going home to face his wife. He does not believe she will understand. The worker says, "Well, you're mostly upset because you're afraid of your wife and what she's going to say about this." Will Reynaldo want to continue to express his worries with this worker?

Enhanced   Blocked

9. Tonda is discussing her need for help with a depression that, she says, started several months ago. She talks about the amount of time she has missed at work and how it has reached a point where she sleeps most of the day. The worker says, "Tell me a little bit about what was going on when this all started." Can Tonda continue to discuss her depression with this worker?

Enhanced   Blocked

10. Persis has been clean for 9 months and comes in to see her case manager about how she is really tempted to use again. Persis asks if there is something more she can do. The worker says, "Oh, come on. It's been 9 months. Get down to business and focus on the good things in your life. It's all up to you whether you use or not." Can Persis continue to explore her temptation to go back on drugs with this worker?

Enhanced   Blocked

CHAPTER 8

Listening and Responding

Introduction

Listening well to others is therapeutic, a healing activity. For that reason, doing so effectively is very important. Although case managers do not practice therapy as we think of it in a counseling setting, they do have many opportunities to listen to others in a way that is therapeutic. At intake and during the course of the relationship as problems and issues arise, the case manager can offer listening as a first and important step in the resolution of people's problems.

Writing in the *American Journal of Psychiatry,* Dr. Stanley W. Jackson (1992, p. 1624) talked about the importance of listening:

> The effective healer in the realm of psychological healing tends to be someone who is interested in talking with and listening to the other person. And these inclinations are grounded in an interest in other people and a curiosity about them. Further such healers have a capacity for caring about and being concerned about others, particularly about those who are ill, troubled, or distressed.

Describing those who seek our help, Jackson wrote, "He seeks to be listened to, to be taken seriously, and to be understood, as crucial aspects of this process." Finally, he talked about the process itself:

> The attentive listening of a concerned and interested healer can, and often does, have a compelling effect on the sufferer. The sufferer often enough responds by telling more about himself, by revealing more. . . . The relationship is deepened—more is said, more is heard, more is understood, more of a sense of being understood is experienced.

Listening to another person in a way that indicates our concern for that person is important in the healing process. **Reflective listening** is a method that allows you to demonstrate such concern and interest in other people.

Defining Reflective Listening

Reflective listening is a term used to describe therapeutic listening and responding—a way of listening that is most helpful to people. This method of listening to others has three purposes.

1. Reflective listening lets people know you have heard their concerns and feelings accurately.
2. Reflective listening creates an opportunity for you to correct any misperceptions.
3. Reflective listening illustrates your acceptance of where the person is at that moment.

When someone talks to you, there are two aspects to which you can listen and respond:

1. The *content* of what the person has said
2. The *feelings* that underlie what the person has said

Responding to feelings is empathic and is, therefore, the most useful kind of response.

When you accurately respond to the feelings the person is experiencing, that person feels heard. Someone is really listening. When clients feel that you truly hear them and that you hear the feelings they are expressing, even when they do not explicitly describe those feelings, they begin to develop trust and rapport with you, making it easier for them to fully talk with you about their problems and issues. As you saw in the discussion of roadblocks to good communication in Chapter 7, there are many responses that are barriers to trust and rapport. You want to provide a safe, accepting environment where people feel free to express themselves. For this reason, it is important to learn how to provide empathic responses that will further a constructive relationship with your clients.

Responding to Feelings

When you let another person know that you have heard the feelings that person is expressing at that moment, you are being empathic. *Empathy* is the ability to hear and experience accurately the underlying feelings and emotions that clients are expressing when they speak. We have empathy when we are able to put ourselves emotionally in their situations and clearly sense what they must be feeling. People may tell you how they feel in actual words, but much of our understanding of other people's emotions comes from their facial expressions, body language, tone of voice, mood, and choice of words. In listening to feelings, we are listening to all of that as well as the words clients speak. In this way, we gain an understanding of the underlying emotions and concerns.

It is important to really listen. Instead, many of us are tempted to think while other people are talking. We think about how wrong they are. We think about what they could do instead. We think about what advice we should give them and how to solve their problem. What we really should be doing is listening for the feeling and determining the degree of that feeling. For instance, a person may be angry or furious or just annoyed. Another person may be a little wistful, sad, or openly depressed. Can you tell the difference when you listen? Of course you can, when you are really listening for the feelings.

Responding to feelings involves the following three steps:

1. Listening carefully to the client
2. Identifying the most prominent feeling you are hearing
3. Constructing a single statement that includes that feeling

This single sentence that includes the other person's feeling is called an empathic response. This is a specific way to practice empathy for clients, to acknowledge clients' feelings and concerns. We use a single sentence because to add more often distracts the person or moves the conversation away from the central concern. A single sentence allows people to know that they have been heard and that they are welcome to continue on the same track; they are reassured.

Reflective listening does not include advice or solutions. A reflective listener does not tell the other person to feel another way or to look at the problem from another perspective. A reflective listener does not judge the message or the feelings. See Figure 8.1 for some examples of correct and incorrect responses in different situations.

FIGURE 8.1 Correct and incorrect responses when listening to feelings

To a person whose new car was damaged in an accident:

Correct: "You must have felt pretty terrible about the accident."
Incorrect: "Before you get upset, you should call your insurance dealer."

To a person who just lost her childcare:

Correct: "I'm sure this has been really difficult!"
Incorrect: "Well, we just need another day care to take your daughter."

To a person distressed over not finding a job:

Correct: "It sounds like you're feeling pretty defeated."
Incorrect: "I'm sorry you couldn't find work. Maybe you'll have better luck tomorrow."

To a person whose daughter has been removed because of child abuse:

Correct: "This isn't going to be easy without her."
Incorrect: "This is what happens when the county feels kids are abused."

In each good example the worker is beginning where the client is in order to be able to move with the client to a solution. In this last exchange, the worker can more easily talk to the client about the steps he or she needs to take to get her children back if she has first acknowledged how difficult this is for the person. Acknowledge first the feelings so the person feels supported and heard and then move to what needs to be done.

In the following dialogue, based on the situation in which a client has wrecked a new car, notice the way the worker stays with the client throughout the exchange:

CLIENT (sighing as he sits down): I wrecked my car yesterday on my way to work.

WORKER: It sounds like you feel pretty bad about it.

CLIENT (sighing again): I do. I guess I should be happy no one was hurt, but I just got the car.

WORKER: It was brand-new and perfect.

CLIENT: I know. I picked it out and ordered it special. It had everything I wanted. I don't know, in a way it was my fault. She ran the stop sign, but I wasn't really paying attention.

WORKER: You're sort of blaming yourself for this.

CLIENT: Oh, I did. I still do. The police said she was clearly in the wrong. But now I'm going through all this unnecessary stuff with insurance and using a loaner car and trying to get a new car.

WORKER: You must feel so disrupted.

CLIENT: Yeah, I do.

Listen now as the same client speaks to a worker who is not trained in reflective listening. Notice how the worker pursues her own agenda and how the client begins to sound defensive and ultimately stops participating:

CLIENT (sighing as he sits down): I wrecked my car yesterday on my way to work.

WORKER: You wrecked your car? How did you do that?

CLIENT (sighing again): I guess I should be happy no one was hurt, but I just got the car.

WORKER: Well, what happened?

CLIENT: The other driver ran the stop sign, but I wasn't really paying attention. In a way, it was my fault.

WORKER: Well, you can't drive and think about 10 other things. When you are driving a car, you have to pay attention to what is going on around you. If you're daydreaming, it doesn't work. You can't do that when you're driving.

CLIENT (sounding defensive): Well, I wasn't really daydreaming or anything. I just didn't notice her. You can't always see everything the other drivers intend to do before they do it.

WORKER: I guess you can't, but sometimes I think there are just too many drivers out there anymore.

CLIENT (nods): But now I'm going through all this unnecessary stuff with insurance and using a loaner car and trying to get a new car.

WORKER: Well, that's all part of it. When you wreck your car you're tied up for months with all the bureaucratic paperwork. And they never give you what you need to buy another one just like it.

CLIENT: Uh-huh.

Let us look at another example of good listening. Notice how the worker in this next example identifies the strongest feeling present in what the client is saying and how the client almost always responds positively to that recognition:

CLIENT *(tears welling up in her eyes):* I never went through this before.

WORKER: It sounds like you are devastated.

CLIENT *(nodding and crying more openly):* My dad died, last week. He...he...well, he had been in dialysis, but he seemed so good Sunday night. Then Monday the nurse called me at work—oh, I guess around 10:00—and suggested I come in, but she didn't say it was an emergency or anything.

WORKER: You must have felt you didn't have to rush, that you had a little bit of time.

CLIENT: Well, no, I went in as soon as I finished up what I was doing, and here he was already in intensive care. His breathing was so labored, so hard for him... *(cries)*.

WORKER: That must have been a shock!

CLIENT: It was! I couldn't believe it. That was my dad lying there. He just fixed my electrical outlet 2 weeks ago. We went out to eat for his birthday. I just couldn't believe he would go now.

WORKER: It was all so sudden.

CLIENT: Oh yes, and then they are talking to my mom and I about how to make him comfortable. He knew me and all, but he couldn't talk. And you know how they say people need permission to die? Well, I told him it was okay to go *(cries)*. I told him he gave his life a good shot and he could go *(whispering)*. And he did *(cries)*.

WORKER: That must have been so painful for you.

Now observe what happens in this situation when the worker's listening skills are inadequate. Notice in this example how the worker brings the conversation around to a more cheerful topic. In this case, the worker may very well be protecting himself from feeling the enormous pain of the client:

CLIENT *(tears welling up in her eyes):* I never went through this before.

WORKER: Like what?

CLIENT *(crying more openly):* My dad died last week. He...he...well, he had been in dialysis, but he seemed so good Sunday night. Then Monday the nurse called me at work—oh, I guess around 10:00—and suggested I come in, but she didn't say it was an emergency or anything.

WORKER: So you went right in!

CLIENT: Well, no, I went in as soon as I finished up what I was doing. I didn't think she meant I had to hurry, and here he was already in intensive care. His breathing was so labored, so hard for him... *(cries)*.

WORKER: Did he have a living will? That would have helped you to know how to handle this.

CLIENT: I don't think he did. I don't know. Mom and I made the decisions. We knew he couldn't go on much longer, and we didn't want him to suffer. I couldn't believe it. That was my dad lying there. He just fixed my electrical outlet 2 weeks ago. We went out to eat for his birthday. I just couldn't believe he would go now.

WORKER: We never really know when death will strike, do we? We just have to look on the bright side, at all the good times we had with people while they were here. We know our parents won't live forever.

CLIENT: It isn't that I thought he would live forever. It was just so sudden or something. We didn't have much warning really. I guess in that respect I can say he didn't suffer, you know, like in a long illness for years and years.

WORKER: See. There's something to be thankful for. There is a silver lining in everything.

CLIENT: Sure.

Did you notice that the client stopped crying? She took the cue from the worker that crying and continuing with the painful story about her father's death was being discouraged. It must have been clear to you as well that a significant part of the story was left out. The first worker received more information about what really happened by saying less. The second worker said more, but cut off part of the story and, therefore, did not receive as much information.

Good listeners stay with people until their emotions are drained off. If a client cries, we know we are helping that person to face and come to terms with intense emotion. People who are supported through this process by skilled listeners heal better than people who have been forced to shut down their feelings in front of the worker and deal with them alone.

Responding to Content

When you respond to the content of what clients say, you are usually doing it to check the accuracy of the information you believe you heard. Listening to content gives you clarity and helps you understand the facts. If the event the person is talking about was traumatic, listening to content helps that person to begin to hear and integrate this experience. We are helping that person to come to terms with what happened and heal. Generally, we listen to content less often than we listen to feelings. Figure 8.2 provides some examples of correct and incorrect responses in terms of listening to content.

Responding to content is a good way to help people who have just been through a traumatic event. As noted, when you repeat the facts of the traumatic event back to the person, the individual can begin to integrate her particular experiences into the whole of her experiences, and her healing is facilitated. In fact, the sooner people

FIGURE 8.2 Correct and incorrect responses in listening to content

> **To a person who is describing losing everything in a fire:**
>
> *Correct:* "So the fire started in the back of your neighbor's house and spread to your back porch?"
> *Incorrect:* "Sounds like a bad fire!"
>
> **To a person who is telling the worker about hearing voices:**
>
> *Correct:* "In other words, these voices have kept you awake at night telling you that there are people watching the house."
> *Incorrect:* "Your hearing voices means you need to see Dr. Preston."
>
> **To an older person whose health has meant she must move into an assisted living facility when she would rather be home:**
>
> *Correct:* "Sounds like you had a wonderful home over on Locust Street and good neighbors."
> *Incorrect:* "Moving here does mean giving up your home and your neighbors."
>
> **To a child in a shelter who is telling about how he saw his dad arrested for assaulting his mother.**
>
> *Correct:* "So the police came to the house and took your dad outside and you and your mother were told to stay in the house."
> *Incorrect:* "Police arrest men that beat women."
>
> In each good example the worker is beginning where the client is in order to be able to help the person tell more and come to terms with what happened. In each correct exchange, the worker elicits more information than would be given when using the incorrect responses and, therefore, will have a better understanding of what happened and how the person feels about it.

begin this process after a traumatic event, the more readily they may be able to heal in the future. Here is an example of a worker who listens to the content of a client's discussion of a traumatic event:

> CLIENT *(looks pale and shaken, and is silent)*
>
> WORKER *(responding first to feeling):* It seems like you've been through something pretty terrifying.
>
> CLIENT *(nods):* I . . . Can I tell you? It was . . . in the parking garage at the hotel. Not late or anything. I heard this person get off the elevator as I . . . (*is silent*).
>
> WORKER: You had gone to the parking garage to get your car.
>
> CLIENT: Yes, and as I was walking toward the car, and it was at some distance, I thought I heard the elevator, and then someone started walking along behind me.
>
> WORKER: The person was walking behind you, but at that point you thought the person was just going to a car.

CLIENT: Right. And just as I was ready to get in my car, . . . I mean, I almost made it, and he grabbed my coat and just pulled, just pulled as hard . . . just pulled me backward.

WORKER: So, in other words, you were grabbed and pulled down from behind.

Notice that the worker rephrases the facts the client gives so that the client hears them again. Through this listening process, it is not uncommon to find that unpleasant memories, which the client has blocked, begin to surface in the supportive atmosphere created by the worker. At this point, some people might say the client is lying or making up things because the story has changed somewhat. More often, however, the client is beginning to remember more of what actually happened because the worker is encouraging and has created a safe environment.

Positive Reasons for Reflective Listening

Some specific therapeutic reasons have been given for employing reflective listening. These are described in the following two sections.

Self-Acceptance

Arnold Beisser (1970), writing about Fritz Perls' ideas, talks about Perls' paradoxical theory of change. The theory asserts that people change only when they are able to accept themselves exactly where they are right now.

Judgments about where clients should be only engages them in defending how they came to the place they are now. This wastes clients' energy and the valuable time you have to work on healing. When you accept where the person is at the moment, the individual can accept herself and move forward. The most healing, and therefore the most therapeutic, practice is reflective listening. By saying "You must be angry," you accept the fact that the client is angry. If you say "You should try to curb your anger and think more positively," the client then has to explain why she has not done so or cannot do so. In this situation, some people say nothing and decide you do not really understand. By accepting, without judging, where the person is now, he can then move on toward something better. People cannot move in a positive direction until they have accepted where they are now.

Drain Off Feeling

In the human service profession, you will meet people in all sorts of life crises and difficulties. Some of the circumstances are very traumatic, and considerable reflective listening will be required on your part if the people are to begin the healing process. A woman who has been raped will start coming to terms with it sooner and heal more readily if she encounters a good reflective listener soon afterward. An older person can prepare for the end of life and feel comfortable about his past life if he can talk about it with a good reflective listener.

Human service workers have been criticized for not listening long enough. Some take only a few stabs at it, and then move into problem solving: Where will this person stay tonight? Who should I call? What facts do I need to open this case? The importance of your role in trauma, in healing, and in helping clients to grow cannot be overestimated. You play that role in large part by practicing good reflective listening.

Points to Remember

Listen Reflectively Long Enough

Do not cut short this important piece of the client's healing process because you feel pressed for time. Be sure you go over the situation thoroughly once and, if you can, review it several times. In cases of violence, reviewing the content several times, along with listening to feelings, helps the victim begin to hear the story and come to terms with it. Reflective listening, in this case as in others, promotes healing.

Introducing Solutions

Do not rush to the solution phase of the interview unless the person seems extremely anxious about what will happen. Even if you have ideas, wait until the emotion has been drained off.

You cannot confront the issues you feel are important if you have not done your reflective listening first. If you do, the client may go along with you, but not as well or with as much involvement as she would if she felt heard and understood. For instance, if you immediately explore your concerns about where the client will stay tonight instead of acknowledging the loss of her home in a fire just hours ago, she will not be as ready to work with you on solutions. Her mind is on her many losses, and her emotions may be ranging from guilt to anger. Listen first. Likewise, if you try to help a young couple with the details of their baby's funeral without listening to their story about the baby and the baby's death, you will be taking care of the details that matter most to you, but the clients are likely to experience you as unfeeling and impersonal. By starting where they are, you can help them to move toward the matters that must be addressed. Most important, you help them to integrate this experience into the whole of their life experiences, making it easier for them to ultimately accept the reality.

If the client would obviously feel comforted to know there are solutions or resources, tell him about these first, but then demonstrate reflective listening at another point in the interview. If the interview is pressed for time, after you give the information that is needed, give at least one reflective listening response. "I know losing your home has been devastating for you" or "This evening must have been heart breaking." The person will leave feeling as if he is understood.

Reflective Listening Does Not Mean You Agree

Just because you say to a client "You must feel very angry" does not mean you think the client should feel angry or should not feel angry. You are simply acknowledging where the person is right now. You have accepted that. This makes it easier for him to accept that. Now he can more easily move from being angry to something better.

You Could Be Wrong

Suppose you say to a client, "It must have made you sad to see your parents go through that." The client responds, "Well, not really. I think I felt more anger than sadness." This is a good exchange. Here you get important, corrected information that allows you to follow the client's concerns more accurately. Reflective listening allows us to clarify and correct things so that we are following where the client is at the moment.

Mind Your Body Language

To facilitate the interview, lean toward the client, look the person in the eye, nod, and look interested and enthused. While the client is talking, do not fiddle with things on your desk, lower your head to write, stare out the window, or glance at your watch. Give body language signals that indicate you are being attentive to what the person is saying. Do not stand over people talking down at them. If the client is in a wheelchair or is a small child, get down to a level where you can make eye contact. Pull up a chair so that you can look at the person directly.

Summary

Good listening is one of the most supportive and healing techniques you will practice in your work with other people. The opportunities to provide solace through good listening skills are numerous and occur in many diverse settings where people come for help. What you give to people in uncertain circumstances and difficult times is the gift of truly being heard and understood. You bring with you the warmth and interest in people that makes them feel valued. Their story is an important story to you. Their anxiety or sad feelings are noted and responded to. Whether or not you are able to effect a positive resolution to people's problems, your listening skills will always provide the support people need to take up the tasks of their lives and go on.

Video Examples

To view the videos that accompany this book, go to CengageBrain.com.

- You can see Keyanna responding to the underlying feelings Michelle has about her situation by watching "The First Interview" online. Two other vignettes demonstrate how this is done: "Developing a Service Plan" and "Helping Tom Solve a Personal Problem."

Exercises

These exercises can also be filled out online at CengageBrain.com.

Exercises I: How Many Feelings Can You Name?

Instructions: In a group of no more than four people, see how many feelings you can name in 10 minutes. Remember that there are many different degrees of the same feeling. After you do the exercise check your list against the list in the appendix.

Exercises II: Finding the Right Feeling

Instructions: When responding to feelings, it is important to know the intensity of the feeling. It is very important to reflect to clients an accurate reading of what they must be feeling. All feelings have varying degrees of intensity. For each word listed here, list other words that mean the same thing but indicate varying degrees of the feeling identified by the original word. The first one is done for you as an example. Check the list of feelings in the appendix to see if there are other words you missed.

HAPPY: *overjoyed, exhilarated, glad, delighted, cheerful, ecstatic, merry, radiant, content, elated, euphoric, ebullient, chipper, bouncy, bright, joyful, pleased*

SAD: _____

CONFUSED: _____

TENSE: _____

LONELY: _____

STUPID: _____

ANGRY: _____

_____

Exercises III: Reflective Listening

Reflective Listening I

Instructions: People communicate words and ideas, and sometimes it seems appropriate to respond to the content of what someone has just said. Behind the words, however, lie the feelings. Often it is most helpful to respond to the feelings.

Following are statements made by people with problems. For each statement, first identify the feeling; write down the word you think best describes how the person might be feeling. Next, write a brief empathic response—a short sentence that includes the feeling. Refer to the sample openers provided in Chapter 7 under the heading "Useful Responses."

1. "When I was in court, the defense attorney really pounded me. You know, like he thought I was lying or didn't believe me or thought I was exaggerating."

    FEELING:

    EMPATHIC RESPONSE:

2. "Those dirty, lousy creeps! Everything was fine in my life, and they really, really ruined everything! I don't care if I go on or not. Why live if someone can just take everything away from you in one night?"

    FEELING:

    EMPATHIC RESPONSE:

3. "I know you said this is temporary housing and all, but I never had a place like this place. I can't stand to think I have to move again sometime, and God knows where I'll go."

    FEELING:

    EMPATHIC RESPONSE:

4. "This whole setup is the pits. He gets to stay in the house after beating me half to death, and I have to go to this cramped little room. Does that make sense?"

   FEELING:

   EMPATHIC RESPONSE:

Instructions Part II: Now go back and respond to the content in each of these vignettes.

Reflective Listening II

Instructions: People communicate words and ideas, and sometimes it seems appropriate to respond to the content of what someone has just said. Behind the words, however, lie the feelings. Often it is most helpful to respond to the feelings.

Following are statements made by people with problems. For each statement, first identify the feeling; write down the word you think best describes how the person might be feeling. Next, write a brief empathic response—a short sentence that includes the feeling. Refer to the sample openers provided in Chapter 7 under the heading "Useful Responses."

1. "Sometimes it kind of makes me sick to think of all the stuff I did when I was drinking. I'd like to go and take it all back, but how do you ever do that?"

   FEELING:

   EMPATHIC RESPONSE:

2. "I just can't go out in the car. All I hear is the screech of tires and the awful thud and scrape of metal. I thought I was dying. I can see it all before me as if it was yesterday."

   FEELING:

   EMPATHIC RESPONSE:

3. "We have a neighborhood problem here! Yes we do! A real big idiot lives in that house. A real nut! He trimmed my own yard with a string trimmer and threw stones all over my car. Ruined the paint!"

FEELING:

EMPATHIC RESPONSE:

4. "I never meant to get pregnant. I know everyone says that, but I didn't! I can't think straight. What about my job and school and all my plans? I feel sick. I feel all the time like I'm going to faint."

FEELING:

EMPATHIC RESPONSE:

Instructions Part II: Now go back and respond to the content in each of these vignettes.

Reflective Listening III

Instructions: People communicate words and ideas, and sometimes it seems appropriate to respond to the content of what someone has just said. Behind the words, however, lie the feelings. Often it is most helpful to respond to the feelings.

Following are statements made by people with problems. For each statement, first identify the feeling; write down the word you think best describes how the person might be feeling. Next, write a brief empathic response—a short sentence that includes the feeling. Refer to the sample openers provided in Chapter 7 under the heading "Useful Responses."

1. "I can tell you now, I just can't go back there. I just feel as if my husband will kill me one of these times."

FEELING:

EMPATHIC RESPONSE:

2. "I can't stand those people! They made fun of that retarded kid night and day. I hope they get theirs!"

FEELING:

EMPATHIC RESPONSE:

3. "I've been clean for 8 months! If you had told me this would happen a year ago, I'd have laughed in your face."

FEELING:

EMPATHIC RESPONSE:

4. "When I was a little kid, my mom and dad got along okay, but now they fight all the time, and my mother says my dad is on drugs and has a girlfriend. Home is like hell."

FEELING:

EMPATHIC RESPONSE:

Instructions Part II: Now go back and respond to the content in each of these vignettes.

Reflective Listening IV

Instructions: People communicate words and ideas, and sometimes it seems appropriate to respond to the content of what someone has just said. Behind the words, however, lie the feelings. Often it is most helpful to respond to the feelings.

Following are statements made by people with problems. For each statement, first identify the feeling; write down the word you think best describes how the person might be feeling. Next, write a brief empathic response—a short sentence that includes the feeling. Refer to the sample openers provided in Chapter 7 under the heading "Useful Responses."

1. "When I took that test, it was really hard. And I guess I was nervous. I mean, I couldn't think of any of the answers."

   FEELING:

   EMPATHIC RESPONSE:

2. "Those guys are lousy! They're always snickering and making fun of other people, especially people who have a disability. They make me sick!"

   FEELING:

   EMPATHIC RESPONSE:

3. "I know Jim said we could be buddies at swim practice, but I'm probably not as good a swimmer as he is. I feel sort of silly trying to swim with him. Maybe he would like to have a better buddy."

   FEELING:

   EMPATHIC RESPONSE:

4. "This whole setup sucks. This other guy gets the tutor, and the teacher tells me to go home and see if my mother can tutor me. She never had this math. Math isn't even her thing. Does that make sense?"

   FEELING:

   EMPATHIC RESPONSE:

Instructions Part II: Now go back and respond to the content in each of these vignettes.

Reflective Listening V

Instructions: People communicate words and ideas, and sometimes it seems appropriate to respond to the content of what someone has just said. Behind the words, however, lie the feelings. Often it is most helpful to respond to the feelings.

Following are statements made by people with problems. For each statement, first identify the feeling; write down the word you think best describes how the person might be feeling. Next, write a brief empathic response—a short sentence that includes the feeling. Refer to the sample openers provided in Chapter 7 under the heading "Useful Responses."

1. "Well, every time I go off my meds, I get kind of crazy. My minister is really putting the pressure on me to quit and let God take over my illness."

   FEELING:

   EMPATHIC RESPONSE:

2. "The people at the halfway house are so nice to me, compared to the way things were with my family."

   FEELING:

   EMPATHIC RESPONSE:

3. "You have some nerve, having the therapist see my son every week for 6 months, and then you refuse to tell me more than 'he's doing better.' How do I know he's doing better?"

   FEELING:

   EMPATHIC RESPONSE:

4. "I've been on the streets since 1972, and I never slept inside a night until now. I don't know, I just can't seem to stay out like I used to without getting this cough."

   FEELING:

   EMPATHIC RESPONSE:

Instructions Part II: Now go back and respond to the content in each of these vignettes.

Reflective Listening VI

Instructions: People communicate words and ideas, and sometimes it seems appropriate to respond to the content of what someone has just said. Behind the words, however, lie the feelings. Often it is most helpful to respond to the feelings.

Following are statements made by people with problems. For each statement, first identify the feeling; write down the word you think best describes how the person might be feeling. Next, write a brief empathic response—a short sentence that includes the feeling. Refer to the sample openers provided in Chapter 7 under the heading "Useful Responses."

1. "I can't believe I was that intoxicated! I just don't believe it. Their gizmo must have been broken or something. I just didn't drink that much and I wouldn't be driving if I had!"

    FEELING:

    EMPATHIC RESPONSE:

2. "You don't expect us to take Alfred into our home, do you? He is very mentally ill—tore up the house several times. I really—well, I know he's my son, but I just can't deal with the way he's been in the past."

    FEELING:

    EMPATHIC RESPONSE:

3. "I can tell you what scares me most. It's being by myself at the house one night and having him come back. I don't know if I can go on living there."

    FEELING:

    EMPATHIC RESPONSE:

4. "I just can't go to class. Not after making a fool of myself the last time. I got every answer wrong when the teacher called on me, and people were making fun. . . . It was terrible!"

    FEELING:

    EMPATHIC RESPONSE:

Instructions Part VI: Now go back and respond to the content in each of these vignettes.

CHAPTER 9

Asking Questions

Introduction

When listening to another person's difficulties, we might find ourselves asking a lot of questions. We usually mean well. In part, we do this to find out more so a solution can be quickly devised. We also do this as a way of filling in the gaps when our reflective listening skills are not strong. It may be easier for the listener to say "Did you have any money?" or "Where did the man say he lived?" than to simply say, "Tell me about what happened."

Sometimes when we listen, we feel nervous about what the other person expects from us. After all, if we are the worker, should we not have all the answers? If we do not have an answer just yet, we can stall for time by asking a lot of questions until we do.

Often, however, the client hears these questions as prying. You may ask questions at a rate the client is not ready to answer. For instance, you may be asking questions further ahead in the story, throwing the client off. Clients may feel pushed to reveal more than they intended or be distracted from the line of reasoning they were following. They may become defensive and their communication guarded if the questions seem to pry or to imply there is only one way to have handled things.

When Questions Are Important

Obviously, at times questions need to be asked. The three times it is important to ask questions are as follows:

1. When you are opening a case or chart for a person and need identifying information (**closed questions**)
2. When you are compiling information for assessment and referral purposes and need facts to do that properly (closed questions)
3. When you are encouraging the person to talk about his or her situation freely to better understand which aspects of it are important to the individual (**open questions**)

Closed Questions

A *closed question* is one that requires a single answer. For example:

>   WORKER: Where do you live?
>   CLIENT: 346 Pine Street.
>   WORKER: How long have you lived there?
>   CLIENT: Oh, about 6 years.
>   WORKER: Have you ever been seen here before?
>   CLIENT: Yes, in 2013.
>   WORKER: And who did you see then?
>   CLIENT: Dr. Langley in outpatient.
>   WORKER: Did she prescribe any medication?
>   CLIENT: Yes, she gave me a prescription for Prozac.
>   WORKER: Do you still take that?
>   CLIENT: No, I stopped a few months ago.

Closed questions are most often used when opening a case or when compiling information for an assessment. In both of these instances, however, clients will need to talk about what has brought them into the service, and you will not rely exclusively on closed questions. Nevertheless, some closed questions are appropriate here to gather the basic information. The person can respond with a simple answer because the questions do not ask for expressions of feelings, descriptions of circumstances, or explanations of problems.

Let's look at another example. Notice that the questions require a single, simple answer.

>   WORKER: You are coming here directly from rehab?
>   CLIENT: Yes
>   WORKER: And you were at Middleton?
>   CLIENT: Yes
>   WORKER: Was it drugs, alcohol?
>   CLIENT: I had a drinking problem
>   WORKER: How long were you at Middleton?
>   CLIENT: About a month
>   WORKER: Do you have a job to go back to?
>   CLIENT: Yes, over at the Farmington Warehouse
>   WORKER: What is your job there?
>   CLIENT: I am in charge of one of the shipping lines

Remember that asking one closed question after another about a person's situation or problem can sound as if the case manager is prying or frantically trying to find a solution, rather than trying to understand what really concerns the client.

Open Questions

Open questions serve the purpose of giving clients more opportunity to talk about what is important to them. By asking open questions, you receive more information about a client's situation. In answering open questions, the client can talk about feelings, underlying causes, supporting circumstances, and personal plans.

Open questions have been shown to put clients at ease. Workers using these questions are not perceived as prying but as expressing real interest in other people or a genuine desire to understand their situation. You can use open questions to obtain examples or elaboration of the problem and to clarify certain aspects of the other person's story.

An open question allows the other person to talk freely about his or her issue. As you listen you find out what is most important to this person about her situation. You are better able to see the other person's point of view. You can better hear her emotions. Open questions give us much more valuable information than closed questions can. For this reason, learning to ask good open questions will enhance your understanding and your ability to establish good rapport.

Often an open question begins with "can" or "could," but there are other ways to start such questions. For example, you might say, "Tell me a little bit more about your divorce." Typically open questions look something like this when listening to the individual talk about his issues:

WORKER: Can you tell me about the night your father left?

CLIENT: Well, my mother had been arguing with him for some time. I could tell he was getting angry. I don't think I really blame her for his leaving. He had done many things to her that she had every right to be angry about. But I guess for him it was the last straw. Anyway, we were having dinner and she began on the same topic of the house. He just put down his fork and got up from the table and walked out of the house.

WORKER: Can you describe your relationship with him after that?

CLIENT: Well, he did come back to the house for his things from time to time. He got an apartment nearby, and I used to stop there on my way home from school. We never stopped seeing each other, and we never talked about my mother. I can't ever remember him asking me how she was or, for that matter, saying anything mean about her.

Would the worker have gained as much information by conducting the same conversation using closed questions? This example demonstrates that the client would have been much less forthcoming:

WORKER: When did your father leave?

CLIENT: I think, August of 2006.

WORKER: Did your parents argue much before he left?

CLIENT: Sure, yeah, a lot of the time.

WORKER: Why did he go?

CLIENT: Well, they disagreed over money and about the house.
WORKER: After he left, where did he go?
CLIENT: He got an apartment near us.
WORKER: Did you ever see him after that?
CLIENT: Yes, pretty much.

Would the client have felt investigated with these short closed questions fired at him, one after the other? Did the worker really understand how the client felt about the divorce and the contact with his father?

In the second example, in which the worker used closed questions, there was room for the worker to assume things that might not be true. For instance, the worker might have assumed that the client blamed his father for leaving. In the first example, in which the worker used open questions, the client's elaboration on the situation demonstrated that this clearly was not the case.

Questions That Make the Other Person Feel Uncomfortable

The Trouble with "Why" Questions

There are three ways that "why questions" can seem offensive or make another person feel uncomfortable. If you ask someone why someone did something or did not do something, you imply that you believe the person should have handled things differently.

- "Why didn't you call the police?"
- "Why did you spend all that money?"
- "Why did you let your children stay out that late?"

Sometimes why questions seem to seek a level of understanding or insight the client may not have. The client has to search for an answer or is pushed to come up with an explanation.

- Why do you think you chose to go to that particular bar?
- Why was it that you and he had that fight on that evening?
- Why is it that each of these incidents happened after she was out with her friends?

At other times why questions can sound like prying, asking for information the client had not intended to discuss.

- Why did you have a fourth child so soon after the third one?
- Why didn't you ask yourself why you chose to do it that way?
- Why are you always avoiding going up to the college to at least get some information?

Asking Multiple Questions

If you fire off a string of questions, the person can feel interrogated. You may sound impatient, and you can confuse the client.

> "Did you see the other person? What did he look like? Did you get a license plate number or some identifying information? How close were you? Did anyone else have information that would identify him? Did the police have any suggestions?"

Too many closed questions can sound as if you are desperately trying to solve the problem for the client. In the example above the worker is getting bits and pieces of information, but probably would have done better to simply have asked, "Tell me about what happened that night?" Then one or two closed questions might have filled in the gaps without pressuring the other person.

Questions That Change the Subject

If the person is talking about how she learned of her mother's death, do not start asking questions about her mother's prearranged funeral. Let the client continue to talk about her mother's death until it seems that she wants to turn to the prearranged funeral. Never ask about something out of curiosity. Do not ask, for example, about the prearranged funeral because you are thinking of getting one for your mother and want to know more about it. Ask questions that stay on the topic the client has selected. If you do ask questions on another topic, make sure the topic is relevant and the questions will actually clarify the person's situation for you.

Here are some examples of questions that change the subject:

- "I heard about the kids before, but where do you work?"
- "So, she died on Saturday, and now you are seeing a lawyer about the will?"
- "That's real neat about how your car looked before the accident. Do you have adequate car insurance?"
- "So, your mother died on Saturday, and you're living in a house by yourself?"

Sometimes questions such as these can disrupt the entire discussion as for instance when the client said, "I got this brand new Mustang with V-6 performance and a six speed transmission. It was ruby red, a really great car." If the worker responds with, "That's real neat about how your car looked before the accident. Do you have adequate car insurance?" the person is likely to respond with something like, "yeah" or "sure." Talking about the car's great features before the accident would end there, possibly leaving the client somewhat annoyed. For the worker, changing the subject cuts off any real understanding of what the person feels he lost.

Implying There Is Only One Answer to Your Question

You can ask questions in a way that implies there is only one acceptable answer.

- "Didn't you go to the police?"
- "Did you tell the other person what you heard?"
- "Did you see to it that he knew what you were thinking?"

Inflicting Your Values on the Client

You can also ask questions based on your own value system. The client, however, may have other values. For example, you may value truthfulness at all costs, whereas your client may come from a group that values group harmony and not hurting another person's feelings. **Questions that imply** that your value system is better are not useful.

- "Did you tell her how you felt before you just walked out?"
- "Did you tell her the complete truth?"
- "Don't you value truth above everything else in this situation?"

Do Not Ask Questions That Make Assumptions

You can word questions in such a way that they make it clear you are assuming you already know the answers.

- "You called the police, right?"
- "You wanted to go to the store, didn't you?"
- "He was being a fool, wasn't he?"

A Formula for Asking Open Questions

Figure 9.1 contains a formula for asking open questions. In the figure, the open question is broken into parts. You can interchange the parts, by choosing one part from each column, to construct good open questions that encourage the other person to feel safe in talking and expressing feelings and opinions. Use this formula in the exercises at the end of this chapter to construct effective open questions that invite others to be open and talk freely with you.

Some Tips for Asking Open Questions

Learning to ask open questions takes practice. It is easy to ask a closed question, such as "Where do you live?" or "How old are you?" When intending to ask an open question, we often start out well and then unwittingly close the question. Some examples of what can happen are presented next.

A Question That Is Not Really Open. "Tell me a little bit about how you got here?" "How you got here" is a closed question and the client can answer, "I came over in the car."

To Open That Question Try This. "Tell me about getting here." The client is more inclined to say more, such as "Well, I came over in the car, but it was scary. I kept looking in my rearview mirror to see if he was following me."

Words That Snap Questions Closed Are How, Why, What, When, Where. You may not be able to avoid using these words in some of your open questions, and using these words in questions is perfectly all right. To leave them out might mean your

FIGURE 9.1 Formula for asking open questions

Openers	Directives	Add-ons/Softeners	Object of the Question
Can you[a]	share	a little bit more about	your husband
	describe	a little bit about	your childhood
	explain	a little more about	your medication
	summarize	something about	what the move was like
	outline	the problems with	the move
	spell out	the larger picture	regarding the move
Could you[a]	talk	a little more about	what your dad said
	give me	a bit more about	your illness
	tell me	something more about	your job
	help me understand	something about	your relationship with your kids
	clarify	a bit	the situation

[a] It is fine to leave the "can you" or "could you" out of the questions. In this case, you would make a request such as "Share a little bit more about the fire."

Digital Download Download from CengageBrain.com

question doesn't make sense. However, when our purpose is to draw people out and make them feel comfortable about talking to us in depth, we want our questions to be as open as possible. When you are about to use one of these words—how, why, what, when, or where—see if you can leave it out for a more open question.

INSTEAD OF: "Tell me a bit about how you found out about your husband's cheating."

ASK: "Tell me about finding out about your husband."

INSTEAD OF: "Can you describe why you left him?"

ASK: "Can you describe leaving him?"

INSTEAD OF: "Could you tell me a little bit about what the doctor said?"

ASK: "Could you tell me a little bit about visiting the doctor?"

INSTEAD OF: "Can you summarize for me when you left?"

ASK: "Can you tell me about leaving?"

INSTEAD OF: "Tell me a bit more about where you were that night."

ASK: "Tell me a bit more about that night."

If we look at one of those pairs of questions carefully, we can see that the two questions are asking for somewhat different information. For example, in responding to the question "Tell me a bit more about where you were that night," the person might tell you where she was and perhaps in addition give you some valuable information beyond that: "I was down by the railroad tracks. I thought I could hop a freight or something like that. I just wanted to get away." When you say "Tell me about that night," you are asking him for much more, and he may be inclined to tell you many more details: "I was feeling terrified. I didn't know who these people were or why they had singled me out. I didn't know what to do or where to go so I went down by the railroad tracks. I think I thought I could jump a freight or something like that. I just wanted to get away. I saw it starting to get light. No one came, and finally I got up and snuck home." In this last example we know a lot more about the night in question, and we have some idea about the client's perceptions and feelings as well.

Open Questions Complement Reflective Listening. Open questions and reflective listening complement each other and go a long way toward putting the other person at ease. Here is an example of how you would combine both open questions and reflective listening in a good interview.

WORKER: Tell me what brought you in today.

CLIENT: My pastor, Reverend Perkins, sent me down after I had a bad time in church. He thought maybe I was depressed.

WORKER: Can you tell me a little bit about what happened in church?

CLIENT: Well, see they have this call to commitment every Sunday. It is for people who want to join the church so, you know, new people go up. But I always feel that my commitment isn't good enough so I go up every Sunday and he didn't think I needed to do that. He asked me after church to talk to him about it and that is when I just broke down.

WORKER: It sounds like you haven't felt really good about each of those commitments you made.

CLIENT: No. Right. I just think well I sinned all week and I need to recommit, but it is never good enough.

WORKER: Tell me a little bit about breaking down.

CLIENT: Oh . . . (begins to weep) I just tried to tell him how important it is to be right with the Lord. I don't want anything to happen to me and I tried to tell him that. I feel as if God is mad at me or something and I need to do this every week to be sure . . . (*her voice trails off*).

WORKER: So you feel that if you don't make a recommitment each week something bad will happen to you.

CLIENT: I do feel that way, but see the way Reverend Perkins says, the call for commitment is for people who want to join the church and . . . he didn't say that but I think he thinks I am making a fool of myself. You know, everybody sees me do that every Sunday.

WORKER: You must feel really caught between wanting to please God and not wanting to go against your pastor.

CLIENT: Yeah. I don't know what to do.

WORKER: Can you tell me a little bit more about your church?

In this exchange the worker stays right with the client as the client explains why she came in for help. The worker combines open questions with reflective listening to draw out the feelings and circumstances of this person's problem. We begin to know a lot about this person: her fears, her beliefs, what is important to her.

Summary

Asking questions helps us to understand our clients and the issues that are bothering them. Asking too many questions, however, can give the impression that we are desperately seeking some sort of solution or prying. The problems people bring to us are theirs. They need someone to listen to their concerns and sort out the best way to approach a solution. We help people do that by listening and asking open questions that encourage them to tell us more about what has brought them to us for help. It helps them to be able to talk about their problems and to hear their concerns reflected back to them.

Combine open questions with reflective listening to feelings and to content in order to create a safe environment for clients to talk to you and begin to work on problem solving. Ask closed questions sparingly, but remember that an occasional closed question is perfectly acceptable; For the most part, confine them to times such as when you are opening a case or you need information to make a proper referral.

Video Examples

To view the videos that accompany this book, go to CengageBrain.com.

- You can see Keyanna use open questions to learn more about Michelle's situation by watching "The First Interview." Danica uses open questions to work with Alison on her service plan in "Developing a Service Plan" and "Helping Tom Solve a Personal Problem."

Exercises

These exercises can also be filled out online at CengageBrain.com.

Exercises I: What Is Wrong with These Questions?

Instructions: Read the questions that follow and decide what makes them bad questions. In writing your criticism, look for questions that assume there is only one answer, inflict values on the individual, make the person defensive, make assumptions, cut off discussion, or change the subject.

1. A woman is telling a worker why she has come to the shelter tonight. Right in the middle of her gripping tale about what was going on at home only a few hours before, the worker says, "How long has this been going on?"

2. A worker has listened to a young mother talk about how she dropped out of school and got pregnant and has no skills. Finally the worker interrupts to ask, "Did you have to get pregnant? Didn't you know about birth control?"

3. A man calls and says he is depressed. He has felt depressed for some time and is now thinking of suicide. The worker asks, "Where is your wife? Are you divorced?"

4. A man is telling you about the night he witnessed a murder. The victim was his brother-in-law, and although he was never very close to him, he feels that maybe he could have stopped his death in some way. The worker asks, "Why don't you just go and ask the police?"

5. A woman has come into temporary shelter with a lot of debts. She has been out looking for work today and is discouraged about not finding anything yet. She sits down tiredly in the worker's office and talks about what her day was like. The worker asks, "Did you have to get so many debts?"

6. A man wants to know if his wife is all right after she has been raped. He is sitting with a worker in the waiting room while his wife is being seen in the emergency room. The worker answers his question with one of her own: "How much does your wife mean to you?"

7. A patient in a partial hospitalization program for the chronically mentally ill tells the worker that when the group went to the mall, one of the patients took a pair of socks without paying for them. The worker asks, "You told someone right away, didn't you?"

8. A woman is telling about the time her coworkers waste when the supervisors are out at meetings all day. The worker responds, "Why don't you say something?"

9. A woman tells a worker about a long and difficult marriage she has endured. She mentions abuse, both verbal and physical, and talks about her own failing health in recent months. The worker asks, "Why can't you just bring yourself to divorce him?"

10. A man is trying to sort out whether or not to leave his employer. He feels that the small company is poorly run and that he could do a better job if he went out on his own. On the other hand, he likes his employer, and he feels sorry for him and the mess he's made of his business. He knows that if he leaves, things will really fall apart. The worker asks, "Don't you value loyalty?"

Exercises II: Which Question Is Better?

Instructions: Read the following questions and decide which of them are better than others. Place a check mark next to those you think are good questions, and then explain why you think they are better than the ones you did not check.

- ☐ 1. The worker to a woman in the hospital waiting room whose baby just died of pneumonia: "How old was your baby?"
- ☐ 2. The worker to a woman who is grieving after her husband died in a hunting accident: "Could you tell me about your husband?"
- ☐ 3. The worker to a teenage boy who is afraid of failing a math course and losing an opportunity to get a scholarship: "Can you tell me a little bit about this math course?"
- ☐ 4. The worker to a young woman who has just discovered her best friend and her boyfriend have been seeing each other behind her back: "Can you tell me something about your best friend?"
- ☐ 5. The worker to an elderly woman whose dog of 15 years has died: "Couldn't you get another one?"
- ☐ 6. The worker to a man who is requesting food for his family after running out of unemployment compensation and being unable to find a job: "Can you describe the sort of work you would be looking for?"
- ☐ 7. The worker to a woman in a shelter who has been out searching unsuccessfully for a house or apartment for herself and her two children: "Where all did you look?"

☐ 8. The worker to a single mother who has been referred for parenting skills training: "Could you tell me something about the problems you have been having with Johnnie?"

☐ 9. The worker to a man with intellectual disabilities whose mother, with whom he has always lived, died unexpectedly: "What did your mother die of?"

☐ 10. The worker to a woman who was accosted and assaulted in her neighborhood and is afraid of calling the police: "Can you tell me a little bit about what happened tonight?"

Exercises III: Opening Closed Questions

Opening Closed Questions I

Instructions: Following are some vignettes in which the worker asks closed questions. Write an open question you think might work better in each situation, and be prepared to tell why you think the closed question is not useful.

1. A human service worker in the emergency room is talking to a man who was hit on the head before he was robbed. He seems to be having trouble getting the story out, but he wants to tell the worker everything that happened. The worker has been with the man a long time. She thinks that it is late and that the man ought to get to bed and rest now. The worker cuts off the discussion with, "Aren't you tired, Mr. Jones?" What open question would you have asked Mr. Jones to help him wrap up his story?

2. The human service worker is trying to learn what happened that resulted in Mrs. Peters being without housing. Mrs. Peters says she has been "on the street a while now." The worker asks, "Have you been on the street for 2 years, 3 years?" What open question would you have asked to learn more about what happened to Mrs. Peters to make her homeless?

3. The human service worker is on the phone with a woman, the victim of child abuse. The woman tells how she has felt recently, how she needed to call, and then sighs and says, "Oh, I don't know how to begin." The worker asks, "Did your father do this to you?" What open question would you have asked to help the woman start telling the story in her own way?

4. An older man has just lost his job after repeated warnings to come to work sober and seek help for his alcohol addiction. He has decided that he should get help now. "Too little, too late," he says with resignation. "I should have been here 6 months ago." The worker asks, "Why did you ever let it get to this?"

5. A child is talking to a youth worker while he waits for his mother to get a place to stay. "We've lived in 16 places," he announces, "and I'm only 7." The worker says, "What school did you go to last?" What open question would you have asked to help the child talk about what all this moving has been like for him?

6. A man calls a hot line and tells the mental health worker he wants to die. The volunteer asks, "Does this have to do with being abused as a child?" The man is startled and says, "Why, uh, no. Not really." The worker asks, "Well, what's the problem?" What open question would you have asked to help the man talk about what was troubling him?

Opening Closed Questions II

Instructions: Put yourself in the place of the worker in the following vignettes, and decide what question you would ask in each situation. Write an open question that you think might work better than the one asked by the worker, and be prepared to tell why you think the closed question is not useful.

1. A worker is interviewing a man in the food bank. He tells the worker that he and his children have not eaten for 24 hours and that he has spent most of that time getting referred around town until he finally got a voucher to come to you for food. The worker asks, "Why don't you have any food?" What would you ask?

2. A woman is referred to the social service department in a large hospital after having a stroke. She is somewhat incapacitated and has had a lot of therapy while hospitalized. Now she is going home and needs therapy at home. The worker asks, "What kind of therapy do you want?" What would you ask?

3. A 16-year-old girl was brought in by her parents after they caught her and some of her friends huffing glue and gasoline. The girl is reluctant to talk and seems a little petulant about being brought in. The worker asks, "Huffing glue. So tell me, you did it to get high, to be one of the crowd, to be smart? What would you ask?"

4. A man and woman have been referred by the county Children and Youth Services for parenting skills training. They are poor and have had their four children removed from the home. They have been told the children will be returned when they complete the course and demonstrate they can use the skills they learned in supervised visitations. The worker asks, "Are your children good kids?" What would you ask?

5. An elderly woman has been having trouble caring for herself in her own home. Twice now, in the middle of the night, she has called an ambulance and has been taken to the hospital for chest pains. When her heart is checked, she is found to be in good health, if a little frail. The worker who is looking into what could be going on asks, "Are you afraid to stay at home alone?" What would you ask?

6. A young woman and her baby have been given a voucher for temporary shelter after she lost the apartment in which she was living. She was evicted for back rent, and her rent fell into arrears only when she was laid off several months ago. She has worked, but she cannot earn quite what she was making before. The worker doing the intake interview asks, "What kind of work have you been doing?" What would you ask?

Exercises IV: Try Asking Questions

Instructions: Look at the case histories that follow and, for each one, write four closed questions and four open questions that you might ask the client.

1. Annette came to your office needing her prescription filled. She was in Marywood Hospital, a private mental hospital, and was discharged on Tuesday. She was given prescriptions, but has no money to fill them. She has no job and probably is eligible for prescriptions paid for by the county. You open a case on her.

Your Closed Questions to Open Her Case Are:

1.

2.

3.

4.

Your Open Questions to Learn More about Her Are:

1.

2.

3.

4.

2. Marie was a client of a partial hospitalization program. She was loud and demanding, but she often felt hurt upon learning that others were afraid of her or reacted to her as if she were angry. As a result of an encounter in the partial program, she is sent to you, her new case manager, to see if there are ways to help her that might work better. You need to understand more clearly what has happened from her perspective and what sort of program she might fit into.

Your Closed Questions to Become Acquainted with Her Case Are:

1.

2.

3.

4.

Your Open Questions to Learn More about Her Problems and Desires for Treatment Are:

1.

2.

3.

4.

CHAPTER 10

Bringing Up Difficult Issues

Introduction

There will be times when you have a concern about something the person has said or done. You may be concerned for your client's well-being, or you do not want your client to do something harmful or continue to behave or think in ways that are destructive. Perhaps, from your perspective, there are some other avenues the person might consider or pursue in order to resolve her problems.

Confrontation

The term *confrontation* is generally thought to mean bringing something into the open that needs to be addressed in a harsh, aggressive manner. Confrontation as such is used to address issues that the client may be avoiding or denying. There are different schools of thought about whether or not this kind of confrontation is useful and what confrontation techniques to use with different populations. Some (Bratter, 2008, 2011) advocate a harsh, reality based method that forces people, particularly teens engaged in self-destructive behavior, to consider their actions. Polcin (2009), writing about individuals with substance abuse issues, argues that some workers feel these tactics are best used to "break down denial with the use of argumentation or even personal attacks" (p. 505). There are those who feel that the effectiveness of this kind of confrontation is not well known due to a lack of empirical research. For these workers confrontation would consist of "nonjudgmental feedback" (Polcin, 2009, p. 505). In other words, the best way to confront others—and whether or not confrontation works is not clearly known—leading workers to "rely on their own experience and intuition when using confrontation" (Polcin, 2003, p. 179).

On the other hand, there are those who argue that, "confrontational counseling has been associated with a high dropout rate and relatively poor outcomes" (Miller and

Rollnick, 2002, p. 7). They argue for a different approach that is more supportive and empathic and contend that harsh confrontation techniques cause the other person to become defensive and resistant to what the worker has to say.

Exchanging Views

We are not counselors, but case managers, and using confrontation, which most researchers agree is a therapeutic tool used in counseling, a strategic intervention in the counseling process, is not something we would do as case managers. Clearly when and how to confront and who can benefit from confrontation are best left to counselors and therapists with years of training.

In this chapter we will look at other ways to bring up issues that need to be considered in a way that the other person can hear, experience as an effort to be supportive, and can use in resolving his problems. We want to think of this technique in four ways:

1. We are collaborating with an equal on how to solve a problem. We have ideas or concerns and other people do as well. Both are equally valid for consideration.
2. We are making a contribution to the resolution of the problem but the other is free to reject the contribution as not useful to him or her at this time.
3. We are making our contribution without a personal need to have the problems resolved the way we think is best. Therefore, we are matter-of-fact in bringing up our point of view.
4. We are offering our ideas tentatively because we don't have all the answers and we are not always right. We stand open to being corrected or hearing another perspective.

What this means is matter-of-factly bringing something out in the open to gain a better understanding and perhaps to help people make meaningful changes or take important new steps. When you bring up your point of view, you are holding up the reality as you see it for the other person to consider. The client is in no way obligated to see things your way, but now both points of view are known and considered. Many opportunities to grow and make constructive changes will be discovered when you engage in a collaborative exchange of views.

The decision to bring unspoken issues into the discussion is another strategic decision. This chapter examines when bringing up these issues might be a useful tool to help you and your client explore differences and resolve possible conflicts.

When to Initiate an Exchange of Views

Discrepancies

There are times when a person will communicate two different messages. Pointing this out can help the person see the discrepancies and can offer an opportunity to look at the situation and at themselves in another way. Some examples of discrepancies follow.

The Person Says One Thing but Does Another. Dalia tells you that she really wants to go to the job-training program and that getting a job is a top priority for her, but she does not register for the classes. On the other hand, she has numerous excuses for not registering, some of which do not seem entirely believable.

Here is what you might say to Dalia: "There have been three job fairs in the last 6 months and we had the applications for a couple of job readiness workshops, but I'm getting the feeling that this isn't the direction you want to go right now. Can you help me out with this?" Clearly the worker is not sure about Dalia's intentions and is asking Dalia to share her view on the matter.

The Client Has One Perception of Events or Circumstances, and You Have Another. Harold thinks you are uncaring and self-involved. He got this idea because you did not come to work the Friday after Thanksgiving even though the office was open. He was off work that day, and he wanted to make an appointment with you so that he would not have to miss work at another time. Your perception is different. To you it was reasonable to be off work the Friday after Thanksgiving because there was only a skeleton staff working that day. You also needed to take a day off before the end of the year or you would have lost some of your accumulated time. Clients rarely come in on this date, and there was a crisis team to cover any crisis that might have come up. To Harold you seem uncaring, while to you your actions seem reasonable.

The Client Tells You One Thing, but the Client's Body Language Sends a Very Different Message. Andrea tells you that she is "fine," that she feels "okay," and that "everything is all right." She looks, however, as if the opposite is true. She speaks in a monotone, looks at the floor as she speaks, and appears depressed and disheveled. These are clues that the spoken message and the unspoken message do not match.

The Client Purports to Hold Certain Values, but the Client's Behavior Violates Those Values. Paul tells you he "likes everyone" and "accepts" everyone. He tells you ethnic differences are unimportant to him and he finds them enriching. In one of his meetings with you, he tells a decidedly racist joke that obviously denigrates a minority group.

All of the examples discussed here are situations that contain discrepancies that deserve to be addressed. Doing so will help to clarify the issues and help you and the other person come to understand one another's point of view. Often ignoring discrepancies interferes with understanding between you and your client because of conflicting perceptions.

Other Reasons to Begin an Exchange of Views

There are other reasons besides discrepancies for **exchanging viewpoints**. You can do this to bring out in the open behavior or communications that seem to interfere with clients meeting their goals. Following are some examples of such situations.

The Client Has Unrealistic Expectations for You. Marcy expects that you will drop everything to see her or to take her phone calls. She does not want to see anyone else in the agency and does not think she should have to see anyone else at night. You are her case manager and she wants you to be there when she needs you.

The Client Has Unrealistic Expectations for Him- or Herself. Miguel has been in a partial hospitalization program for a number of months and has been sick for about 4 years. Stress seems to trigger his psychotic symptoms, and regulating his medication is difficult. He is very good at cleaning and janitorial tasks around the center, and there is a good supervised janitorial program for clients in which they hold a regular job and clean actual establishments. Miguel can have a job there but he is set on going to work at the highway department and getting a job "driving a steamroller." He applies for the job repeatedly but gets no response.

The Client Asks for Assistance, but Actions Indicate the Client Is Not Interested. Serena asks you to help her find suitable housing so she will not have to stay at the shelter any longer. You have some leads she could pursue, but she breaks appointments, calling in to say she was detained and will reschedule. She does not follow up on the leads you give her, and the two apartments she went to see that appeared suitable she turned down for minor problems, refusing to live there.

The Client's Behavior Is Contradictory. Art comes in to group and tells the group he will stop drinking. He never misses AA meetings, gets a good job, and begins to help others stop drinking. Later you learn that he is actually drinking in spite of what he says in group and at AA meetings and that he goes to AA on Tuesday and Thursday and to his favorite bar on Friday and Saturday nights. Art's behavior is contradictory in another way. While he talks to newcomers in the group about how helpful it is to stay in group and how wonderful the agency is, he has been denigrating a certain member of the staff outside the building where he goes to smoke during the break.

The Client Is Engaging in Behavior That Appears Destructive. Claudia has been cited by child welfare several times for leaving her two children under the age of five alone during the evening while she goes out with friends. She tells you that she does not want to lose her children and that she does take care to see that they are in bed before she goes out. However, her behavior is viewed as endangering her children's welfare and there is every possibility that the child welfare agency will remove the children if Claudia's neighbors report her again.

The Client Is Looking for Solutions to a Problem. Marcus needs a place to stay after being in prison for 4 years. He seems to be floundering about where to look and where to start. For now he is "crashing" on the couch at his brother's apartment, but the apartment is small and with the brother's wife and two children there Marcus doesn't think this is a good solution. He tells you he has talked to "some people who know a guy who has some apartments over on 19th Street."

Using I-Messages to Initiate an Exchange of Views

Because the viewpoint or the perceived problem is yours and the observations are your own, your exchange should begin with or include a reference to you. The term used by Dr. Thomas Gordon for these statements is *I-messages* because they contain the words *I* and *me*. The exchange is not helpful, as we have seen, if statements contain the accusatory *you*. Figure 10.1 shows some examples of correct and incorrect I-messages to demonstrate the difference between them. The first example consists of messages to a client who was late on Tuesday; note the use of *I* in the correct version and the use of *you* in the incorrect version. The second example demonstrates a worker's concern for what her client is about to do.

FIGURE 10.1 Examples of I-messages

To a client who was late on Tuesday:

Correct: "I'm concerned about when we got started on Tuesday morning. Starting late got my day behind more than I wanted, and I spent a lot of time trying to catch up. Could we look at your scheduling and mine and see if there is a way we can start on time?"

Incorrect: "You were late on Tuesday, and you held me up. My whole day was behind, and I spent a lot of time trying to catch up. Can you get your times straight so this doesn't happen again?"

I-messages broken into the four parts:

Correct: (1) "I'm concerned about when we got started on Tuesday morning. (2) Starting late got my day behind more than I wanted, and I spent a lot of time trying to catch up. (3 & 4) Could we look at your scheduling and mine and see if there is a way we can start on time?"

To the client who is distressed over having to go before the District Justice:

Correct: "I guess it just seems to me that you could get in more trouble if you follow through on your plan to give the District Justice a hard time about this. I'm thinking it might cause him to be even tougher on you. Let's look at this and see if there is some other way to handle this."

Incorrect: "If you go out there and give the District Justice a hard time, all you are going to do is get yourself in a lot of trouble. My advice is to cool down and just go in there and listen to what they have to say."

I-messages broken into the four parts:

Correct: (1) "I guess it just seems to me that you could get in more trouble if you follow through on your plan to give the District Justice a hard time about this. (2) I'm thinking it might cause him to be even tougher on you. (3 & 4) Let's look at this and see if there is some other way to handle this."

A complete I-message usually contains four parts:

1. Your concerns/feelings/observations about the situation expressed in a nonblaming, matter-of-fact manner
2. The tangible outcome you *believe* you see as a result or the possible consequences for the client
3. Your suggestions or thought concerning all of this, offered tentatively as a contribution to resolving or improving a situation
4. An invitation to collaborate

The second part of Figure 10.1 provides more examples of messages given to clients, with each message broken into the four parts. Compare the correct and incorrect messages. Note the following about the incorrect ones: They begin immediately with the accusatory "you" rather than "I," and they contain no invitation to the client to collaborate on a solution (in the second example, the worker gives advice instead).

The Rules for Offering Your Point of View

There are ways and times to talk with people about the issues that concern you. An important goal is to do so in a way that allows people to hear you and make use of what you have said (Strong and Zeman, 2010). We all benefit from the feedback of others, but the manner in which it is given often interferes with our ability to accept and use that feedback.

The following text discusses rules for making I-messages less threatening and more acceptable to the listener. Figure 10.2 contains examples of correct and incorrect messages for each rule. As you read about each rule, examine the sample messages in Figure 10.2 under the heading for that rule. Note that in the correct messages the speaker emphasizes "I" and "me," taking responsibility for the observations, ideas, and concerns, whereas in the incorrect messages, the emphasis is on "you."

Empathy. Inviting another person to consider your point of view needs to originate from the empathy you feel for the other person and their struggles with problems (Strong and Zeman, 2010; Leaman, 1978). Part of being empathic is being able to sense how the other person will feel when you begin to express another point of view. Perhaps what you have to say is something that is painful to hear, hard to consider under the circumstances the client is experiencing. Leaman (1978) writes that when we express our concerns what we say "must emerge from a genuine empathic concern" (p. 630). Leaman goes on to point out that when we put our views out for consideration it should be "an act of caring" for the other person and motivated by "feelings of caring and commitment" (p. 631).

If your concern invites others to examine themselves, their behavior and their choices, this can be difficult. What impact will your bringing up your viewpoint for consideration have on your relationship with the client? Before you bring up your ideas and concerns, consider how this will affect the other person and the relationship you share.

FIGURE 10.2 Examples of I-messages based on rules for confrontation

Empathy

To a person having trouble keeping his food stamps:

Correct: "It sounds like this has been so frustrating for you. I'm just thinking that maybe we could make it a little less frustrating if we called down there and asked them what they want you to do to keep your food stamps. What do you think?"

Incorrect: "We can just call down there and find out what they want you to do about keeping your food stamps."

To a person who leaves her children at night while she goes out with friends:

Correct: "I wonder if your leaving the kids in the evening when you go out with friends isn't seen the same way by the neighbors and Child Welfare. I'm thinking that they probably think that leaving the kids like that is neglectful and that is why they call Child Welfare. What are your thoughts about why the neighbors are doing this?"

Incorrect: "Leaving kids alone at night is neglectful and that is why the neighbors call Child Welfare and that is why Child Welfare is threatening to take the kids away."

Timing

To a person having trouble getting a job in her field:

Correct: (Said during the third contact) "You know I'm wondering if part of the reason you are having trouble getting a job might have to do with the interview itself and how others are viewing you. I'm thinking maybe we should look at ways to help you come across as a serious candidate. What do you think?"

Incorrect: (Said during the first contact right after the client has expressed a concern that she is always turned down for jobs in her field. The worker doesn't know her well enough to judge whether she is serious or not.) "You know I'm wondering if part of the reason you are having trouble getting a job might have to do with the interview itself and how others are viewing you. I'm thinking maybe we should look at ways to help you come across as a serious candidate. What do you think?"

Be Matter of Fact

To a person whose goals are unrealistic for the present:

Correct: "Here's what I am thinking about this degree in biology. It just seems to me that some of your goals are a bit further down the road for you. I'm wondering if we could look at some preliminary steps for you to take first to help you get ready. What do you think?"

Incorrect: "You're not ready to undertake a degree program."

(continued)

FIGURE 10.2 *(continued)*

Be Tentative

To a person who may not be seeing all of the issues with his mother:

Correct: "I guess I'm wondering about this problem you're having with your mother. I could be wrong, but when you describe the way she talks to you, it sounds to me as if she is angry for some reason. Do you think we should look at that? What do you think?"

Incorrect: "It seems to me your mother is angry at you."

Focus on Tangible Behavior or Communication

To a client who says she is fine but looks depressed:

Correct: "Allison, it seems as if you are feeling pretty blue about things. Things have been pretty rough lately. I'm thinking maybe we should talk about your mood and how you are feeling. What are your thoughts about this?"

Incorrect: "You look depressed. What's wrong?"
or
"I see you as depressed. I'll set up an appointment for you to see the doctor."

Take Full Responsibility for Your Observations

To a person who needs housing but is doing little to obtain it:

Correct: "I think what is bothering me right now is that it *seems* as if the sessions we have together to get you better housing aren't as important to you as I first thought. What I mean is that to me it seems you have some other more important priorities in your life right now. I might be wrong. I'm basing this on the fact that you never went to see the three apartments that were available to you. Can we talk about where you are right now with housing and where we should go from here?"

Incorrect: "Because you never went to see those apartments, I can see you don't care about housing."
or
"The way it appears, housing certainly isn't a high priority for you. You never followed through."

Refrain from Accusing

To a person who has unrealistic expectations for you:

Correct: "I'm getting the feeling that you would like me to be available to you whenever you call or come in including evenings and weekends. I could be wrong about this. It is just the way it looks to me right now. Can we talk about what your expectations are?"

Incorrect: "I think you think I should be here for you 24 hours a day and that is unrealistic."

Refrain from Anger

To a person who is having trouble maintaining sobriety:

Correct: "I'm really concerned that you are drinking when you are away from the program and not talking about these relapses in the group. This is pretty important to me and part of that is probably because I think it makes you seem untrustworthy to others in the group. How do you see this?"

FIGURE 10.2 *(continued)*

Incorrect: "You can't come to group and lie to people about your drinking. You've been drinking outside group and you're lying about it when you don't bring it up. I put my time and my energy into these groups and into helping people like you and you do this behind my back?"

Refrain from Judgment

To a person who needs permanent housing but is not pursuing it:

Correct: "Can we take another look at your priorities and see where housing for you and your children fits in? I was under the impression that this was pretty high on your list, but you haven't kept the four appointments we had to discuss it. Maybe we need to look at this differently. Let's talk about this."

Incorrect: "Housing is not your top priority. I thought it was."

Do Not Give the Client a Solution

To a person having trouble remembering appointments:

Correct: I need to see if we can get this appointment problem cleared up. My schedule gets all off when you forget to come in. Let's see if there is a way to resolve this."

or

"There probably are some different ways we could approach this appointment problem. I have some thoughts, and you probably do too."

Incorrect: "Go get an appointment book. Write all of our appointments in the book, and that way you won't forget."

or

"You should get an alarm clock that works and have your landlady call you up every morning. That way you can't miss."

Always Collaborate

Correct: "How can we look at this differently?"

or

"What do you think we can we do to change this?"

or

"Is there something we should be doing differently?"

or

"How can we resolve this?"

or

"Let's look at this together."

Incorrect: "You need to do things differently."

or

"You need to change things."

or

"I think you need to figure out how to handle this thing."

or

"You need to find a solution here."

or

"You better take a good look at this yourself."

Digital Download — Download from CengageBrain.com

Timing. When you listen to the other person you may begin to formulate ideas or concerns that you feel need to be expressed. It is important to know when to do this. There are two factors that are important. First, the relationship must be developed before you introduce your points. There needs to be rapport and a sense of trust that exists for the client with regard to you as his or her case manager (Strong and Zeman, 2010). Never begin an exchange of views before you have established good rapport with the other person. Leaman (1978) writes that "rapport is essential" and stipulates that the client needs to see the case manager as "a helpful person" first (p. 632).

The second factor involves when in the developing relationship would you introduce your views. Although you may be able to see what issues are involved and wish to introduce a different point of view during the first interview, that is rarely effective (Leaman 1978; Polcin, 2003). The relationship needs to be well established before you invite an exchange of views with the other person.

Be Matter of Fact. Do not become excited or judgmental or petulant. Refrain from using words or phrases that are derogatory or denigrate people and the choices they have made.

Be Tentative. This is particularly important. You could be right or you could be wrong in your observations, but this is not a competition to see who is right and who is wrong. What you are putting forth for consideration might not work from the client's perspective. For that reason, it is not helpful to present your ideas as though they are true or right.

What does tentative sound like? Use phrases such as, "I'm just wondering if . . . ," "I'm thinking that maybe . . . ," Do you think that one thing we should consider is . . . ," I could be wrong but let me throw out this idea and see what you think."

Focus on Tangible Behavior or Communication. *Tangible* refers to what you can observe or hear. Sometimes when we bring something up for discussion, we tend to be vague about what the actual problem is. We might generalize or just describe our feelings about it. This is not enough information for the client to use to make a meaningful change. For example, a worker might say, "I don't know. I just think you let people walk all over you. I could be wrong about that but it seems a little like that. What do you think?" This is a nice I-message, but it is vague. It would be better to say, "I don't know. I just think when Curt took the rent money from you and never paid you back twice and then Jean asked for a loan for her car and you gave it to her it was like they felt they could take advantage of you or maybe that they feel they can use you. I could be wrong about that but it seems a little like that to me. What do you think?" In this second instance the other person gets a lot more valuable information.

Take Full Responsibility for Your Observations. If you recognize that what you observed is what *you* observed and that it is perfectly all right for your observations to be incorrect or different from another's observation, it will be easier for you to take responsibility for your observations. If you are wrong, the perception can be corrected, particularly if you have been very tentative and nonjudgmental.

Refrain from Accusing. It may be tempting to blame or accuse the client for the situation. Doing that is not empathic or an act of caring. It is an attack on the other person's self-esteem. Refrain from doing that because it prevents the person from hearing you. Accusatory exchanges are not between equals because the accuser assumes he is right. Accusations sound like this: "You brought this on yourself when you decided to write those letters." Or in another instance, "You said these things to the group. You set this up so that we can't trust you. This is your doing."

Refrain from Anger. Sometimes it is tempting to use hostile confrontation to punish a client who has made you angry. In these situations, you might use public humiliation or denigrate the person. In the relationship you hold with clients you have a degree of power. Your words can wound the other, diminish your relationship, and cause the other person to feel defensive. Again, the client will not hear the important message.

Refrain from Judgment. No matter how much empathy you feel you have for the other person you really can never know entirely what it is like to walk in their shoes, to face the obstacles they have faced, to have grappled with the life circumstances they have had. Judgments are generally impossible if you are genuinely tentative when you give feedback. Give your viewpoint carefully and listen carefully to the client's response (Strong and Zeman, 2010). Polcin (2009) refers to using "explicit but nonjudgmental feedback about the potential consequences" (p. 505) (of behaviors).

Refrain from Producing a Solution. Because of your position with regard to the client, who is already having problems, any solution you give could be seen as imperative. Even the words "should" and "ought" sound like imperatives to the client and are best avoided. What we really want is for people to develop their own solutions. If we have a contribution to make to that process there are ways to do that as noted above. What we don't want to do is take over or offer our solution as the only one or the best one.

Always Collaborate. When offering feedback or your point of view at the end ask the other person what she thinks about what you have said (Strong and Zeman, 2010; Polcin, 2009). When you have laid out your point of view ask, "What do you think?" or "How do you see it?" This establishes the fact that you see this as a shared search for an answer or a positive change of direction.

Asking Permission to Share Ideas

There will be times when you will want to offer information or suggestions. For example, suppose you are working with a man who wants to start seeing his children who live with their mother. You have some ideas about how he might go about that. Rather than giving the solution, ask permission to share some ideas (Miller and Rollnick, 2002). You might say, "I have a couple of ideas that might be helpful to you, but I want to be sure it is all right with you to share these now." Or you could say, "There are some things my clients have done in the past that worked well for them. Would you mind if I shared a few of these with you?" In this way, solutions and advice are given only with the client's permission, leaving the client in charge of his situation and free to reject the offer of ideas.

The opposite approach would be to simply give the advice or hand out solutions. You could even start your message with *I*. You might say, "I have found it is better if you stay away from the children until there is a court order for you to see them." You may feel this is an I-message, but you have given a solution without permission. There is the very real possibility that you could make the client defensive, arguing against the very thing you see as a good solution. On the other hand, your client may appear to agree but actually feels resistance to being told what would be best, a resistance he does not express. It is better to ask permission to share the idea first before plunging in. When giving your ideas, do so tentatively and ask for feedback from the client.

For example, Naoko was working with Paul on housing. Paul, who suffered from schizophrenia and had a problem with alcohol, was not happy with the place he lived. Most of the people there had drinking problems as well, and Paul felt they tempted him to drink more and skip his medications. On the other hand, Paul told Naoko that these people were accepting of his illness, friendly, and often very helpful. Naoko had some ideas about where Paul could move where he might feel secure and have friends, but not be with the people who had problems with alcohol. Before Naoko gave these ideas she said, "You know I was thinking of a couple of places that might work for you if they have an opening. Would you mind if I told you about them?" In this way, Naoko made it clear that the ultimate decision was Paul's and she was only offering suggestions.

Asking permission to share ideas is a good way to proceed when you may be handling a new case in an urgent situation. There is a problem that needs to be solved immediately and the people involved are looking to you for ideas about how to solve it. Even though this is the first encounter with this person, if you ask permission to share your ideas about how to address the issue, it will be more collaborative. For example, you might have to do a crisis intervention with a person who is having a mental health crisis. The family is distressed, looking to you and your agency for what to do next. Here you might say to them, "There are a few ways we can handle this. Would you mind if I lay these out for you and we look at them together?"

It is always best to have more than one idea to share with clients so that they feel there is a choice. Emphasize that the choice is theirs to make and that they would know best which of these ideas, if any, would work for them.

Advocacy: Confronting Collaterals

There may be times when someone is interfering with the client's treatment, your ability to interact effectively with the client, or your client's progress. When this happens you need to speak about what is best for the client. You are advocating for your client when you stand up for that person's best interests. For instance, a night-shift nurse supervisor in the emergency room took it upon herself to keep the interview room open. Even though the room was there for workers to interview victims of

FIGURE 10.3 Sample messages for confronting collaterals

To the nurse in the emergency room who is trying to clear the interview room:

Correct: "I need you to help me complete this interview. I expect to need this room for about 45 minutes, and then I will have all the necessary information."

or

"I can see you want the room. Could you give us another 45 minutes to complete the interview? This must be done before the patient leaves."

Incorrect: "Oh dear, we'll only be a minute, and I need this information too. May we stay awhile longer?

or

"I thought we could use this room any time. What seems to be your problem?"

---

domestic violence, violent crime, or rape, the nurse would barge in, in the middle of the interview, and try to clear the room. Such situations generally include something someone is doing that:

1. Adversely affects the client
2. Adversely affects your work with the client

In situations like this, you need a firmer message. The message would:

1. Not sound tentative
2. Be pleasant, but firm (smile, but mean what you say)
3. Contain an implied or explicit request for help
4. Thank the other person if they comply

Examples of correct and incorrect messages to the nurse in the emergency room who is trying to clear the interview room are shown in Figure 10.3.

In dealing with other people whom we see as making our work more difficult, it is tempting to throw out the rules and simply show our annoyance or exasperation. The problem with that approach lies in the fact that we need to work with other people and the agencies they represent. In this field, we must be able to communicate well with one another if we expect to help the people we serve learn better ways of communicating. Your anger directed toward the nurse in the emergency room can affect relations between your agency and the entire emergency room staff. If this is an important part of your work, such strained relations will affect patient care. Remaining firm, but diplomatic, often prevents such problems.

On Not Becoming Overbearing

It is a little tricky to stay where the client is and still express your own concern. Sometimes a technically correct I-message is really about your agenda and is not sensitive to the client and where the individual is with the problem at the moment. Such an I-message comes across as intrusive.

For instance, a woman is suddenly widowed. Her husband died in an accident on Tuesday night. You went to the home as part of the crisis team the night it happened because police said she was extremely upset. Tonight you are doing a follow-up visit. When you talked to her the first time, you learned that she is the second oldest of five children. Her brothers and sisters do not live nearby, and she made no move to call them in spite of your suggestion that she do so and your offer to do it for her. You feel that family can be very supportive at a time like this. You have reached this conclusion because you and your family are close and supportive. In this situation, you might send I-messages like those that follow. The parts that are italicized actually express a view or opinion belonging to the worker and do not leave any room for the client's perception.

- "I will honor your request; however, *I feel you may be avoiding a source of real help.*"
- "*I'm uncomfortable that* you don't want your family to be aware of your husband's death. *Family support can be very comforting, and I'm sure that they will not be inconvenienced.*"
- "*I'm not clear* about why you want to keep this from your family. I feel that they would want to know."
- "*It seems to me* that going through this alone *will be very rough for you.*"
- "*I feel that talking to your relatives will be very helpful.*"
- "*I have a problem with you wanting to do this alone.*"

Suppose it turns out that some years ago this woman was in trouble. She was a rebellious teenager and left school and ran away from home. Her parents seemed not to care, and when she attempted to return home at age 19, they told her she had caused them enough grief and she was not welcome there. She moved here, went to college, got a master's degree, and married a local dentist. She feels better off without her family who has never offered her support in the past. She does not tell you all this because she just met you and she does not know you well enough to go into all the reasons why she left home and is estranged from her family.

There are five ways we can make our I-messages ineffective.

1. Using the words "but" or "however" reverse what we have just said supporting the client's point of view—"I do see your point but . . ."
2. Expressing how we see the situation, using *I* correctly, but never inviting the client to describe how she sees the situation "What are your thoughts about this?"
3. Suggesting a solution but not asking the client for his solution. "What ideas do you have that we could use to solve this?"
4. Coming across as the way we view the situation is the only way to view the situation. "I feel this is not something you can handle alone so we need to call someone for you now."
5. Failing to consider that there may be extenuating circumstances that you are not privy to.

Think about such possibilities very carefully when you frame an I-message. Be sure that while you speak your concern you leave plenty of room for the fact that you do not know everything and that you could be very wrong. Sounding tentative helps: "I could be wrong." Or "I am not necessarily right about this."

Follow-up

When people respond to your ideas or views use your reflective listening responses and open questions to learn more and move the conversation forward productively. For example, you might say, "So in other words you really do not want to call your family after so many years of having no contact with them." The person can agree, if that is the case and will probably move on to tell you more about her situation.

Summary

Providing our own point of view must be done carefully and with considerable tact. Our goal is to introduce another perspective, and we want the person to be able to hear and use what we have to say. This process also involves listening carefully to what the other person says in response. It is often helpful to ask permission to introduce our ideas. Once we begin to express our view of things, it is best to sound tentative and to invite people to respond to the ideas we have raised. We do not want to convey a know-it-all attitude that imposes on the client the solution and viewpoints we think are best.

Video Examples

To view the videos that accompany this book, go to CengageBrain.com.

- In "Helping Tom Solve a Personal Problem," you can see Danica bring up some difficult issues with Tom as they sort out what Tom needs to do to complete his program.

Exercises

These exercises can also be filled out online at CengageBrain.com.

Exercises I: What Is Wrong Here?

Instructions: Look at the following worker communications, and identify what is wrong with the way each one is expressed.

1. To a person who is drinking and taking tranquilizers: "That's a risky thing to do!"

2. To a person who is driving without a driver's license: "You're just doing this to tempt fate."

3. To a person who is always forgetting to take his insulin: "I'm sick of these so-called lapses of memory. You must want to feel sick most of the time!"

4. To a person who bounced three checks in 3 months because she cannot seem to balance her checkbook: "Go take an accounting course, for heaven's sake!"

5. To the person who has lamented not spending enough time with his son: "Children are important. They grow up fast. You only have so long to spend with them when they are kids. You should keep that in mind."

6. To the person who had trouble completing a high school equivalency exam and is now talking of becoming a doctor: "You need to be more realistic about what you can and can't do. Think of some other career."

7. To the woman who has completed 10 weeks in a rape victim support group and is still unable to work or leave the house much, but who says she is fine and getting over it: "It doesn't seem to me like you're getting over it. If you wanted to get better, you would force yourself to go out more."

8. To the man who complains about his neighbors but spends time on his porch yelling at the children, which starts neighborhood feuds and tensions: "You're always yelling at them. Of course they fight with you!"

9. To the woman who has been in a wheelchair for several months following an accident in spite of her doctor's feelings that she could now be up walking with crutches: "You need to get out of that chair and practice walking. The doctor says you don't need the wheelchair."

10. To a child who says the other kids do not like him, but who is always hitting the other children and calling them provocative names: "You're half the problem, you know. Stop yelling and hitting everyone, and they'll like you better."

Exercise II: Constructing a Better Response

Instructions: Now go back to the vignettes above and construct a tentative *I-message* that invites **collaboration**.

Exercises III: Expressing Your Concern

Expressing Your Concern I

Instructions: In each of the vignettes that follow, you have a problem—a concern about something affecting the client. For each of these situations, construct an I-message from you to the client. Be sure to follow the rules for bringing up your point of view. Make certain you sound tentative and ask for collaboration. Use several sentences to soften and put forth your ideas.

1. A woman, who has been a prostitute, recently discovered she is HIV+. She is currently staying in a shelter where you see her. Several nights she comes in drunk and tells you, "Hey, it doesn't hurt as much this way." The next day you

approach her with an I-message expressing your concern and initiate an exchange of views.

2. A woman you have been working with tells you her husband is really a dear. He has done many wonderful things for her, and she is feeling guilty about calling you, but he does keep her confined to the house and slaps her a lot. You use an I-message to express your concern and initiate an exchange of views.

3. A man with two children needs temporary shelter. His oldest, a daughter, is old enough to drop out of school; and in the course of placing him, you learn that he has encouraged her to do just that. He tells you he needs someone at home to look after the place, now that they have one, and to see that the younger child is taken care of. You use an I-message to express your concern and initiate an exchange of views.

4. A woman, recently diagnosed with ovarian cancer, is using a prescription medication her doctor gave her to help her with the anxiety of facing the perpetrator in court. Lately you feel she has been abusing her medication. Her speech seems slurred, and you often see her slip one of the pills into her mouth. You use an I-message to express your concern and initiate an exchange of views.

5. A man has been on pain medication for a number of months, prescribed by his doctor for intractable low back pain. Following surgery he continues to feel a need for the medication although he is walking well and is pain free. When his doctor began to reduce the prescriptions for the pain medicine you learned that your client began to purchase heroin on the street. You use an I-message to express your concern and initiate an exchange of views.

6. A woman has not come out of her house since she suffered a major injury at work. Although her doctors say she will be able to return to work if she goes to rehabilitation, she refuses to go and cites her concern for her fragile recovery. You have talked to her many times by phone and invited her to attend support groups at the rehabilitation center where you work and to see a counselor, but she never comes, and you are becoming aware that she is terribly fearful. You use an I-message to express your concern and initiate an exchange of views.

Expressing Your Concern II

Instructions: In each of the vignettes that follow, you have a problem—a concern about something affecting the client. For each of these situations, construct an I-message from you to the client. Be sure to follow the rules for confrontation. Make certain you sound tentative and ask for collaboration. Rather than a single sentence, try using several sentences to soften and put forth your ideas.

1. A man who has been sitting by his wife's side since she slipped into a coma is weary and has neither eaten nor slept for over 24 hours. You approach him with an I-message expressing your concern and initiate an exchange of views.

2. A woman who is refusing to take medication that would prevent her from having a psychotic episode comes to you and says she is not sure what to do. She does not feel well, but she would like to be able to handle things without medication. You use an I-message to express your concern and initiate an exchange of views.

3. A man whose wife just left has told you he wants to give up his job and simply leave the area, having no further contact with either his ex-wife or his children. You are concerned that he has not had time to think this through. You use an I-message to express your concern and initiate an exchange of views.

4. A man on your caseload, addicted to heroin, has been arrested for selling heroin to a minor. You go to see him at the county prison where he tells you that selling heroin was the only way he could support his own habit. You use an I-message to express your concern and initiate an exchange of views.

5. A woman staying in the shelter where you work has left her baby in the care of others repeatedly and gone out. She says she is going to the store or to look for an apartment or a job, but others let the baby lie in the crib and cry. You have had to feed and change the baby on several occasions. You use an I-message to express your concern and initiate an exchange of views.

6. A man is waiting for his Social Security disability check to start. He has a serious heart condition and has been told he should not be out in extremely cold or hot weather. You stop by on a home visit and discover he is out on a cold day shoving piles of snow off the driveway. He tells you it is not that cold and this is not "shoveling." You use an I-message to express your concern and initiate an exchange of views.

Exercises IV: Expressing a Stronger Message

Instructions: In each vignette that follows, you have a problem with the behavior or actions of someone; this person's behavior is affecting the goals of your work with the client or is adversely affecting the client. For each vignette, construct a firmer message that explicitly or implicitly requests this person's help.

1. You are interviewing a man who appears to be quite delusional in the hospital emergency room. The new security officer at the hospital does not seem to understand that the behavior is part of an illness, and he keeps entering the room and asking, as if the patient cannot hear, "Is he giving you any trouble? Do you want me to take care of him?" Your message expresses your need to continue the interview and your need for privacy.

2. You have been working with a man who was beaten and robbed. Because of the injuries, he has been unable to work. His employer calls you several times, saying he thinks the man is simply "freaked out" and needs to get over it. The boss tells you that he has told the man this on several occasions, and says that the man just yells at him. You need the boss to understand the severity of the situation, and you feel it would be helpful if he did not keep calling the victim with his negative opinions. Your message expresses your need for the boss to work with you and the client more constructively.

3. You have been working with a child in temporary housing. You have discovered the child is very artistic, and you have found an artist who is willing to volunteer time to teach the child on Saturday mornings. The mother of the child is upset and tells you that it is impossible "the kid has any talent" and that "anyway, he's got chores on Saturday morning." Your message expresses your need to see the child's potential fully realized.

4. You are interviewing a rape victim when her boyfriend barges into the room and demands, "What's going on in here?" Your message expresses your need to continue the interview.

   Note: Do not allow another person in the interview room with the client until you and the client have decided *privately* whether that person should be there. In other words, do not discuss, in front of the boyfriend, whether the boyfriend or anyone else can stay. Do not exchange glances with the client. In such situations, never ask the woman, *in front of the man*, if it is all right for the man to stay during the interview. Lead the man outside when you give him your message. Later, when you and the woman are alone, you can ask her whether she would like to have him present, but *always make it appear that the decision to have him wait outside is yours*. It is possible that she is afraid of him and will feel compelled to agree to his staying if she is asked about it while he is in the room. If she is fearful or embarrassed, the quality of the interview will be compromised.

5. An elderly woman is trying to decide what to do about her need for help. The decision is between staying in her own home with assistance, or selling her home and entering a nursing home. She is very torn. You have arranged for help, which seems to be working well, and you visit her each week. During your visits, the woman discusses with you her options. The decision is a difficult one for her. When you visit her, a woman who lives next door invariably appears and offers her advice and expresses her doubts that the woman should be living alone. Your message to the neighbor expresses your feeling that her behavior is not helpful.

CHAPTER 11

Addressing and Disarming Anger

Introduction

People do become angry. They express anger and hostility in ways we might find quite unpleasant. We can expect that there will be times when the people with whom we are working will forcefully express their anger. As professionals, it is helpful to view the anger as a clue to other underlying issues or as a clue to problems that need to be resolved. Using the anger to help us better understand the other person is better than reacting to it defensively or personally. When people are angry, it is not about you. It is about frustrations and concerns in their own lives. If you are an effective, reflective listener, you will hear these underlying causes and feelings, and you will respond in a manner that disarms rather than provokes the anger.

Common Reasons for Anger

When clients are angry, it is often because of one of the common reasons listed here:

- *The client is angry about something the agency has done.* The agency in which you work will have policies and regulations that you must follow. Sometimes the agency is bound by state and federal laws as well. These laws work better for some clients than for others. Clients who feel that the agency's policies have caused them to be treated unfairly or with insensitivity to their particular circumstances may react angrily.
- *The client is angry about something you have said or done.* As noted earlier, there will be times when the client or the client's friends and relations will have a problem with something you have said or done. Without your intending that it should

happen, a client may completely misunderstand what you have said or may misread your intentions. On the other hand, you may not always be completely tuned in to where the client is at any particular moment and may unwittingly say or do something the client finds upsetting.
- *The client is fearful.* Many people are frightened by the turn their lives have taken. The changes that have occurred that brought them to your agency may make them feel as though their lives are out of control. They may attempt to reassert control through the use of anger, and they may lash out at you because you are the safest target or the closest person at the moment.
- *The client is exhausted.* Some clients you see will be exhausted. These people may have been grappling alone with an issue or problem for a long time, or the circumstances they are facing now may be taking all their energy. They sense that they may not be able to carry on, which may cause them to direct anger at you.
- *The client feels overwhelmed.* Other people feel overwhelmed by problems. They may feel that they cannot handle all that is facing them. Sometimes they feel the extent of the burden is unfair, and so they lash out at you.
- *The client is confused.* Some people are confused by policies, circumstances, others' reactions to them, or the steps they must take to right a difficulty. Rather than admit to feeling confused, some clients become angry and blame the system or you or your agency.
- *The client feels a need for attention.* Some people feel insignificant and demeaned. It may have nothing to do with you, and it may very much relate to a lifetime of living in the margins or having one's problems or contributions trivialized. These people need to feel valued and worthwhile. The problem for you is that your best efforts may not always be good enough. Sometimes such people are extremely tuned in to slights and suspected rejection. They may become angry with you for reasons that you feel are unfair or unwarranted. As always, you are the professional person and need to speak to the condition of the client in a professional manner.

People become angry for many reasons. Knowing how to disarm anger is important. It will enable you to move toward a more meaningful dialogue and a better resolution.

Digital Download Download from CengageBrain.com

Why Disarming Anger Is Important

You cannot be as effective in your work if you are dealing with a client who is angry. The client cannot be expected to move the relationship to another level; but you, as the professional, can be expected to practice the techniques that will allow the relationship to move beyond the anger. The major reasons for disarming anger are as follows:

- *Eliminates an obstacle to true understanding.* Disarming anger diffuses the anger, making it less of an obstacle to true understanding. People who are angry cannot really hear each other. If you are genuinely interested in why the client is

reacting in this manner, you need to reduce the anger so that you can better understand what is fueling these strong emotions.
- *Shows clients you respect their message.* Disarming anger shows the other person that you respect the message even if the way it is expressed is not helpful. By moving to another level beyond the anger, you can indicate to angry clients that their concerns are important to you even when you are having trouble with the way they are addressing these concerns.
- *Enables you to understand the problem.* Disarming anger allows you to become aware of the actual problem. Only when you have disarmed the anger can you and the client actually address the underlying concern. As clients feel heard and understood, they are more likely to begin to collaborate with you in looking at their problems and the solutions.
- *Allows you to practice empathy.* Disarming anger allows you to practice empathy, seeing the situation as the other person is seeing it. Disarming anger is an important part of establishing rapport. If you become angry yourself, you are caught up in your own feelings and needs at the moment. On the other hand, if you think about the reason the person is angry and you speak to that situation or to those feelings, you are responding empathically. This lets the client understand that you are not going to engage in an angry exchange, but you are going to respect the client's concerns and feelings.
- *Focuses work on solving the problem.* Disarming anger focuses on solving the issues and problems, and not on who is to blame. Disarming anger techniques do not allow for exchanges of blame. Angry clients may hope for such an exchange with you wherein they blame you and you defend yourself, often by blaming them in return. The purpose of disarming anger is to fix those things that legitimately need to be fixed.

Many people sound angrier than they mean to. They are often anticipating the angry response of the other person. As human service workers, we read anger as a signal that the client's needs have not been met, and we focus on resolution of the problem that has caused the angry emotions, regardless of whether we think the client's anger is legitimate.

Avoiding the Number-One Mistake

Countless times human service workers encounter people who are openly angry. Many of those workers choose to take that anger personally. Taking anger personally is the number-one mistake when dealing with an angry person. It is a foolish mistake to make.

As noted earlier, people become angry for a number of reasons. Some of these reasons have nothing to do with the worker specifically. Other times the anger may be caused by something the worker or the agency has done, and the anger may be rude and denigrating. Nevertheless, beyond disarming the anger, it is important that when you encounter an angry client, you refrain from taking the anger personally.

It is not about you and how you feel. It is always about the client and your professional response to the client. A worker who chooses to take the anger of a client personally might end up in a conversation something like this.

CLIENT: Where the hell were *you* on Tuesday?

WORKER: What do you mean?

CLIENT: Where the hell were you? I came in to get a voucher for food, and you weren't here.

WORKER: Why are you shouting at me? I wasn't here, but you don't have to shout.

CLIENT: I do have to shout! You say to come in here for a voucher, and I did that, and you were not even here. Where the hell were you?

WORKER: Look, Mr. Peters, I don't have to tell you where I was. If you came in and I wasn't here, why didn't you tell someone else what you needed? I'm not the only person who can help you.

CLIENT: I get so damn tired of the way you guys act like prima donnas. Who the hell gave you the right to tell all of us when to come and when to go? You say come in, I come in, like a fool, and *you* decide you'll just go someplace else.

WORKER: Well, if that's the way you feel, you certainly don't need my help. I've spent quite a lot of time with you, may I remind you? You have gotten a lot from this agency. I'm not sure I'm going to put up with this shouting at me.

CLIENT: Well, what are you going to do about it? I can tell you that you are a piss-poor caseworker if I want to. I can't do much else around here, but I *can* do that!

WORKER: You're an idiot. Go out and get the voucher from Mrs. Charles, bring it back here, and I'll sign it (*begins reading papers on her desk*).

In this example, the relationship is damaged, and there is an unsatisfactory resolution. Bitter feelings remain for both the worker and the client.

There is a better way to handle situations like this one. This chapter will explain how to use a four-step process to deal with anger. The central question you want to ask yourself is this: Can I feel empathic toward this angry person and hear the pain behind all this anger—or am I likely to get into a power struggle with this angry person to show I won't be pushed around? Empathy is the professional response. Power is the unprofessional response.

Erroneous Expectations for Perfect Communication: Another Reality Check

Some human service workers have the erroneous expectation that their clients will give them no trouble. In their view, clients not only will never get angry, but they will follow suggestions, be appreciative, and never raise doubts, criticisms, or resistance. This sort of thinking is a trap, and workers who fall into it often become exasperated or punitive with clients who become angry. Workers who are susceptible to the trap have envisioned themselves as helpful people dispensing good works to people in need. The reality that some people will not want their help, will criticize the help they receive, will show no appreciation or will become angry has not entered their thinking when they chose this field.

We all have had bad times in our lives, and we look back on those times later and think, "I wasn't myself then." These times may have been isolated incidents, or they may have been prolonged periods when we were under a lot of stress. The people who seek our help are under a lot of stress. In addition, many of them have problems precisely because they have trouble communicating easily with others. Anger and other forms of negative communication may be all they learned.

Expect anger, disarm it, and treat it matter-of-factly. In this way you will not allow a client's anger to obstruct your work with the client, nor will you carry completely unrealistic notions in your head that clients won't or shouldn't get angry. They will get angry, but you will know what to do.

For example, Jane was a worker in a home for three individuals with mental illness. Kip had a bipolar diagnosis and was doing well. In fact, maintained on his medication, he was pleasant and cooperative. He was working at a local supermarket and seemed about ready to move to an apartment of his own. Then it was discovered that the medication he was taking, Lithium, was adversely affecting his liver. Liver function tests came back showing this deterioration. Doctors immediately removed Kip from the Lithium and placed him on an alternative medication.

Almost at once Kip's personality turned irritable and angry. He accused Jane of spying on him, and he became erratic about going to work. When the residents in the home went shopping for their groceries, he either sat in the van with his arms folded, refusing to get out, or he created scenes in the supermarket about things he wanted to buy that would have shattered the careful budget he and the others had constructed.

His outbursts in public were embarrassing to Jane, and in the home she often endured a lot of his anger. Jane's approach was twofold. She actively advocated for a reexamination of Kip's medications, and she was firm with Kip but never angry. Many times she told him she understood that he was not feeling like himself. She refused to take anything he said personally. On more than one occasion, his accusations actually made her laugh, and Kip laughed with her, recognizing momentarily how silly his accusations were.

Jane's superiors, and particularly the treating psychiatrist, all believed that Kip could have become dangerous had Jane not steadfastly refused to escalate the situation or take it personally.

The Four-Step Process

In his book *Feeling Good*, David Burns (1980) suggests a four-step process for disarming anger. The material in this section is adapted from his book. First, we will look at the individual steps, and then we will look at how these steps work in actual practice.

1. *Be appreciative.* It is frightening enough to tell people you are angry about something they have said or done. You will put clients at ease if you can say something like "I appreciate your coming to me with this" or "It is helpful to know how you are feeling about this" or "Thank you for bringing this up."

2. *Ask for more information.* A client who is upset may be skimming the surface of an issue. To understand the issue better, ask for particulars. Do not grill the person or sound defensive. You might say, "Can you tell me when this happened?" or "I guess I'm not sure when this happened. Can you help me out?" or "How often did this happen?"
3. *Find something with which you agree.* Never make up something just to sound agreeable, but see if there is not some little piece of what the client has said with which you can agree. You might say, "I think we have probably done this in the past." or "I can see where you would look at it that way." or "There probably was a lapse the day you are referring to."
4. *Begin to focus on a solution.* Focusing on a solution should involve *collaboration* whenever possible. Remember, however, that the client owns the problem; the extent of your assistance is a conscious and strategic decision that you must make.

Digital Download Download from CengageBrain.com

Begin with your objective point of view. Listen to the client's point of view. Then decide if you will make changes or leave things as they are. You might, after explaining your point of view, say, "I can see where you thought that. I think from now on I'd like to write you ahead of time." or "I can see where you thought that. Right now we are really short-staffed, so writing to you ahead of time isn't really an option. But I'm glad you brought this up. In the future, we will take another look at it." Own your perceptions and own your decision regarding the problem. Use "I," not "you." Let us return to the situation we looked at earlier and see a more effective approach for handling the problem.

CLIENT: Where the hell were *you* on Tuesday?

WORKER: I'm not sure I know what you mean.

CLIENT: Where the hell were you? I came in to get a voucher for food, and you weren't here.

WORKER: What time were you here on Tuesday?

CLIENT: Oh, about 2:00. You say to come in here for a voucher, and I did that, and you were not even here. Where the hell were you?

WORKER (USING ACTIVE LISTENING FIRST): Well, this must have been really inconvenient for you. I appreciate your telling me about this. Did you ask anyone else to help you?

CLIENT: No, I didn't. I didn't know I could.

WORKER: Sure you can. I can see where you would think I didn't care about your getting food when you expected me to be here.

CLIENT: I know. We didn't have any dinner Tuesday night. Just potato chips and cheese—oh, and there was a little milk left.

WORKER: I'm really sorry that happened. Let me see that you get the food you need today, and let me explain better than I did the other day how this works. If I or any of the other workers tell you to come in for a voucher, then you can come right to the office, and whoever is doing the intakes can take your information and see that you get the food you need. This shouldn't have happened this way, and we don't want it to happen again.

CLIENT: Thanks.

As is rarely the case, all the elements of the four-step process are present in this exchange. In this example, the worker asks for more information with genuine interest ("What time were you here on Tuesday? Did you ask anyone else to help you?"). She goes on to express appreciation ("I appreciate your telling me about this."). She indicates that she agrees with the way the client viewed the situation ("I can see where you would think I didn't care about your getting food when you expected me to be here."). Finally, she moves on to focus on a solution ("Let me see that you get the food you need today, and let me explain better than I did the other day how this works.").

The worker in this example does some other things that make it clear she is not going to take the client's anger personally. She uses reflective listening ("Well, this must have been really inconvenient for you."), letting the client know that he is being heard and respected. She also takes some responsibility for the mix-up even though she may actually have explained it perfectly ("And let me explain better than I did the other day how this works.").

We might change this vignette just a bit. Perhaps the worker actually did explain to the client on the phone before he came in how the agency works. There are many reasons he might not have heard her: anxiety over trying to make sure his kids would eat that night, anger over having to go to the agency in the first place, uncomfortable feelings of helplessness or inadequacy over his inability to fix his situation on his own, and the stress of not eating and having hungry children at home.

Although the worker may not know specifically what has generated the angry outburst, she is fully aware that there are forces at play in this man's life beyond his need for her to be present when he arrives at the agency. For that reason, she remains respectful throughout the entire exchange, and she moves with genuine interest and concern through the steps of disarming anger. In other words, she does not take his anger personally and feel a need to confront it with anger of her own.

What You Do Not Want to Do

There are a number of things you need to avoid doing. Figure 11.1 contains examples of these things. The incorrect example for each point illustrates what you want to avoid, and the correct example shows you a better way to handle the situation. As you read, refer to the figure and compare the correct and incorrect examples that illustrate each point.

- *Avoid becoming defensive.* Do not fall into the trap of defending yourself. It is okay to have made a mistake or to be wrong. When you defend yourself you indicate that you wish to argue the points the other person is making. It is better

FIGURE 11.1 Examples of What Not to Do

Do Not Become Defensive

To a person who feels the worker did not spend enough time with her:

Correct: "I might have cut the interview a little short."
Incorrect: "We don't have a lot of time around here. I'm doing the best I can.

Do Not Become Sarcastic or Facetious

To a man who works and is frustrated because he needs a later appointment but keeps getting an early morning appointment:

Correct: "I'm glad you brought this up again. We really do need to get this straight."
Incorrect: "Here we go again! Thanks for telling us, again, how inefficient we are."

Do Not Act Superior

To a woman who thinks her daughter should have different services:

Correct: "We ought to look at this more closely. I'm glad you told me about this. You may be right."
Incorrect: "The services have been chosen for your daughter by professionals in the field of child development, and they know what it is she needs."

Do Not Grill the Client

To a man who believes his aunt is being neglected by the agency:

Correct: "Tell me more about what you see happening with us and your aunt. We may need to look at this situation more closely."
Incorrect: "When exactly did we fail to come out to your aunt's house? How often did this take place, and did she ever tell us about this before? We need to know specifics before we can determine if this is a real problem. What other problems did you encounter with us?"

Digital Download — Download from CengageBrain.com

© Cengage Learning®

---

to indicate that you want to hear the points the other person is making. If you begin to defend yourself, it makes the other person angrier, and you lose an opportunity to really resolve the problem.

- *Avoid sounding sarcastic or facetious.* When you thank people for their comments or agree with something they told you, it is possible that you will sound sarcastic or facetious. This is especially true if you are feeling defensive. "Thanks so much for telling me" can sound cynical or it can show a genuine desire to understand. Monitor the tone of voice you use in these encounters to be sure your tone of voice is not escalating the situation.
- *Avoid coming across as superior.* It is all right for you to be wrong in your perceptions or behavior, and it is all right for the client to be wrong too. If you feel especially threatened or angry at your clients, it may be tempting to denigrate them in some way, pointing out how little they actually know about the situation or how little experience they have and how much more knowledgeable you are. This is a good time to show that you and the other person need to work together equally to solve a problem.

- *Avoid grilling the client.* In order to better understand the problem from the client's point of view, you will need to ask questions. However, too many questions can make the other person feel as if she is being cross examined. Avoid grilling the client by asking numerous questions, one after the other, in a doubtful tone of voice. If people are nervous, you will only make their nervousness worse. Most people grill another person in a triumphant attempt to prove the other person wrong. That is not your goal here. Your goal is to genuinely try to understand.

Look for Useful Information

It is hard when someone is angry or irritated with you or your agency to really hear what the issues are. However, you can benefit from the feedback you are receiving if you really hear it. Sometimes the client is bringing you valuable information that will help you to make constructive changes in yourself or in your agency.

In one agency, there were a lot of angry clients calling for help. They all had been discharged from a certain program without follow-up services or with follow-up services that had not been confirmed. This often left clients without much-needed medications or months of waiting to be reinstated. The agency was grateful for the clients' feedback and developed a questionnaire for the receptionist to use when such calls came in. She was instructed to handle each call with a genuine interest in what exactly had happened with each client. Gradually, a picture emerged of precisely what was wrong and how to fix it. In this example, an entire agency benefited from the clients' feedback. A more efficient operation will keep clients from returning with recurring problems and will save money and time for other clients.

When people are angry they are telling you that something is not working, something is wrong. Listening carefully can help you to discover what that is. It may be something you can change or something happening at the agency or something you have done or it could be something the client is doing that interferes with his care. If you can hear it you have an opportunity to address it.

Safety in the Workplace

Agencies where there are clients likely to be assaultive generally have in place policies and procedures to minimize the danger these people might pose for staff and other clients of the agency. Your agency should have policies and procedures related to how to handle situations that could become dangerous. Who do you call? Where do you conduct interviews where others are nearby? How is the furniture positioned in your office and interview rooms to prevent someone from blocking the door and preventing your being able to leave? If a person has been assaultive or combative in the past what is the agency's policy related to how that person is served in the future? There is research that indicates that staff training plays a considerable role in preventing dangerous situations and harmful outcomes. Staff training was shown to increase the confidence of professionals dealing with clients who might become assaultive (Kynoch et al., 2011; The Joint Commission, 2010). When you go to work in an agency, look at the policies that are in place and attend any training that is offered. If none of these is offered at your agency ask that they be developed.

The Importance of Staff Behavior

On rare occasions, people become so angry they seem to be about to lose control. Their demeanor moves from rational expressions of anger to increased belligerence, threats to the safety of others, or actual aggression toward people and objects in the vicinity. Research shows that staff people play an important role in defusing these explosive situations (Kynoch et al., 2011). An even tone of voice, continued reflective listening, and relaxed movement work best.

Lisa, a nurse in a community program for the mentally ill, discovered Phil eating lunch one day with a gun lying by his plate. He had been angry about his medications earlier in the day, but that problem seemed to have been resolved. Instead of quietly approaching Phil and suggesting the gun be left with the nurse until the end of the day, Lisa became hysterical. Rushing about the room, she loudly began to clear out the startled patients, thrusting them through the door. "Call the police, call the police," she kept shouting to other workers. Phil, alarmed by her actions, grabbed his gun and pointed it at her. He began to yell at Lisa, "Just shut up, shut up, before I shoot you. Be quiet." Lisa dashed from the room and cleared the entire building. Police came from every direction. The area was cordoned off, and a standoff ensued into the afternoon.

Lisa's loud, hysterical tone of voice, her panic, and her hurried actions all combined to make Phil agitated. Before long, the situation had escalated. What Lisa should have tried first was asking Phil to come with her to another part of the building. If Phil left the gun at his place, another worker could have secured it. If he brought the gun with him, Lisa might have said matter-of-factly that perhaps it would be better to leave the gun with the staff for the time being. If Phil gave her the gun, she could have taken steps to secure it.

Even if Phil were resistant and wanted to keep his gun, the staff could have asked the clients to bring their lunches into the group rooms for after-lunch groups. If they made this request in a tone of voice that indicated that it was nothing out of the ordinary, clients would have complied. In the meantime, police could have been called to come quietly and help to disarm Phil.

In another situation, Jim, a young mental health case manager, was working with Alex. Alex wanted to go into the hospital, fearing that he was getting sick again and would hurt someone. On that particular day, there were no beds, and Jim's supervisor suggested he help Alex find an alternative to hospitalization until a bed was available. Jim was afraid of Alex, who spent most of that day sitting in the waiting room. Each time Jim explained that no beds were available yet and that an alternative needed to be found, Alex grew more belligerent.

The last time Jim returned to report that there were still no available beds, he did so in what he thought was a very firm manner. Reasoning that Alex seemed about to become uncontrollable, Jim assumed that if he approached him with a firm, superior tone of voice, he could keep Alex from getting any angrier. In fact, Jim's superior tone was harsh and was heard by Alex as denigrating. He began to shout and pound the wall to demonstrate to Jim "just who is in charge here."

Clearly the way in which staff react to a person who appears angry and out of control often plays a role in whether the situation escalates or is defused. Remaining matter of fact, practicing empathic reflective listening, and remaining calm are important in maintaining a controlled situation. If you or your clients are in danger, certainly the first thing to do is to secure the safety of everyone. However, situations that escalate because workers fuel them—inflaming the client's anger by becoming loud, agitated, or angry themselves—can rapidly spiral out of control. Remaining calm and moving deliberately to prevent a dangerous situation from worsening is the responsibility of the professional.

Summary

Disarming anger is an important skill used to preserve your relationships with your clients and to prevent anger from escalating. The goal is to reach an understanding about the problems or concerns that are fueling the anger and to resolve those where possible. Becoming angry yourself can only escalate the situation, making real **problem solving** and collaboration with the client impossible. In previous chapters, you have learned many techniques by which to convey to clients their importance and your interest in what they have to say. Use these to advantage when dealing with a client who is angry. Remain matter of fact, refuse to take the anger as a personal insult, and reflect back the underlying concerns and feelings of the client.

Video Examples

To view the videos that accompany this book, go to CengageBrain.com.

- In "An Angry Consumer," Keyanna practices the skills discussed above when Michelle comes in angry about not being able to fill her prescription. You will see that when Michelle recognizes Keyanna's genuine desire to help, Michelle becomes less distressed.

Exercises

These exercises can also be filled out online at CengageBrain.com.

Exercises I: Initial Responses to Anger

Instructions: In the examples that follow, formulate an initial response to the anger and criticism you hear. On a separate piece of paper construct your answer, looking at the steps for disarming anger, and use those steps that seem appropriate. The four steps are: (1) thank the person for the comments, (2) ask for more information, (3) find some point on which you can agree, and (4) begin to look for a negotiated solution.

1. A man is coming to your agency for assistance after release from a drug rehabilitation center. He wants you to do more for him than you think is wise. You have been very helpful in ways you could, but you have also insisted he do some things for himself because you do not want him to become dependent. He is frustrated, and one day when you suggest to him that he try to call his lawyer himself, he blows up and yells, "This crazy place, sucking up the taxpayers' money—and for what? I get so sick and tired of your trying to make me do everything when that's what you're paid for. A bunch of idiots is what you are! Incompetents! Sure, I can do it myself! If I wanted to do it myself, I wouldn't have come to you, would I?" What is your initial response?

2. A woman who is in your shelter feels neglected. Twice you are interrupted when you are talking to her because of severe emergencies. You apologize both times and continue your discussion with her, but you are short-staffed and things at the agency are unpredictable. The second time this happens, when you are able to get back to her, she cannot remember what she had been saying. That upsets her. She says, "You all sure can find plenty of reasons to avoid talking to me. Every time I sit down to talk about my case, you get up and run off. Now I can't remember where we were. I don't see you running off when you talk to Alice or Cindy. Just seems like every time I need help, well, you have something more important to do." What is your initial response?

3. Do you see differences between the issues the man is having with the agency and the issues the woman is having with the agency? Are you and your agency more to blame in one instance than in the other? How does your response differ as a result?

Exercises II: Practicing Disarming

Instructions: Following are some opening sentences said by angry clients. It is up to you to develop the exchange, including more information regarding what the client is angry about and the responses of the worker. You do not have to use the disarming steps in any particular order. See if you can add some active listening and open questions as you go along. See if you can put yourself in the clients' shoes and empathize with their feelings.

1. CLIENT (*talking loudly, and banging his fist on the desk*): I have a beef to pick with you! You tell me I'm a mental case and then give me medicine that makes me feel like a nut case.

   WORKER:

2. CLIENT (*barging past the receptionist to the worker's office, obviously angry*): My kids and I are hungry! Know what that means—to be hungry? We're hungry and you . . .

   WORKER:

3. CLIENT (*looking at the worker cynically*): So you say the agency is understaffed. Really? Understaffed? Who are all these young girls running around here doing nothing? Is it too much to ask that I get an appointment without waiting 4 weeks?

   WORKER:

4. CLIENT (*Sits down unsmiling, looking past the worker*): I am just furious. I'm tired of trying to get into group on the phone. I talk to someone who says there is an opening for me. That's happened three times now. Every time I come in for group the leader says the group is full. A misunderstanding they tell me. Well, what I want to know is how many of these little misunderstandings do you people have a day? Do you think the rest of us aren't busy? If I ran my life the way you run this agency I would be all screwed up. I want a place in group. I want it now. I don't want to hear excuses or come in next week or call me tomorrow. (Voice rising) I want to go home today with this matter all settled. Am I being clear enough for you?

   WORKER:

CHAPTER 12

Collaborating with People for Change

Introduction

We have been looking at basic communication skills that create an environment in which good rapport and meaningful change are possible. There are, however, additional skills and information that might make the process of helping another person more understandable and give us a more realistic picture of how people grow and change. These enhance what you have already learned so that you give another person more robust support for change. In this chapter we will look at some enhancements that will give you added skill in your work.

In 2002, William Miller and Stephen Rollnick wrote a book that changed the way many people approached their clients. The book, *Motivational Interviewing,* was written primarily for therapists working with individuals in the field of addictions. Their ideas about respecting and collaborating with clients even under very trying circumstances speak to the values expressed throughout this textbook. Although their book is written for therapists there are numerous ideas and skills that case managers can find helpful. For that reason, we will discuss some of these ideas case managers can use to enhance their work with clients.

What Is Change?

Change has different meanings for different people. It can mean immediate changes to their painful circumstances, or it can mean long-term changes that affect the person's future well-being and sense of competence. You may begin by looking first at the near term changes such as abstinence through admission to a rehab program, relief of severe anxiety through the use of medication, or a change in an older person's unsafe living arrangements. But at some point you and the person will address how to prevent the same problems from occurring again and how to develop a worthwhile future.

When change involves giving up a way of being, an old habit, or accepting one's diagnosis, it can be difficult. People do not just change, as you can well imagine, even when they have suffered a setback or uncomfortable crisis, asked for help, or said they want to change. Fear, uncertainty, stubbornness, denial, lack of confidence or a lack of hope, or even the inability to envision the future can be significant obstacles to change. Change takes **commitment** and hard work. Your role is to support that hard work and commitment.

Stages of Change

Because change does not happen quickly it is better to view it as a process. In order to help us better understand how change takes place Diclemente and Valesquez (2002) looked at change as a process. Their model breaks that change process down into five stages a person might go through as he works toward changing. The model was originally developed for a program to help expectant mothers stop smoking. However, we have found that the model is useful any time a person needs to make changes that are difficult. One person may need to stop smoking; another may need to take her medication; another may have to lose weight for medical reasons; a fourth person might want to upgrade his skills and return to school; and another, in a program for parents who have physically abused their children, may want to learn better ways to parent. When you meet a person for the first time, he or she may be at any one of the different points in the change process. Knowing where the person is in the change process helps you to be more effective. Let us look at the stages.

Stage One: Precontemplation

People in this stage are really not thinking about change and therefore are not ready to change. They may not see a need to change, or they may have tried in the past and been unsuccessful.

In this stage you might reflect back to the person the fact that she is not feeling ready to change. This might be a time to invite her to look with you at her behavior. Possibly you would use an I-message expressing your concern for how this behavior or lack of behavior adversely affects her. Invite her to give you feedback.

Listen to why she is reluctant to change. Explore obstacles she sees. Be matter of fact and accepting of where the person is now without passing judgment or arguing. In addition, make it clear that whether or not the person decides to change is their decision to make. You will not pressure one way or the other.

Stage Two: Contemplation

In this stage a person is willing to explore but not yet willing to commit to change. What you might be hearing is **ambivalence** about making any changes. Ambivalence is normal when we seriously contemplate changing from something familiar to something unfamiliar. In this stage he is still not ready to change.

Again, accept that he is not ready and reflect that back to him. "You aren't really feeling good about doing this." When people feel ambivalence there is some fear that you will pressure them or force them to make changes they are not ready to make. In this stage it is a good idea, therefore, to reassure the person that the decisions to make any changes are up to them.

Talk over the problem with him. To support a person's contemplation, talk through the risks and the benefits involved in change at this point. Ask him if he sees any good points to the way things are now and what bad points he might see about the current situation. How does he see the problem? What might be some good solutions? What does he risk if he changes, and what rewards might he reap? **Collaboration** is important in this stage as you give information and learn from your client.

You may encounter people who will never move out of this stage or remain in this stage for a long time. How long and whether or not they change is up to them so it is unwise for you to feel responsible if a person does not move beyond the contemplation stage.

Stage Three: Developing a Plan

At this stage the person is ready to change but needs a plan. This is a collaborative endeavor as you seek to develop a plan with her and overcome the obstacles she may see ahead. From this collaborative effort you and she aim to find the best plan of action.

She may have tried to change previously and in talking with you finds that she learned some valuable lessons from those experiences. Now is a good time to brainstorm. Without judging the ideas talk about the options she sees and ideas you have. Eventually a plan presents itself.

Sometimes the initial plan is only a baby-step toward a much bigger goal. That may be what she can handle at the moment. Support her efforts. If she indicates she will try something you believe may not work, use a very tentative I-message to bring up your own concerns without discouraging her.

Stage Four: Implementing the Plan

In this stage people will actually change their thinking or behavior or their old habits, but think of this as practice. The new way of thinking and doing things still feels new. During this trial run at changing things, it is normal to be uncertain. Some people will miss the way things used to be. Use reflective listening, as you listen without judgment to their concerns. Sometimes a person putting a plan into action will not be entirely committed or may feel awkward and inadequate. Your **encouragement** and support can facilitate the work this person is doing.

Stage Five: Maintaining the Changes

In this final stage the person is seeking to make the changes a permanent part of her life. People take varying amounts of time to make the changes permanent, and some people never quite accomplish that. What obstacles are in the way of long-term success? Are there still things that will support the original behavior, such as triggers? Does the environment support a change?

Your support in the way of follow-up and encouragement are crucial at this stage. Acknowledge the successes. Work through collaboratively new obstacles that might crop up. Things can happen in this stage that you are not expecting.

- A person might test herself. Can she revert to the old ways just a little bit and still be okay?
- He may have discovered that the change cost him personally. Perhaps he lost friends or he had to move. Sometimes the costs aren't visible at first.
- She begins to doubt she can do this long term or sustain the effort it takes.
- He may have a sudden urge to go back to the old ways, to drink again, to throw out the medicine and go without, to return to a destructive relationship.

Your support and encouragement are crucial at this point.

Relapse

A person may find that he cannot sustain the new behavior or way of being over a long period of time. And so he may return to his addiction, another may stop his medications, another may return to an abusive relationship, while another may resume abusing her children. These relapses usually occur gradually. There is an initial slipup, which is often followed increasingly by the old behaviors.

Look at the attempts your client has made rather than focusing on the failure. Talk about how she successfully maintained the changes over a period of time. What was learned? What can be done differently next time? Talk through the relapse, using your reflective listening skills, to help the person make sense of what happened and why.

In some cases people can learn how to recognize a relapse, self-managing their disorder or addictions. Education about what to do and how to prevent relapse can help. Advance directives can help if the person is likely to have a relapse in her mental disorder. An advance directive allows the client to stipulate how she wants the next relapse handled.

Look together at how the relapse happened: identify what is likely to bring about a relapse, note the early signals of a relapse, prepare a prevention plan, and plan for how a relapse should be handled if it occurs again.

A Case History

Let us follow Claudette, whose situation illustrates the stages of change. Claudette came seeking help for a compulsion to shoplift. After numerous years in therapy and several short stints in the county jail, she still felt the urge to steal. She confided that she thought she had an "addiction" because getting things out of the store without paying for them was a challenge and there was a thrill when this was accomplished successfully. When she came to the agency, she was in danger of losing her children and felt the need to try once again to stop her negative behavior.

Claudette was beyond the precontemplation stage (stage 1). She had thought about changing before she came to the agency (stage 2, Contemplation). When she came in, the case manager took the social history and the two of them talked over the problem at some length. Claudette blamed an older sister and her friends for her "bad habit," which started when they were all teenagers. Throughout the interview the case manager assured Claudette that what was decided would be her decision. They talked a little bit about the negative consequences of Claudette's stealing, but they also examined the thrill she said she experienced each time she stole successfully. The case manager made several suggestions with Claudette's permission and Claudette considered them, including a 12-step program, but Claudette seemed unsure. The case manager suggested Claudette think about the suggestions and also what she thought would work well and return to talk further.

Claudette made an appointment the next week. She had once again shoplifted, and she was concerned that she would, in time, get caught and lose her children because shoplifting was a violation of her parole. Claudette and the worker devised a plan together (stage 3), which included Claudette attending the 12-step program every evening for the next month and seeking a sponsor there. In addition, Claudette suggested that she would not go shopping unless someone was with her who would be distressed if the person found her shoplifting again. A likely person was her aunt, and when contacted, the aunt was willing to accompany Claudette for the next month when she went shopping (Stage 4, Implementing the plan).

Claudette did really well and was into her fourth month in which she had abstained from shoplifting. (Stage 5, Maintaining the changes) She reported that she liked the people she had met at the 12-step program and liked her sponsor. In addition, she had been shopping without her aunt on several occasions without any desire to shoplift. Her case manager encouraged her pointing out the strength and determination Claudette was using to change her behavior.

Just before Claudette's time in the 12-step program was completed, however, Claudette called to say she had shoplifted on a Sunday afternoon when her aunt had been busy at church and her sponsor was out of town. Out of milk, Claudette had gone to the store, promising herself that she would only buy the milk and return home, but she had also stolen four pairs of children's socks.

Claudette had relapsed. The case manager spent time talking to Claudette about how the relapse happened, what she thought had triggered it, and what she might have done differently. With Claudette's permission, the case manager brought together the people who were most concerned and wanted to help, Claudette's aunt and her sponsor. The initial discussion led by the case manager centered on what could be learned from the relapse. What had Claudette learned about herself? What more needed to be addressed in planning that was missed in the original plan? The case manager focused on the success of the last few months and on Claudette's strengths and ingenuity, which she used to refrain from shoplifting.

Once again a plan was devised, and Claudette actively contributed. This time it was agreed that she would call her sponsor or her aunt if she felt a compulsion to shoplift. It was agreed that Claudette and her sponsor would look for someone from the 12-step program to act as a backup sponsor if Claudette found herself feeling a need to shoplift with no support around. She would continue in the 12-step program every weeknight for another 3 months, and she and her case manager would be in contact once a week. With the sponsor acting as a peer counselor, Claudette began to make important changes. At the end of the month she had not shoplifted, and her nights in the 12-step program were reduced to three nights a week (stage 5, Maintaining the changes).

Now the case manager began to work with Claudette about her future, one Claudette saw as productive and beneficial. Together they looked at the sort of work Claudette might do, the education she would need to do that work, and the need for better parenting skills. Claudette wanted to get off welfare, and she wanted to be involved in other *healthy* activities that would take her away from the compulsion to shoplift. In the next few months, she started school to become a nursing assistant and was enrolled in a parenting workshop. After a year there had been no further shoplifting, and Claudette seemed to be on the road to real recovery. Her grades were good, and she confided that most of her time was spent parenting or studying. She and her peer counselor remained friends, and now Claudette was beginning to help another person with a similar problem.

In this case study, the case manager used all the skills he knew would be important. He stayed where Claudette was during the process, even when she relapsed. He collaborated and at times coached her. He worked through the stages with her and accepted her ambivalence. Claudette never had to listen to scolding or denigration, and she never had to use time and energy defending herself to her case manager. Instead the focus remained on where Claudette was at the moment and where she would like to go from there.

The question becomes: If change is a process and the person is working with a case manager, what role does the case manager play in supporting change for that person? Below we look at some specific ideas and skills that help you do that.

Understanding Ambivalence and Resistance

We help people change by accepting their ambivalence and understanding their reluctance to follow through.

Accept Ambivalence

We have talked about ambivalence so let us look at it more closely. It is important for your own realistic expectations and the therapeutic nature of your relationship with clients that you accept ambivalence as normal. Miller and Rollnick (2002) point out that ambivalence is a normal part of change. They write, "(people) often recognize the risks, costs, and harm involved in their behavior. Yet for a variety of reasons they are also quite attached and attracted to their addictive behavior" (p. 14). We see people whose behaviors are destructive, but they are ambivalent about changing those behaviors. Miller and Rollnick write, "Passing through ambivalence is a natural phase in the process of change" (Miller and Rollnick, 2002, p. 14).

Resistance Misunderstood

People often move toward their recovery or change at different rates due to their ambivalence. Most people do not readily change without experiencing some ambivalence, some conflict about it, or some degree of pessimism. These thoughts and feelings are a normal part of changing. Some people stay with their ambivalence longer than others, and some people never can move beyond it.

Case managers can make two mistakes. They could mistakenly characterize this ambivalence and the behaviors ambivalence elicits as resistance. Resistance suggests oppositional behavior, people who are uncooperative and difficult. It is somehow the client's fault that she does not follow through with our directives or even plans we worked out with them. We often see this resistance to our planning as a negative part of her problems, a behavioral problem she uses that obstructs getting better or changing her life for the better. Some of that may be true, but we need to back away from these ideas and consider other possibilities.

The second mistake is thinking it is the case manager who is responsible for the client making changes. The point to remember is that it is up to the client to change. Some people change and others do not. Some people change rapidly and others take years to effect a change. *The client is the one responsible for whether or not change takes place.*

Sometimes things move too quickly for a person. Change may be coming in unexpected and very uncomfortable ways, and the old, safer ways of being and doing are giving way to the unknown. The individual may have excuses, procrastinate, find reasons to avoid you, or just ignore plans you and she made together. It may be that the goals in the action plan you both developed looked good on paper, but now she is finding it enormously difficult to implement the plan the way you both constructed it.

According to Miller and Rollnick (2002), we should see resistance as a "signal to respond differently." (p. 40). In other words, what goals we worked out with the person or the way we are supporting the person's path to change needs a different approach. Often untrained workers argue with the individual, pointing out the work put into the plan for change or the person's initial participation, and why the plan is a good one. That worker is laboring under the false belief that by arguing he can make the other person see things the way he does. But that other person may have a very different perspective, and plans that seemed reasonable when the two of them sat down together in the office no longer seem reasonable when the client attempts to put the plan into practice in real life.

According to Miller and Rollnick (2002), writing about the role of the worker, "One does not directly oppose resistance but rather, rolls or flows with it" (p. 40). They go on to write, "Reluctance and ambivalence are not opposed (by the worker) but are acknowledged to be natural and understandable" (p. 40).

Here are some ideas you can use when the plan is not working out well or the client is having trouble following through with the plan.

- *Coming Alongside*. Rather than arguing start where the client is. As Miller and Rollnick (2002) point out that there is always a "degree of discrepancy between status and goal" or between the way things are now and the way the client would like them to be. They write that the discrepancy exists "between what is happening at present and what one values for the future" (p. 10). Reflect back to people matter-of-factly their feelings about not following the plan, not making the change.

    WORKER: "So in other words your ex-husband called you and was having trouble on his job and you felt you were the best person to help him with that so you went back to him."

    WORKER: "It sounds like you really felt an urge to eat those desserts because all your friends were celebrating and now you feel sort of down on yourself."

    WORKER: "So there was a party for your father's 75th birthday and you felt that one drink or even two would not mess up your sobriety."

- *Discuss the Reluctance You See*. Talk openly about what you perceive to be reluctance, using an I-message and making it clear that you could be wrong or have misunderstood.

    WORKER: "I can see this is really hard for you."

    WORKER: "To me is seems like you really feel torn about this."

    WORKER: " At a party where everyone is drinking it is hard to be different."

- *Allow People to Continue as Before*. Do this collaboratively and respectfully. This is not a cunning technique to trick people into changing their destructive behavior. Simply reflect back to the other person what you feel he or she sees as the reasons not to change. Often this allows the other person to argue for change.

    WORKER: "So for right now you would like to stay with him."

    WORKER: "It sounds like you are feeling that for the time being you want to eat the way you did before, at least for awhile."

- *The Decision Is Theirs*. Point out unequivocally that the ultimate decision for what takes place belongs to the client. Neither you nor anyone else intends to force someone to do something he or she is not ready to do or feels uncomfortable about.

    WORKER: " You are in charge of what you decide to do."

    WORKER: " How ever you want to proceed from here is your decision."

- *Ambivalence Is Normal*. Acknowledge that feeling uncertain, wanting to reexamine things, stop and back up, and all the things people do when they are not sure about changing are normal. Ambivalence comes before real change. The person needs to hear that from you and know that you accept where he or she is at the moment.

WORKER: "It really can be hard to make these changes. Let's talk about how this is going and what you're going through."

WORKER: "It really isn't always easy to make changes like this. Maybe we can slow down and take another look at this together."

Encouragement

We support people in their efforts to change and in their setbacks with the strategic use of encouragement. In the stages of change above we looked at the importance of encouragement, particularly where a person's environment doesn't provide supports for change. Here we look at this important skill more closely.

When Clients Are Discouraged

Part of the case manager's work in the relationship is to provide encouragement to people who may feel very discouraged. Many of the people we see are discouraged as a result of their circumstances, disease, or illness, or because of what has happened to them. Here are some symptoms commonly seen in discouraged people:

- The situation seems to overwhelm them.
- They have low regard for their capabilities or put themselves down.
- They are unwilling or unable to take responsibility.
- They describe their circumstances or other people as totally domineering and overwhelming.
- They have no trouble discussing their problems but are unable to focus on solutions to those problems.
- Their goals seem impossible or unrealistic to them.
- They set impossible standards for themselves.
- They sprinkle their conversation with negative global statements such as "*Everybody* hates me" or "*Nobody* ever calls here" or "*Everything* is rotten" or "Bad things *always* happen to me."

How Case Managers Motivate and Encourage

Case managers often provide the encouragement people need to reach beyond their current situations. As noted above, case managers encourage people to look for and try alternative ways of doing things, but often thinking about change is frightening to people. These are skills you would employ beginning with your first contact with the client and they are part of your strategic assessment skills. The following sections discuss these techniques.

Begin Where the Client Is. The first step in encouraging others is to accept people exactly where they are in the change process. This starts with the very first contact with clients. The worker does not denigrate the person for needing help or coerce the person to be better or work harder. Arnold R. Beisser (1970) talked about Frederick

Perls, the father of Gestalt therapy, and what Beisser called Perls's "paradoxical theory of change." Beisser defines the theory this way: "Change occurs when one becomes what he is, not when he tries to become what he is not" (p. 77). Perls believed that forcing people to be different never accomplished real or lasting change. People first have to come to terms with who and where they are at the moment and find some degree of acceptance in that.

If you start immediately telling people how to do things differently, how their behavior has been the root of the problem, or how erroneous their thinking has been, they will find it important to argue and defend themselves—to explain why they have done those things or thought that way. Valuable time is wasted and rapport is lost as people move into a defensive position rather than a collaborative one with you.

One case manger related that she had begun to work with a person who had battled numerous psychotic episodes. This woman had a severe mental disorder that had required emergency hospitalizations in the past, but once she was stabilized on medication and doing well, she would stop the medication and try her best to go on as if she did not have a disorder to address. This is much like the diabetic who knows he is diabetic but goes on drinking sodas or eating cookies anyway in an attempt to ignore his disease. The case manager said it was not until she and the client worked together, moving carefully through the stages of change, that the woman was able to accept the facts of her illness. She began to move beyond denial, to thinking of small steps that would help her take better care of herself, and eventually she achieved some of her long-range goals. Until the woman accepted her disorder, no change or recovery was possible. As of this writing, this woman is getting married and holds a meaningful job.

As a case manager, you may be the first person to really listen to people entering the system. Use this opportunity to bring acceptance and understanding to the way things are now. Let people be exactly where they are at the moment they first see you, without remonstrance from you. In other words, convey to the person that it is perfectly all right to be where the person is now, given what has gone before. Once the person's footing is secure he or she can see about moving to another, better spot of the person's choosing. Beisser (1970, p. 77) wrote, "The premise is that one must stand in one place in order to have firm footing to move and that it is difficult or impossible to move without that footing." To really provide useful assistance to people and encourage them to grow into their true potential, start exactly where they are, not where you think they should be.

See the Person's Strengths. Case managers who use encouragement as a tool see people as basically capable and wanting to take as much responsibility for their lives as possible. Our clients come with varying degrees of independence and abilities. Seeing these accurately is part of your assessment skills. Some people are independent when we meet them, some can grow into full independence, and others will always need some help. In this wide spectrum of abilities and needs, you will undoubtedly see the strengths of each individual and point those strengths out to the person. Hearing it from you is often all it takes for a person to stop and reconsider. Then together you can work to help the individuals maximize and take pride in those strengths. For example, you might say, "I think that what you told her really explained the situation and sounded like you have this well organized." Or you might say, "I was impressed by your ability to finish the course even when you hated the instructor. That's real perseverance."

Accurately Assess the Person's Obstacles. Sometimes people appear overwhelmed and either unable or unwilling to take responsibility. Try to understand accurately how real these obstacles are. For example, an inability to use public transportation may be a small obstacle that is easily overcome for a depressed college graduate but a very real obstacle for an individual with moderate intellectual disabilities.

A person may not follow through on projects or goals the two of you set together. The reason may have more to do with the projects than with the individual. It may be that the two of you set a goal that is too complicated for now. Perhaps the goal looked safe when the client sat with you in the office, but now it looks terrifying. Explore this with the individual. Don't assume that people who do not follow through are being obstinate or uncooperative.

Appreciate Every Effort. Suppose your client resolves to try something new, and the first efforts are not very successful. To encourage this person to try again, focus on the efforts, not on the results. The attempt to grow or to change is more important than whether or not it worked the first few times. Point out that the attempt gave valuable experience or information and that it creates a basis for improving.

Paula, a case manager for people with intellectual disabilities, wanted to help Bart learn how to get to his new job. Bart had been hired to collect the market carts in the parking lot of a large supermarket and bring them into the store. The job gave Bart something useful to do each day and a small income. However, the store was nearly 6 miles from Bart's group home. Paula initially walked Bart to the bus stop and showed him that he needed to catch the number 12 bus. Bart insisted he could do this, but he ended up on the number 2 bus several times. He was downcast when he called from the store. He had been most of the morning riding buses even taking the number 12 going the opposite direction. When Paula got to the store Bart was down on himself, calling himself "stupid" and talking of giving up the job. Paula, however, told Bart he had done a good job of striking out independently to get to work and she would help him fix the mistake. Her upbeat message was encouraging. She and Bart caught the number 12 together the next morning and Paula got off before the store so that Bart could get the rest of the way on his own. They wrote the number 12 in big letters on a paper, which Bart carried with him each morning until he got the hang of it. Bart had the idea of putting a chalk x on the sidewalk in an out-of-the-way place so he could see it and reassure himself he was at the right bus stop. From then on Bart was able to get to work independently. Paula focused on Bart's efforts not on the mistake, which allowed him to start over and learn his bus route.

Never Lose Sight of Potential. Focus on people's potential. You know all about their deficits or weaknesses. So do they, but they have far less certainty about their potential for growth and change. Look at clients' strengths, their past experiences, and their accomplishments, however small. Use these to guide you in planning the small steps they can take toward positive change.

FIGURE 12.1 Encouragement vs. discouragement

The Encourager	The Discourager
1. Says you can	1. Says you can't
2. Starts where the client is	2. Starts where the case manager is
3. Works at the client's pace	3. Work's at the case manager's pace
4. Helps people do their best	4. Wants people to compete
5. Believes people are capable	5. Believes he or she as case manager is more capable
6. Talks directly to people	6. Talks down to people
7. Is supportive of others	7. Is critical of others
8. Is enthusiastic about other's potential	8. Has unrealistic expectations for other people
9. Let's people develop their own personal standards	9. Sets personal standards for people
10. Sees resources and strengths in people	10. Sees people as mainly weak and incompetent
11. Is interested in the smallest accomplishments	11. Is interested in the smallest mistakes
12. Believes people can change	12. Believes people never change

Source: Based on D. Dinkmoyer and L. E. Losoncy, *The Encouragement Book: Becoming a Positive Person* (Englewood Cliffs, NJ: Prentice-Hall, 1980).

Digital Download Download from CengageBrain.com

Figure 12.1 summarizes the differences between a human service professional who provides encouragement and one who provides **discouragement**.

Recovery Tools

Another way we help people change is by employing the 10 fundamental components of the **Recovery Model** (See Appendix A). Below is an abbreviated list of these components. The full list can be found in Appendix A.

- *Self-direction:* Individuals take charge of how they will recover, including making significant choices and decisions.
- *Individualized and person centered:* Consideration is given to the very unique attributes each person brings to recovery, including culture, preferences, personal experience, and a collection of strengths and needs.
- *Empowerment:* People participate to the fullest extent possible in decisions that will affect them and are given the information they need to participate fully. They speak for themselves and control the elements in their own recovery.
- *Holistic:* Recovery addresses a person's whole life, looking at more than the issues that caused the person to seek help. A person's role in the community, preference for where to live, what work to pursue, and what spiritual needs are not being met, are all considerations.

- *Nonlinear:* Recovery does not happen in predictable steps but involves growth and occasional setbacks.
- *Strengths-based:* Recovery is based on the strengths people bring to the process, including their interests, talents, accomplishments, coping skills, and much more.
- *Peer support:* Mutual support and encouragement are provided by peers who share experiences and information and even advice.
- *Respect:* People are accepted and appreciated, their rights are protected, and discrimination and stigma are eliminated wherever possible.
- *Responsibility:* Individuals are personally responsible for their own self-care and path to recovery, relying on their own courage and energy to do so.
- *Hope:* Recovery involves striving for a better future, and people are encouraged to envision themselves overcoming obstacles and moving forward.

These components are principles we follow to support the client's efforts. Let us examine several of these in more detail.

Self-determination

Self-determination means that to the extent the individual is able, he or she drives the recovery or change process. Self-determination involves the person's right to direct his care and to determine the elements that will go into the service plan. In some locations clients are given the resources to pay for the care they deem important to their recovery. For example, Gladys thought she would recover better if she had acupuncture along with her medications and counseling. Arrangements were made to make the resources available for Gladys's acupuncture, and Gladys attributed her seven sessions of acupuncture with augmenting and enhancing her road to recovery.

Within the limits of what is reasonable and the person's capacity for self-determination, people have the freedom to choose what is best for them and the authority to develop meaningful plans directed toward their own recovery. Their vision of recovery prevails, not the case manager's.

Peer Support

Another recovery component is peer support. Some time ago, when people with problems were viewed as incompetent and hapless, asking one of these people to help another was seen as ridiculous. Today peer support has become a significant part of a person's recovery. Peer counselors greatly diminish the feeling of being alone with one's problems.

Sometimes a person will recover if given more sustained time and support, but the case manager does not have the time to devote to this one person. Sometimes a client will benefit from the help of another who has "been there." Peer support by others who have waged a similar struggle and recovered can have an enormous impact on recovery.

One agency director offered this analogy: "A person is in a hole and can't get out. The public walks by, but they don't know what to do so they call the experts. The experts come and bring a ladder, but it isn't enough. Then the peer support counselor comes along and gets in the hole too. The peer shows the person in the hole how to get out, how to use the ladder effectively because they've done it before themselves. That is the beauty of peer support." Today peer counselors are viewed as having something valuable to contribute to the team, and they are used in a wide variety of settings.

Communication Skills That Facilitate Change

There are some communication techniques, many of which come from *Motivational Interviewing*. These enhance your ability to support people as they make positive changes. Here we will look at a few of these.

Make It Safe to Explore

Using reflective listening from the start is important in setting a climate in which exploration of the problem is safe. You might start an interview the first time you meet the client with questions that show your interest in the person's point of view:

- "Tell me a little bit about how you see this situation."
- "Tell me a little bit about what brought you in today."
- "Let's talk about what you feel is most important here?"
- "Tell me what you feel we should address first?"

You might continue with something like this:

- "So it is really hard right now to see a way out of this?"
- "So what you feel is most important is to contact your ex-wife to see what she knows about this?"

These are simply considerate ways to address people in any interview, but the questions indicate that you value clients and their agenda for themselves. It respectfully allows others to set the priorities. The safety comes from your acceptance without judgment of the other person's goals and concerns.

Steer around Initial Worries

Sometimes clients come in believing that we will prescribe the best course of action. Betts was sure the case manager was going to recommend a dietitian who would put her on a stringent diet. Bill came in convinced that the case manager would tell him he had to stop drinking before he could be helped. Others come in convinced that there are just too many obstacles to surmount in order to solve their problems. Perhaps they feel there isn't enough money, too little time, too many people will be disappointed or feel betrayed, or solving the problem will entail giving up too much. A good case manager might say something like this:

- "Wow, you're really moving ahead of things. Today I just want to know how you see the problem and what you think about it."
- "You're way ahead of me here. I just want to know what you think about your situation."

Articulate Self-determination

When it comes to changing a situation or behavior, state the obvious:

- "In the end, what happens here is entirely up to you."
- "You are free to choose either alternative."
- "You will be the one to decide what happens here."

For many people, this sense of personal autonomy is relieving and takes away one reason for taking an adversarial stance with the case manager. When you have one agenda for the person and he has another or when there are two alternatives and you like one and he is torn between the two you can set up tension and communication barriers. Say, for instance, that Pete wants to stop drinking and he also feels as if doing so is a lost cause because he tried it before and it didn't work. If you are arguing for him to change and he is arguing with you about why this won't work, it makes the relationship far less effective for Pete. You are not helping him change at this point but simply arguing for your point of view. Let Pete decide which alternative to choose and whether or not this is the time to change.

Help People Talk about Change

Arguing with you about why they should not change convinces people that they should not change. For example, if you tell Pete why he needs to change he is probably going to tell you why that is nearly impossible and convince himself that change is nearly impossible. The opposite is also true. When people tell you why they should change, they are very likely convincing themselves that change is possible and can be done. If Linda explains to you why staying in an abusive relationship has numerous disadvantages and dangers to herself and her children she is more likely to consider leaving seriously. If you argue with Linda telling her all the reasons she needs to leave she is very likely to explain why leaving is not possible. What you want to do, therefore, is engage people in what Miller and Rollnick (2002) refer to as **"change talk."** Here you are constructing the interview so that they tell you why change would be beneficial. Using your skills, ask open questions about change. Rather than asking why a person doesn't change, ask instead about how that person sees change occurring:

- "Tell me a little bit about how you would like things to be."
- "What are some of your ideas about how you might bring this about?"
- "What do you think might be a good place to start?"
- "How would things be for you if you did leave."
- "Tell me something about what you have considered doing to change the situation."
- "Let's look ahead about 3 years and tell me where would you think you would like to be?"
- "Say you are coming in here about 3 years from now. Tell me a little bit about how things would be different."

When clients respond to questions like these, be sure to reflect the responses back to them. This way, clients hear the responses twice.

- "So, in other words, you thought you could cut down gradually by smoking one cigarette less each day for a month?"
- "What you really want is to be free of your symptoms and go back to school."
- "So staying there has some real dangers associated with it."
- "In other words, if you left you could go back to school and continue your nursing degree."

With comments like these, you place people in a position in which they are the ones arguing for and articulating change. It is the client who is describing how change would be beneficial.

Discuss Discrepancies

People become ambivalent because discrepancies exist between the way things are and the way they would like things to be. The more a person deals with these discrepancies, the more likely it is that the person will move toward change. The idea here is to allow the person to discuss why it would be an advantage to change.

For example, Trudy did not want to take her medication because she felt it labeled her as mentally ill. On the other hand, she recognized that she not only felt better on her medication, but she was also more productive. Consequently, she went on and off her medication with predictable swings in mood. Here is what a good case manager might ask:

- "Tell me a little bit about the problems in the way you are taking your medication now."

Here the case manager elicits from the client herself what might be impractical or negative about her use of her medication. This is more effective than the case manager lecturing the client about how to take her medications. A good case manager might also say:

- "Tell me about how things would be if you took the medication consistently."

Again, the case manager is allowing the client to talk about the advantages, rather than listing them for her. A case manager with poor insight and skills might be tempted to ask instead:

- "Well, why don't you take the medication the way it was prescribed? It would make you feel better. It would no doubt help you go to work and get things done."
- "Do you really think this is not a problem for your health? You are staying in bed, sleeping all the time, neglecting your family and your hygiene. How do you think this isn't a problem?"

Open questions about what Trudy values or would like to see in her life 4 or 5 years from now also illustrate the difference between the way things are and the way the client would like things to be. For example:

- "Let's put aside how hard it might be to change and tell me how you would like things to be in 5 years."
- "Tell me something about the values you have that make you consider staying on the medication."

Again, reflect the responses to such questions back to the person, so they are heard twice.

It may be easier for people to change when they can see a clear difference between the way things are and the way they want them to be. Ask questions that will help your clients see that discrepancy. Look for ways to allow people to discuss the disadvantages of their present situation and the advantages that could be had if things were different:

- "How would things be different if you decided to do this?"
- "Maybe we could look at how this change would make things better."
- "Could we look at how things might be a bit better if you do this?"

In these questions, the case manager is asking the client to discuss disadvantages.

The case manager might also ask:

- "Tell me about the concerns you have right now if you make the change."
- "Describe some of the main reasons for not making a change like this?"
- "Let's look at not taking the medication. I'm wondering if you were to stay off the medication, how would things be then?"
- "Tell me a little bit about the worries you have about your situation?"

In these questions, the case manager is setting the stage for the client to recognize why it would be advantageous to make changes. It is not the case manager telling the client why changing would be a good thing. It is the client telling the case manager. In addition, it is the client who spells out what the disadvantages are in leaving things as they are.

As always, reflect back what clients tell you so they hear it twice.

Allow People to Express Ambivalence

If people feel safe with you, they will talk more openly. For that reason, you will be more likely to hear people talk about their ambivalent feelings. This is particularly true if you use good communication techniques that value or appreciate the client. When people tell you about their ambivalence, you know you are doing something right. Focus, using reflective listening, on what the person is concerned about.

For example, Jose knows he should diet, but he tells you that doing so would mean a damper on family get-togethers, denying himself the food he loves, and eating food that seems boring to him. On the other hand, he wants to live a long life, see his children grow up, and knows that his present weight has caused his diabetes and high blood pressure. Through reflective listening, the case manager accepts these concerns as making sense within the context of Jose's life. A good case manager might say:

- "So it would be hard to eat food without much taste, particularly at family gatherings where there is lots of good food."

Here the case manager accepts the client's concerns. A poorly skilled case manager might be tempted to say:

- "You don't have too many options if you want to reduce your blood pressure."

In this response, the case manager argues for the going-on-a-diet side while the client is forced into taking the no-diet side of the discussion.

Bring Out Confidence

Consider a lack of confidence part of normal ambivalence. Some people will be more confident that they can pull off a change than others will be. Ask for expressions of optimism and confidence:

- "Tell me about your personal strengths that will help you succeed here."
- "If you decide to change, what do you think will work for you?"
- "Tell me about the confidence you have that you can do this."

These questions allow the other person to express explicitly what he or she has that will contribute to success.

Generally people have made some changes or have some accomplishments in their past. Talk about that:

- "You said you wanted to drop out of school, but in the end you didn't. You graduated. Tell me how you did it."

Ask for details so that people have to talk about their successes and will hear how they overcame obstacles.

Pima did not want to finish high school. She was having some trouble with English, she liked being home with her mother doing housework, and she did not have hopes of going on to college. But Pima stayed in school and graduated. The case manager explored this situation to highlight Pima's strengths.

- "Tell me about deciding to stay in school."
- "You managed to keep going? Tell me about that."
- "Tell me about some of the reasons you were so successful in completing high school?"
- "Tell me a little bit about the strengths you used to see it through?"

Of course, you would not ask all these questions or ask them in rapid-fire order. Everything the person says you would reflect back and explore. "So in other words you took it one day at a time but you felt like you were sort of muddling along." These kinds of questions help people to talk in some detail about previous successes and to tell the case manager about those successes, rather than the other way around. Talking about their own success tells people why they can probably succeed in making an important change in their lives now.

You might also ask people to list for you the positive strengths and characteristics they see in themselves that will help them get through a change in their lives. Again, by using reflective listening between questions, you can encourage people to tell you what strengths they have, rather than your telling them, although occasionally you may have to begin the list of strengths for the person. "Sounds like you see yourself as pretty stubborn," might be one way to feed back what the person has told you.

Facilitate Commitment

Another important approach is to elicit intentions to change from the client. A case manager might ask:

- "Can you describe something you would be willing to try at this point?"
- "Tell me about your intentions to just get started."

Generally, at some point, people will develop a plan they believe will work for them. If you ask permission to share that plan with others, it often strengthens a person's resolve to follow through. For example, are there family members the person wouldn't mind inviting to hear about the decisions that have been made? Are there other workers who might be invited to hear about the change? Always ask permission to do this, make sure it is the client who does the telling, and accept any hesitance to share the plan with others.

Reflect the Opposite Side

Sometimes a person is reluctant to make a change. In this situation, it can be helpful if you carefully take the opposite side. That is, you reflect the reluctance back without sarcasm. This is not a cynical ploy to trick your client. The client is not an adversary but a collaborator. For example, Trudy did not want to take medication, even though she had been too depressed to work. The case manager said:

- "You really feel that taking medication is a sign of weakness and will label you a 'mental patient,' and it seems better to you to be depressed and unable to work at this time rather than take the medicine."

It is important to hear the matter-of-fact tone the worker is using, rather than a sarcastic, facetious tone. The case manager has reflected back the reluctance and even added the other side of the situation, what it will be like without the medication. In this case, the client responded by saying:

- "I suppose I should take the medication if it would really get me back to work."

Here the case manager's empathic reflection of her client's reluctance, along with highlighting the obvious consequences, caused the client to rethink her reluctance in a new light. Now it is the client who argues for positive change. In another example the worker said, "I can see that you find it really exciting to go to the track and make those bets. You have a lot of friends there and when you win you really feel great. It sounds like you want to continue to do that for now even though it is costing you a lot of money." In this case the client responded with, "I don't know. It does seem like I could find something else to do." The client did not sound convinced but he was beginning to look at things from another perspective. The worker went on to say, "tell me a little bit about those things." The client began to talk about his other interests. At no time did the worker resort to arguing with him about not gambling.

Trapping the Client

The client expresses ambivalence. She is not sure she wants to change, to stop smoking or to return to school or to take her medication. The case manager tells her why she should want to change. The client tells the case manager all the reasons why she shouldn't want to change or why she can't change. She talks about how it is too late to change. She admits this behavior has some negative consequences, but she focuses on all the positive rewards she gains from it. In the process, she convinces herself that she does not want to change and that she probably couldn't if she tried. The case manager concludes the situation is hopeless. The client is resistant, in denial. The client has already made up her mind and doesn't want to change, and so the case manager stops really trying.

Another person is an alcoholic, and the case manager uses the aggressive form of confrontation with him. The case manager readily tears the person down in an effort to get him to "take a really good look" at himself. The case manager tells him he could change if he wanted to, and that the problem is he is too lazy, too self-centered, and too much of a "cry-baby." Most of us would lose trust in a situation that produced that kind of confrontation. When the client drifts away, the case manager decides that the client is unmotivated and hopeless and labels the client "resistant."

The problem in the two scenarios just described is that arguing with people about changing and confronting clients with what bad people they are generally won't help them make constructive changes. People stay and work on their situations and problems more often when they feel the environment is safe, they are in control of their own situation, and someone will listen to their vacillation on the way to changing (Miller and Rollnick, 2002).

In the two instances described earlier, well-intentioned case managers used techniques of dubious value and unwittingly further trapped the individuals in their negative situations. People change when case managers use the attitudes and skills you learned in the previous chapters and practice the techniques noted above. Therefore, we will look next at how to enhance these skills and use them strategically.

From Adversarial to Collaborative

Brainstorming

Spend some time with the person who is having trouble considering how to bring about desired changes. Ask the person to list as many ideas as she can think of for making the changes. You add some ideas too. As the ideas are proposed, list them without judgment.

Jenny and her case manager were brainstorming about what to do about Jenny's abusive marital situation. In the course of this brainstorming, Jenny offered several ideas that her case manager felt were impractical and even dangerous. For instance, Jenny suggested that she and her husband continue to live together and seek counseling. The case manager knew from experience that in most such cases, the mere mention of abuse before a counselor, provided the abuser went for counseling, could result in more abuse after the counseling session was over. Nevertheless, the case manager said nothing until all the options were on the list and no one could think of any other. Then, while going over the various options, the case manager asked permission to share her experiences with this particular course of action and Jenny agreed. As a result of the discussion, Jenny decided to go to counseling for herself and ask her husband to join a group for abusive men. She wasn't sure her husband would go to such a group, but she felt this was the best way to start.

In brainstorming, if all the options are laid out, usually a solution appears. Sometimes several solutions appear, or a good one is developed by combining elements of several different ideas.

Offering Information and Advice

In one case management unit, the clients were told what they needed to do. Phillip resented being told by his case manager that he needed to go to the welfare office; he had hoped that a change in his medication might mean he could get a job. While he felt he might need welfare at the moment until he "got on his feet," he resented never being able to talk about what he hoped would come next.

As a case manager, you should offer your own ideas only:

1. After you have fully listened to and explored those of the client
2. After first asking for permission to do so
3. When you believe it can help the person or you need to warn the person about a threat to his or her safety
4. After making it clear it is up the person whether or not he uses the advice or not

It is arrogant to approach people as if we have all the answers for their lives. It is respectful and collaborative to offer ideas after we have heard theirs. Our purpose is to have people work together with us, not be put off by us.

Summarizing

Miller and Rollnick (2002), in their book *Motivational Interviewing*, discuss the importance of a good summary. If you are skilled at reflective listening, you can do summarizing well. Miller and Rollnick suggest three types of summaries that reflect back to people important points that they need to hear again. If your summary hits most of the important points, clients will move again to discussions of how the changes can take place.

Collecting Summaries. In collecting summaries, you simply pull together the important points you have heard thus far. These should be short so they do not interfere with the person's train of thought. You can end these with the question "What else?" to keep the person going on the same topic. Here is an example by a case manager:

- "Let me just go over what you have said so far. You are interested in quitting drugs, particularly marijuana, and you think if you stay away from school you would be less likely to have access to it. You feel pretty sure your friends won't understand and may even make fun of you if you tell them that you want to quit. What else?"

Linking Summaries. In a linking summary, you bring together and link important points, some of which were made earlier in the discussion. These summaries are often used to clarify where the person is feeling ambivalent. The case manager gives both sides expressed so far and allows the client to explore them at the same time.

- "Let's see. You have decided that you want to stay away from school to avoid your friends and access to drugs. You feel this will help you to stop using drugs. On the other hand, you also think it would be a good idea to get your high school diploma, and you are hoping to use it to go to the community college for a degree in math."
- "So you want to stay away from school in order to avoid your friends who sell drugs and you want to get your high school diploma."
- "It seems like a good idea for you to stay away from school to avoid your friends who are selling and using drugs. At the same time you feel you need to get your high school diploma."

Notice that the case manager is using reflective listening and is matter-of-factly acknowledging the fact that there is ambivalence about what to do in this situation. In other words, the client sees the situation in two different ways and the case manager is simply reflecting those two views.

The case manager in the previous examples uses the words "on the other hand" or "and" rather than "but," and there is a reason for this. When you use the word *but*, you tend to negate all that was said before it. You want to acknowledge both sides of the situation as equally important.

Transition Summaries. In transition summaries, you are preparing the client to move forward, whether toward another interview with you or one with the provider where the client will be referred. Your transition summary might go something like this:

- "Let me see if I can summarize where we are before we stop today. You are thinking about dropping out of school in order to avoid drugs and the friends you have who push them. You feel that these friends would not understand your plan to stop using drugs and would not support your plans to stay in school and attend the community college. You feel you should get your high school diploma, and you are willing to explore how to do that with the counselor you will be seeing next week."

Summaries help people clarify, organize, and start again. These summaries are one of the ways the case manager shows acceptance and prepares the way to move forward. Always use a transition summary at the end of your time with a person so that you and the client can move toward the next contact you have together.

Steering around Obstacles

Just because people argue against change does not mean they are dead set against it. Some case managers are tempted to give up at this point. They hear the person giving reasons the change can't happen or won't work, and they decide the person is right.

Maybe the client is right, but before you jump to that conclusion, it is important to shift the discussion to a point where the client can talk about what change would be like, how it might happen, and what things would be like if there were no obstacles and the client got there finally. People often come to us with no hope and little vision. They have lost faith in themselves and lost their hope for the future. The transfusion, if there is to be one, comes from the case manager who uses reflective listening and asks the right questions to help people begin to envision change.

This is another way of steering around obstacles. Here is an example of how such an exchange might sound:

CLIENT: I can't see any way out of this mess.

WORKER: Right now it looks pretty hopeless to you.

CLIENT: Well, yes.

WORKER: Tell me a little bit about what brought you here.

CLIENT: My wife left, I don't know where she is, and I feel like I'll never get my marriage back together again.

WORKER: You'd like to get your marriage on track again?

CLIENT: Yeah, but I'm pretty sure she won't go for that. She was pretty mad when she left. Said she was through and I haven't heard from her since then. I tried to talk to her once but she changed her phone number.

WORKER: Tell me about your marriage.

CLIENT: (talks about a troubled relationship and his wife's complaints that he is rarely home and is not interested in the children).

WORKER: When you talk about getting the marriage back on track, what would your marriage look like then?

CLIENT: (describes a better relationship, including his being more attentive).

WORKER: I know things look bleak now, but let's take a minute and try to think of all the ways you might be able to bring this about. What ideas do you have for starters?

Some case managers might listen during an exchange like this—using good reflective listening and even asking good open questions—and conclude that the client is right and the marriage situation is hopeless. In this example, however, the case manager goes directly to the better vision. (You'd like to get your marriage on track again.) Later she asks for information about what his on-track marriage might look like. This brings the client to a point where he can begin to envision change. Finally, the case manager suggests they look at options for change anyway.

These are just a few of the ways case managers begin to gently steer clients toward a positive vision of the future. When the case manager remains hopeful and encouraging, the client receives a transfusion of these positive attitudes from the case manager and begins to look in a new direction. When the case manager assumes that all is hopeless because that is what the client says, it is the case manager who has taken the transfusion, and it is not a very constructive or helpful transfer. Nor is it the reason the client sought the case manager's help—so that they could both feel equally bad about change. Case managers are responsible for maintaining a positive, but realistic, attitude and helping clients explore solutions to their situations in a hopeful way.

Summary

Nearly every person who comes to us for help is looking for something better, some sort of improvement. The request for help, however uncertain the person might be, indicates some sense that things could possibly be healthier or more appropriate, more useful or more comfortable. Change and recovery are difficult processes and proceed at different rates for different people.

Many people look at their mental or emotional illness, their addiction, their current marital situation, their criminal history, or their self-destructive behaviors and choices and decide to change. Some of them do make changes in the direction of a more stable, healthy life on their own. Some do not. When people reach out for help, it is often case managers who facilitate the move toward the change they are contemplating.

Just because people reach out for our help, however, does not mean that they are entirely prepared to change their situation or their behavior. People come to us who are thinking about possibly changing or who are mandated to come by their employer or the courts. People come to us because they know they should change but really like things the way they are. They are, in a word, ambivalent. There is a conflict for them between what is presently knowable and what they think they should be doing or even what they want to do, if they thought they could. Some people have lived so long with their illness, situation, or addiction that they feel hopeless. Others just want a little coaching to get on another track in life. Still others come to us for help again and again but never make the change.

Coaching, collaborating, and planning are ongoing processes as people grow and see themselves as doing more, accomplishing more. In the work you do with people, the first plan may open up to a second more independent plan that paves the way to an even more accomplished third plan. Some plans move quickly. Other times clients may become stuck and things move slowly or erratically. Expect that clients will move at the different rates and some may never completely resolve their problems.

Seeing people as ultimately responsible for the direction their lives take and viewing ourselves as a resource, a coach, an encourager, and a source of support for change makes our work more effective and gives people more opportunity to recover successfully.

Exercises

These exercises can also be filled out online at CengageBrain.com.

Exercises: Helping People Change

Instructions: Looking at the techniques you learned in this chapter, decide what techniques and skills would be best to use in each of these situations.

1. Kelly has talked and talked about getting a divorce from her husband who abuses her occasionally and has always been remote and uninvolved in their marriage. You feel strongly that she should leave because you see the abuse escalating. (a) What are some interviewing techniques you might use to help Kelly decide what she should do from here? (b) How responsible is Kelly for making a change and how responsible is the case manager?

2. Marlo wants to stop drinking and get her children back. They were removed from her home a year ago after Marlo had her fourth DUI and caused an accident on a city street. She has been in jail and is now in a halfway house while on parole. (a) What are some interviewing techniques you might use to help Marlo decide what to do now? (b) What can Marlo do to help herself? How do you get Marlo to begin to talk about what she can do?

3. Ben has been sober for 2 years and is now working consistently on a job he likes. He divorced his wife during the time he was in rehab because she could not give up her drinking and seemed to "want to pull me down." Now she has returned, intoxicated, and begging him to take her back. They talked into the night about "old times" and both of them got drunk. (a) What are some interviewing techniques you could use to help Ben at this juncture? (b) Who makes the final decision about what to do about Ben's ex-wife? (c) Construct a response you might give Ben that accurately and matter-of-factly reflects his ambivalence.

4. Elmer does not want to be abusive to his wife and wants his marriage to resume. His wife left and is staying in a shelter where she has indicated that she will not come home until Elmer gets some help. Elmer admits to you that abusing his wife was "probably wrong" but seems to resist any ideas about how to go about changing his behavior. (a) What are some interviewing techniques you might use to help Elmer at this point? (b) How responsible is the case manager for Elmer changing his behavior? (c) Should the case manager be actively involved in getting Elmer to change and the wife to come back home? Yes? How active? No?

5. Elise is using alcohol to self-medicate. She suffers from debilitating depression, but since adolescence she has turned to alcohol and street drugs to alleviate the symptoms of this depression rather than take her medication. Her reasons for doing so are that the medication is too expensive, coming in every time she needs a prescription is "a pain," and she can go to the liquor store and buy cheap liquor without the inconvenience. Lately she seems to be thinking that maybe using alcohol is not the best idea, but she is defensive about why she has used alcohol in the past. (a) What are some interviewing techniques you can use at this point to help Elise? (b) Does the case manager have an obligation to explain the negative effects of alcohol on depression or is there a way to get Elise to describe these? How would you get Elise to do that?

6. Christy has a diagnosis of schizophrenia. She believes, when she is ill, that she is related to the president of the United States and she has made her way to Washington on several occasions. Both times Capitol police have arrested her and then had her hospitalized in Washington. Christy has returned home now and is talking to her case manager. She realizes when she is on her medication consistently that she is not related to the president, but she tells her case manager that she gets an overwhelming desire to see the president to warn him about "all the bad people out there." She goes on to give an elaborate description of how she is related to the president through her "earth mother." (a) Should the case manager explain to Christy how she is not related to the president? (b) Are there other interviewing techniques that the case manager could use? (c) Is the case manager responsible for making sure that Christy does not return to Washington? If yes, to what extent?

CHAPTER 13

Case Management Principles: Optional Review

Introduction

In the first half of this book we have discussed the principles of case management. We have considered ethical and legal principles and how these support best practice for case managers. We have considered ways to communicate with people that are helpful and support them as they make important transitions or attempt to maintain the situation they find best for them. In this chapter we pull these together showing how they work in combination to create the kind of environment in which people seeking your help will be able to thrive and succeed.

Combining Skills and Attitudes

Let's begin by looking at the skills and attitudes you learned in the previous chapters. In the first sections of the book, you learned the importance of respect for the person and your responsibility for caring for people who are emotionally vulnerable. The following skills and attitudes were stressed:

- *Acceptance* of clients where they are at the moment they reach out for help is one of the most important and facilitating attitudes. Acceptance of where the person is in the change process as he or she attempts to change or grow.
- *Self-determination* requires an attitude of respect for the person's choices and the ability to honor the person's choices.
- *Informed consent* engages clients in the process of making decisions about their services and treatment with sound information. It involves information, often provided by case managers, that helps people make the decisions that direct their lives.

- *Not insisting on your solutions* is yet another way that the client's choices and views are honored. The case manager is aware of boundaries and allows other people to decide for themselves the best solutions.
- *Individualized planning* requires that we see each person as a separate and unique individual. Planning with people involves seeing their unique strengths and needs accurately.
- **Child and Adolescent Service System Program (CASSP)**, *the recovery and resiliency models, and self-determination concepts,* all of which come from federal and state governments, mandate that we consider the person's strengths, ideas, and choices and approach people with hope and vision.
- *Cultural competence* requires case managers to understand and relate appropriately to cultures other than their own, to learn about those cultures with which case managers interact often and to respect differences in lifestyle and customs.
- *Nonjudgmental attitude* is the attitude that allows us to listen and accept what the other person brings. It is this attitude that allows us to be where the client is without a need to pass judgment as to whether the client is good, stupid, smart, or a bad person.
- *Motivating and encouraging* are important in instilling hope for something better, faith in oneself, and the possibility of meaningful change. Encouraging people requires us to see the other person's strengths, their goals, and support their attempts to reach those goals.
- *Collaboration* is yet another way to show respect for people, involving them in the process of planning. Collaboration involves seeing client's contributions as equal to yours in the planning process. In collaborating we actively work with people to support their vision and ideas.
- *Seeing others as separate people* is another attitude that commits the case manager to keeping the boundaries between who is the client and who is the case manager clear and free of personal emotional intrusions.

Digital Download Download from CengageBrain.com

The skills and attitudes listed above create an environment wherein others feel safe. These principle attitudes demonstrate respect for others, and foster their self-determination. Being mindful of these principles and practicing the skills associated with them helps to foster trust and gives people the structure within which to change.

Next we learned communication skills—the very skills that demonstrate these beneficial attitudes toward other people. Let's review what we know:

- *Reflective listening* demonstrates considerable acceptance for where people are at the moment and shows our interest in what they are thinking. Our respect in turn fosters self-acceptance and enables people to move forward.
- *Open questions* are another way to show we want to go where the other person thinks is most important, that we are interested in their viewpoint and want to hear what that person feels is most relevant.
- *I-messages* are meant to introduce our concerns or thoughts in a nonintrusive manner. If we do this well it is clear that these ideas are owned by us and are simply

an invitation to the other person to explore with us. Our "I-messages" are very tentative because we always consider the possibility that we might be wrong.
- *Disarming anger.* Even here, when the person is angry, we invite criticism, refuse to be defensive, and appreciate the information—all ways to accept and respect the concerns of the person without needing to respond in kind.

Digital Download Download from CengageBrain.com

These communication skills are tools that give us an effective way to demonstrate regard for others. All of us feel better in the presence of someone who respects us, is not judging us, and is hopeful about our situation. We begin to relax, to talk, to explore ideas, to be ourselves. In this type of situation, it becomes safe to think differently about the future. This is exactly why you learned these attitudes and skills: so that the people who seek your help will trust you and your intentions and, therefore, be able to work on changing their situations.

Practice

You are now at the point where you can practice holding entire conversations with clients using all the skills you have learned. At first, this will seem awkward and mechanical to you. You will hesitate as you try to decide whether to use a reflective listening response or an open question. You will feel that you are engaging in a dialogue that is halting and unsatisfactory. With practice, the skills become second nature. Your responses will be smooth and sound unrehearsed and genuine. The following exercises are designed to help you begin to put all the skills together in the same conversation.

The vignettes and exercises below are designed for practice. These can be written exercises, but there are other ways to use these. Working with another person as the client, you can record these and bring them to class for others to critique or for a grade. These exercises can also be used in the classroom for role play, wherein you practice combining what you have learned into one interview.

Exercises

These exercises can also be filled out online at CengageBrain.com.

Exercise I

Instructions: Follow the instructions for each question based on the communication skills you have learned in class.

1. You have been called to an elderly woman's home. The family is upset that she is refusing to leave the home for an evaluation at the hospital even though she can no longer walk, is incontinent, and cannot get to the kitchen to fix meals for herself. She lives alone. When you arrive, you find her daughter and son-in-law

exasperated from their attempts to convince her to go to the hospital for a medical workup, as her doctor has recommended. You enter the room, and after introducing yourself and finding a place to sit down near her, you ask her an open question. What is your open question?

2. The woman tells you why she is not going to the hospital. "I don't want to," she says, "Why should I? I'm an old lady. I've lived a full life. I want to die right here. Harry [her husband] died down at the hospital. He thought he was coming home, and he never did. He didn't want to go down there, and we all made him go. We thought he would get well, come back home. He never did. I'm certainly not going through that." Choose a word that describes the feeling you think the woman is expressing, and then give your empathic response.

   FEELING:

   EMPATHIC RESPONSE:

3. You are quite concerned about this woman being at home alone. As you sit there, you can see how extremely frail and weak she is. After a while, you express a very gentle I-message or concern that *you* feel regarding her being home alone. What is your comment?

4. In response, the woman answers, "You know, I never said I wouldn't go anywhere! I just said I wouldn't go to that godforsaken hospital where Harry died." Write an opening sentence that shows you are beginning to collaborate on a solution that might work for her and still accomplish what is needed. Use the word *we*.

Exercise II

Instructions: Follow the instructions for each question based on the communication skills you have learned in class.

1. You are seeing a 40-year-old man for the first time. He has a diagnosis of schizophrenia. This is a chronic condition that started when he was 19 and went away to college. Today he has come in to see if you can help him with a better place to live. He tells you he just wants to "live somewhere else, that's all." You sense that he is upset. You ask him an open question. What is your open question?

2. In response, the man looks away, and then slowly begins to talk about the kids in the neighborhood. He says they make fun of him and try to stop him when he walks home at night. "I get sort of sick when I see them. The other night I stayed out in the alley most of the night. It just seems like they are out to get me. They throw bottles at me and call me 'psycho.' I don't know. I just stayed out so they wouldn't get me." He looks sad and embarrassed. Choose a word that describes the feeling you think the man is expressing, and then give your empathic response.

   FEELING:

   EMPATHIC RESPONSE:

3. You are quite concerned about the stress of this situation for him, knowing stress can trigger a relapse. You are not clear what else he does during the day or what friends and family he might have for support. You ask him an open question about his activities. What is your open question?

4. The man tells you that he is enrolled in Goodwill during the day and that he has been taking the bus on his own to the center where he works in the retail shop. "When I have to take the bus, I stand on the corner, and they always see me. Or when I get home and get off the bus, they seem like they are waiting for me." Choose a word that describes the feeling you think the man is expressing, and then give your empathic response.

   FEELING:

   EMPATHIC RESPONSE:

5. You still want to determine if this man has any supports other than Goodwill. Ask him an open question about that?

6. In response, the man answers, "My family got funny on me years ago. I have a sister. She lives in Lemoyne, and she always has me come over at Christmas. She is the only one that bothers with me. The family always acts nice at first, but I can tell I get on their nerves. I don't stay real long." Choose a word that

describes the feeling you think the man is expressing, and then give your empathic response.

FEELING:

EMPATHIC RESPONSE:

7. The man continues, "But I have this friend from Goodwill—Paul. He told me I could come to his house if things get bad. He didn't want me to stay in the alley like I did." Respond to the content of this statement.

8. The man goes on to say, "Yeah, Paul is a real good friend. The other day when I ran out of Geodon, he lent me some of his until I got my prescription filled down at the county." You are concerned about his use of another patient's prescription. How do you express your concern using an I-message?

9. The man says, "Well Cindy, down at Paul's apartment building, she works for the county, and she got upset too, like you. She said Paul ran out of his stuff too soon to get a refill, and they had to see the doctor especially to get him more pills." Respond to the content of his comment.

10. Now that you know what supports are in place, you need to know what he thinks about finding another place to live. Ask the man an open question to learn what ideas he has about finding another place, and to begin collaborating on a solution. What would be your open question?

11. The man answers by saying, "Well, I was thinking I could get into the program Paul is in. It's part of Goodwill. They told me I was doing good and didn't need to be in that apartment building, but I know a lot of the people there, and I could, well, feel better if I lived there—or down near there somewhere." He is referring to a personal care boarding home that is sponsored by Goodwill. How would you indicate you want to collaborate with him on the issue of a new place to live?

12. He seems relieved and you tell him you will check into his moving and get back to him in the next few days. Then you summarize what took place in the interview today. What is your transition summary here?

Exercise III

Instructions: Follow the instructions for each question based on the communication skills you have learned in class.

1. Mrs. Sylvestri is worried about her son, Manuel. She tells you that he seems to have trouble concentrating in school and that the teachers have told her he is "a daydreamer." He is falling further and further behind, and she is wondering if her son is reacting to the recent death of his father in a construction accident. She says, "I don't know if you saw it on TV or not. He was working on that sewer project down in Carsonville, and the walls of the trench fell in on him. All his buddies were there, trying to dig him out. It was on the news." You respond with active listening. Choose a word that describes the feeling you think the woman is expressing, and then give your empathic response.

   FEELING:

   EMPATHIC RESPONSE:

2. Mrs. Sylvestri seems grateful for your understanding and goes on to say that Manuel saw the newscast of the accident at a friend's house. "He was over there playing about the time the news came on," she said. "I hadn't gone to get him because I had so much to take care of, the hospital and then the funeral home and calling his brother to decide what we were going to do. And Manny saw it on the news. The neighbors didn't know, not yet." What is an open question about her husband you might ask her?

3. She tells you he was a good father and a good provider, that Manuel is their only son, and that they were unable to have other children. They came together from Puerto Rico 9 years ago and worked hard to buy a small house in Carsonville and serve the community. "Now he is gone, just when Manny is about a teenager, and . . ." Give a reflective listening response. Choose a word that describes the feeling you think the woman is expressing, and then give your empathic response.

   FEELING:

   EMPATHIC RESPONSE:

4. Mrs. Sylvestri asks you if you can help with Manny. She tells you what a "good kid" he is and how much he helps around the house. "He has cousins, all his uncles and aunts, they live in Carsonville or around, so he has family, and me, but he won't pay attention in school. He hurts, you know?" Ask her an open question that will help you to know what she might have in mind for Manuel.

5. Mrs. Sylvestri responds with the fact that the school wants him tested for attention deficit disorder, but she does not think he is "mental or anything." Instead, she believes he is grieving and says, "I don't know what you do for a kid who just lost his father like Manny did. I don't know." Ask permission and then using an "I-message," suggest children's Arbor House that specializes in children and grief.

6. She looks interested. "Are there other kids, that many other kids who have problems like Manny? So sad to think. What do they do there?" You describe the group sessions and the activities they have for the children and the age range of children Arbor House accepts. Then you move toward collaborating with her on a way to engage in these services. Demonstrate how you go about doing that.

7. You know that there are also services for parents who are widowed and feel Mrs. Sylvestri might benefit from these. You ask permission to share what you know with her.

8. Mrs. Sylvestri tells you she would like to get started with the programs at Arbor House. After you make an appointment for her to be seen for the first visit there on Tuesday evening and you give her directions you ask her if she has any other concerns. How do you ask that?

9. When Mrs. Sylvestri says she would like to just start with Arbor House you give a transition summary before you terminate the session. Demonstrate how you do your transition summary.

Exercise IV

1. Kevin comes in asking for help. He had been addicted to pain medication in the past and went through a period of rehabilitation, NA meetings, and abstinence from pain medication. Today he tells you that following extensive dental surgery he was prescribed Valium and Vicodin (hydrocodone). He completed the prescriptions and was able to wrangle a prescription for each of these from his family doctor by telling him that he had been given nothing for pain and the dentist was out of town (which was true). "I can see I am getting back into that addiction thing," he tells you. Choose a word that describes the feeling you think Kevin is expressing, and then give your empathic response.

   FEELING:

   EMPATHIC RESPONSE:

2. Kevin tells you he does not think he will ever be able to stop using drugs and says, "I never should have come in here. It's a waste of your time. I'm a loser, at least in this life, and nothing is ever going to change for me." He looks resigned. You steer him around his initial worries. What do you say to steer around his initial worries?

3. In order to help Kevin think about how his life would be without using medications in this way you ask him to put aside his worries about changing and tell you what some advantages there are to not being addicted. How do you go about eliciting this "change talk?"

4. Kevin tells you how much more productive he was when he was not taking pain medications and how he has been off work for a week now because of the surgery. "I have the time to be off and at first I really missed it, but then as I got more on the drugs I just kind of got back in my old rut of sleeping and looking for the next high. The high is . . . I just like the feeling. Everything is fine then." You reflect back his ambivalence, articulating both sides of his drug use. Demonstrate how you matter-of-factly reflect both sides of the argument.

5. He agrees that there are two sides to his drug use. He likes the way it makes him feel, but it makes him a less responsible person and he regrets that. You,

however, want to focus on the time he has had recently free of drug use. You give a response intended to bring out his confidence. Demonstrate how you do that.

6. Kevin talks about how he threw himself into his work, took two college courses to augment his skills at work, and managed to work up to crew chief. As he talks about what he accomplished he looks somewhat pleased, even surprised. Give a reflective response to content.

7. Kevin smiles and nods. You then ask him to tell you about the strengths he used in order to get to that point. What are his personal strengths and characteristics that helped him achieve this? Demonstrate your open question.

8. It seems to you that Kevin would like to find a way back to abstinence and become productive again. You give a reflective response.

9. Kevin agrees so you continue by asking him if he would be willing to make a small commitment to get started. You want to know what he sees as the first step he might take. Demonstrate your request here.

10. Kevin says he would be willing to get his medications and leave them with the agency and try withdrawing on his own for a few days and then return to work. Once a good plan about how to do this is decided upon you give a transition summary. Demonstrate your summary.

Exercise V

Instructions: Try this exercise as often as you need to in order to become more proficient in the communication skills. Find a partner with whom to practice the skills. And record your conversations.

1. First, you be the worker and let your partner play the part of a person who is seeking help for a personal problem or condition. Play the recording back. Critique your efforts with your partner, looking at ways you might improve your responses.

2. Ask your partner at the end of the dialogue how your responses felt to him. Did he feel reassured, understood, supported? Did your partner feel genuine interest and concern on your part?

3. Next, reverse roles, letting your partner be the worker while you take the role of the client. Again, record the conversation and go over the recording together, looking for ways to improve your partner's skills.

4. Give your partner feedback about how her responses felt to you.

5. Once you have made a good recording of you as the worker responding to the client, hand it in to your instructor or play it for the class. Be prepared to receive additional pointers for improvement. At this stage in your training, every word can be examined. Soon, however, you will begin to know exactly what to say.

CHAPTER 14

Documenting Initial Inquiries

Introduction

In many agencies, phone inquiries are logged on the computer and kept on file there. The "New Referral or Inquiry" form used in this text contains the kind of information the individual would be asked to provide during an initial call to the agency, particularly if the person were asking for an appointment. Learning to gather this information properly is important to ensure that the person's situation is handled well from the beginning.

This first form, for our purposes, is extremely basic. The form is used to take information from people calling *on the phone* to obtain services. Some agencies do more extensive phone intakes. Others ask a few questions on the phone and have the client fill in forms in the waiting room before ever seeing a case manager for the first time. This particular form was designed to give you practice in noting general information about the client and practice in capturing succinctly the presenting problem. You are to ascertain what the problem is, in brief, and set up an intake appointment in which the caller will be seen in person for a more in-depth history and evaluation. Remember to use all the communication skills you have learned in previous chapters to make the person feel at ease while describing the problem. Asking for help is difficult, but it is less so if the phone worker is empathic and accepting.

PLEASE NOTE

You cannot accept a case and fill out this entire form on the word of another person. The other person can refer the client to your agency, but the client must call in order for the intake to be valid. Exceptions to this rule exist when (1) the person is incapable of calling due to a mental health emergency (other forms are used for this situation); (2) the person needing services is a child (a parent or guardian can call for the child); (3) the caller is very infirm, has an intellectual disability, or is a frail older person (a family member or friend can call for that person); or (4) the person needs an interpreter to communicate.

Walk-ins

Some case management units report that many people do not call but come to the office in person. Usually, this same form, or one like it, would be filled out by reception personnel. In some places a very similar form is given to the client to fill out in the waiting room. Generally walk-ins, whenever possible are seen that same day. People who feel a need to come in seeking help usually feel their situation is urgent. Seeing the person for a full intake can prevent an emergency room visit later or a crisis in the person's life that might escalate if there were no intervention. For our purposes we will use a phone intake form, but it is a good idea to be aware that people do not always use the phone and may come in when they feel they need help.

Guidelines for Filling Out Forms

The following guidelines apply to filling out all forms:

1. Use black ink. Never use a pencil to fill out a form.
2. Do not use correction liquid on forms; it is not permissible.
3. Be sure to sign and date any form you complete or note you write.
4. N/A (Not Applicable) is generally used where there is no answer or the question does not apply to the client.

Steps for Filling Out the New Referral or Inquiry Form

A form for phone inquiries is found in the Appendix C. It is titled "New Referral or Inquiry." Use the following step-by-step process when filling out this form. Make a photocopy of the blank form in the Appendix C at the back of this book, and fill it out by following these steps:

Step 1. Place the person's name, sex, date of birth, and address at the top of the form in the spaces provided.

Step 2. Place a home phone number on the form, and the work number of the client if the client is working.

Step 3. The designated client will either:

   a. Be a minor and have a parent or guardian, in which case you circle or underline "parent" on the form and write in the name of the parent, or
   b. Be an adult with a spouse, in which case you underline or circle "spouse" on the form and write in the name of the spouse, or
   c. Be neither of these, in which case you write N/A in big letters on that line.

Step 4. If the person is employed, place the name of the employer on that line. If the person is not employed, place N/A on that line.

Step 5. If the person is in school, place the name of the school (complete with what kind of school—college, elementary school, high school) on that line. If the person is not in school, place N/A on that line.

Step 6. The individual will either be:

   a. A self-referral, meaning the person found out about your agency through the phone book or a friend and called in on his or her own. If that is the case, write "self" on that line. Most calls are self-referrals; or

   b. Referred by a doctor or other professional. In that case, place that person's name on the line. You are asking the caller, "Who referred you to our services?" Answers might be Dr. Graham Smith or Attorney William Burns. When writing a doctor's name be sure to indicate if it is an MD or a PhD.

Step 7. Under the section marked "Chief Complaint," always tell why the person called *today*. Do not say the person called today because her husband is abusing her. The husband may be abusing her, but what made her go to the phone today?

Here are some reasons people might give for calling today:

- Today the person decided she cannot go on.
- This morning, his employer insisted he get help.
- What happened last night was the last straw.
- She saw a medical doctor within the last 48 hours who told her she needs counseling.
- He just had a fight with his spouse and is afraid of what he might do.
- He just hit his child.
- This morning he started to think about going back on drugs again.
- She thinks she will hit her child and has called for help to stop herself.

In filling out the "Chief Complaint" section, capture why the person called on this date and not on some other day. Begin this section with "Client (or the person's name) called today because..."

Capturing the Highlights of the Chief Complaint

You have a small space and can use very few sentences to describe why the person called today and not some other day. In thinking about what the caller has told you about why he or she called, choose the most important points. The details will be noted when you take the social history. Here are some examples:

> John Haulik called today because his employer requested he seek help for drug problem (crack). In last 2 weeks, he has missed or been late to work every day. Sleeping on the job and cited for safety violations. Client sounded distressed and anxious to begin treatment.

> Jane Wilson called today after a serious fight with her husband involving physical abuse. Jane states husband has been verbally abusive in past, but not physically. Client is hospitalized and looking for alternative safe living arrangements upon discharge. Client sounds depressed, but cooperative.

Following are some guidelines to keep in mind for capturing the highlights:

- *Keep the reasons from being too complicated.* Do not make these first cases psychiatric emergency situations that need either immediate attention or a commitment. In other words, in this first exercise, do not create clients who are hearing voices, are contemplating suicide, have made a suicide attempt, or are a danger to others.
- *Be very specific.* Do not use general descriptions such as "her husband beats her" or "she lives with an alcoholic" or "he has been having a hard time at work." Tell when the last beating was, what the most recent problem with the alcoholic husband was, and what the most recent problem was for the client at work.
- *Keep the reason for the call brief.* Do not include a lot of background information, as that will be acquired when you do the social history at the time of the evaluation. Give just enough background information to let the next worker know the context of the client's problem. For example:

    - Angelica called today because she was severely beaten by her husband on Tuesday during an argument over dinner. Client was hospitalized and is seeking alternative shelter. There is a history of domestic violence, which the client feels has worsened in the last 9 months.
    - Horace called today because his employer warned him that without evidence of treatment for problems with alcohol, employment may be terminated or suspended. He admits to drinking while on the job today. There is a pattern of binge drinking followed by missed work and problems with coworkers.

Steps for Completing the New Referral or Inquiry Form

The following steps complete the process described in the previous section, which ended with Step 7.

Step 8. Under "Previous Treatment," keep the notes brief—just note when, where (and with whom if you know that), and for what. Keep from being too wordy in this section. For example:

Incorrect: Susan saw Dr. Piper at the Waldenham Clinic, 432 Muench Street. She started to see him in June of 2013 after her first son was born and continued to see him for 6 months. He was treating her for postpartum depression.
Correct: Seen in June of 2013 for 6 months by Dr. Piper, Waldenham Clinic, for postpartum depression.

Step 9. The intake is "taken by" you. *This is the first place your name is to appear on this form!* Put the date of the intake next to your name.

Step 10. Under "Disposition," note the name of the person to whom you refer the new client for intake and the date of the intake appointment. In many settings, the person who handles the phone inquiries is not the same person who sees the clients when they come in for their first appointments. For training purposes, we will assume that you will be doing both the phone inquiry and the client intake, in which case you would write your own name, along with the date of the intake appointment, on that line.

Step 11. Under "Verification Sent," write "Yes" and the date. The date you use here is the date you send out the verification form, usually the same day on which you take the phone inquiry.

Figure 14.1 shows a new referral or inquiry form that has been filled out correctly. Look at the form to see how the worker filled in each element.

FIGURE 14.1 Sample new referral or inquiry form

Wildwood Case Management Unit
New Referral or Inquiry

Client *Karen Elaine Markley*   Sex *F*   DOB *4/9/82*

Address *1234 Pleasant St.*

Anytown, PA   ZIP *01234*

Home telephone *555-555-5555*   Wk telephone *555-555-5555*

Parent or *Spouse*   *John H. Markley*

Employer *Evansville Township*

School *NA*

Referred by *Walter E. Carmichael, M.D.*

Chief complaint and/or description of problem

Client called today because she suffered an incapacitating anxiety attack at work. A coworker took her to Dr. Carmichael who referred client. She states that she suffered first attack 4 years ago following birth of son. These attacks are increasing in frequency, particularly at work. Client sounded distressed and was tearful at times. She seems eager to begin treatment.

Previous evaluation, services, or treatment *Was seen by Dr. Allen Peters, Winston Clinic, from 9/2009 to 7/2010 for general anxiety and agoraphobia*

Taken by *Marcia Andrews*   Date *9/13/2014*

Disposition *Referred to Marcia Andrews for intake 9/27/2014*

Verification sent *Yes (9/13/2014)*

Digital Download — Download from CengageBrain.com

After a person has inquired about services from your agency, it is important to bring the person in for a more thorough history and evaluation of the problem if the person is seeking services. You will set up an appointment for the caller on the phone at the time of the call or soon after you hang up. The next step is to send a letter verifying or confirming this appointment.

Evaluating the Client's Motivation and Mood

Complete your note with a single sentence that indicates how the caller seemed to you. For example, you can mention how the client sounded. Did the caller seem depressed, glad to have reached you, relieved to be getting help, guilty over what has happened? Did the person seem eager to engage in services, skeptical that you can help, cynical about complying with forced treatment, or cooperative?

Here are some examples of sentences that might summarize how the caller seemed to the phone worker:

- Curt expressed a desire to begin treatment immediately and seemed angry.
- Marci seemed depressed by the circumstances but motivated to follow through with services.
- Pete expressed skepticism that anyone could help him but seemed motivated to seek help.
- Aisha was tearful and seemed depressed during the interview.
- Harold seemed agitated by these recent developments and somewhat unwilling to follow agency procedure.

Steps for Preparing the Verification of Appointment Form

Not all agencies use verification letters. Those that do, do so to confirm for people the appointments that were made with them for an initial intake in the office. In this way, the agency hopes to cut down on the number of missed appointments and the number of hours reserved for people that are not used by them because they fail to show up. (Blank copies of the forms referred to in this section can be found in the Appendix C.)

Today many agencies have more clients than they can see easily. Long waiting lists are the result. An agency cannot afford to waste an hour on a person who does not come in for a scheduled appointment. Although it may give the individual worker a much-needed break and time to catch up on paperwork, it is an hour for which the agency will not always be reimbursed because no services are given. For that reason, some agencies send out a verification letter to remind people of the appointments that are reserved for them.

Following is a step-by-step procedure for filling out the verification form. A blank verification form can be found in the Appendix C at the back of this book. Make a photocopy of that form and follow these steps to fill it out:

1. On the verification form, be sure that the date you send it out is the same date you said you sent it out on your new referral or inquiry form.
2. Be sure to address the person by name.
3. Fill in the date, time, staff, and location of the interview. The date is the date you listed under "Disposition" on the inquiry form. You can decide on a time. The staff person will be you. The interview will take place at the Wildwood Center.
4. Sign your name. Your signature should line up precisely under "Sincerely" and over "Case manager." Do not sign out to the right.

Figure 14.2 contains a sample verification of appointment form with all the information added. Look at the form to see how the worker addressed each element.

The next step in the process is when the person actually comes to the agency for a more thorough evaluation of his or her situation. This is sometimes called an intake appointment. In this chapter, we have practiced phone intakes; in Chapter 15, we will turn to the first appointment and examine how to prepare to meet the individual for the first time.

FIGURE 14.2 Sample verification of appointment form

Wildwood Case Management Unit

Verification of Appointment

Date 9/13/2014

Dear *Mrs. Markley:*

This letter is to inform or remind you that you have an appointment scheduled:

Date: _____9/27/2014_____

Time: _____2:30 p.m._____

Staff: _____Marcia Andrews_____

Location: _____Wildwood Center_____

Please contact me if you have any questions or if you need to reschedule.

Sincerely,

Marcia Andrews

Case manager

Summary

Taking an intake from a person on the phone requires two important skills. The first is skillful communication, something you have been working on previously. Using the skills you have acquired, you will be able to draw the caller out and learn the reasons for the call. In addition, you will need good observation skills. Your task, during this first phone call, is to assess the needs of the person calling and assess the degree of distress the person is experiencing.

On your "New Referral or Inquiry" form, you will need to be able to document not only what the person shared with you on the phone but also the way that person sounded, how motivated the person seemed, whether the person's conversation with you seemed reasonable, and how distressed the individual seemed to be. As you practice your communication skills, begin also to practice listening to the tone of voice and the underlying emotions the client may not express directly.

Exercises

These exercises can also be filled out online at CengageBrain.com.

Exercises I: Intake of a Middle-Aged Adult

Instructions: Using the blank form in the Appendix C titled "New Referral or Inquiry," develop a client intake. You may use any problem that would ordinarily come to the attention of a social service agency. Your client should be an adult, at this point, calling on his or her own to seek services.

In developing your client and your client's problem, read the American Psychiatric Association's *Diagnostic and Statistical Manual of Mental Disorders* (APA, 2013) and look at books and articles on specific problems (such as domestic violence, alcoholism, divorce, depression, addiction). Look at the chapters in the companion textbook, *Fundamentals for Practice with High Risk Populations* (Summers, 2002), for information on how your client might be feeling and what issues or problems the client might be facing when he or she makes the first call to your agency.

If time permits, do several adult intakes with each client having a different reason for calling.

Practice sending a verification letter to your hypothetical client.

Exercises II: Intake of a Child

Instructions: Using the blank form in the Appendix C titled "New Referral or Inquiry," develop a client intake for a child, a person under 16 years of age. In this case, a parent or guardian would be calling on behalf of the child. A doctor, a school counselor, or a teacher may have referred the parents to you, or the parents may have felt they needed help and sought your services without a referral. Typical issues confronting children are problems with school, behavioral problems, and adjustment problems to events such as divorce or the death of a parent.

In developing your client and your client's problem, read the *DSM 5* and look at books and articles on specific problems common to children. Look at chapters (particularly those on children's mental health and on developmental disabilities) in the companion textbook, *Fundamentals for Practice with High Risk Populations* (Summers, 2002), for information on how your client and your client's parents might be feeling and what issues or problems the client's parents might be facing when they make the first call to your agency.

Practice sending a verification letter to your hypothetical client.

Exercises III: Intake of an Infirm, Older Person

Instructions: Using the blank form in the Appendix C titled "New Referral or Inquiry," develop a client intake for an older person, a person over 80 years of age. In this case, a child or close friend or neighbor would be calling on behalf of the client. A doctor may have referred the caller to you, or the caller may have felt the client needed help and sought your services without a referral. Typical issues confronting frail, older adults are problems with self-care, independent living problems, untreated medical conditions, malnutrition, and depression and anxiety.

In developing your client and your client's problem, read the *DSM 5* and look at books and articles on specific problems common to older people. Look at the chapter in the companion textbook, *Fundamentals for Practice with High Risk Populations* (Summers, 2002), for information on how your client and the concerned caller might be feeling and what issues or problems the caller might be facing when making the first call to your agency.

Practice sending a verification letter to your hypothetical client.

CHAPTER 15

The First Interview*

Introduction

You have spoken to the person by phone, and you have arranged for the person to come into the office for an interview. This is the **first interview** for the individual. Even if this person was at one time a client of the agency, we will assume the case has been closed for some time.

The general purpose of this interview is to begin the assessment process. In doing that we establish the following basic information about the person:

- Strengths, including external support systems, talents, successes, capabilities, and positive attitudes and events the person defines as a success
- Weaknesses, including gaps in the external support system, lack of experience or information, negative attitudes, and events the person defines as failures
- Current problems that caused this individual to seek help *now*
- Potential problems
- A sense of who this person is

---

*Adapted from *Where to Start and What to Ask*, by Susan Lukas. W. W. Norton & Company, Inc.

Your Role

You have three tasks to accomplish in this first interview. First, listen and convey an accurate understanding of clients' perceptions about themselves and their problems. When you convey this understanding, it does not mean that you necessarily agree with them, but it does mean that you have heard them accurately. To do this well, you need to allow people to proceed in their own words. As they talk to you about what led to their seeking help, you can reflect back their feelings and perceptions about their situation, responding to feelings and to content. In this way, you sort out with the client what is important.

Second, you formulate a professional understanding of what it is the individual is experiencing and what this person will need while being served by your agency.

Finally, strive to establish rapport with clients so that they feel comfortable with you and with your agency. Some people, no matter how hard you try, will never warm to the interviewer, but most people respond positively to a worker who is warm, genuine, and empathetic.

What you are doing is formulating a good assessment. You have sought an accurate and professional understanding of what the person is experiencing, you have begun to note what this person will need, and you have established the rapport that will allow you to move to collaboration with the person around a plan.

The Client's Understanding

In most cases, people have recognized the need for help; but in a few cases, they may feel they do not need to be in your agency. The courts mandate that some individuals seek help or face jail or the permanent removal of their children. In situations in which clients feel forced to come to your agency, you may encounter hostility. In either case, you must indicate that you have heard all of their concerns about being there. You can convey this through your ability to reflect back how clients are feeling about being in the agency.

Even when people believe they need help, they may not be clear about what their problems are or how the agency can help them. They may be clear that the current situation is painful, but unclear about how to describe it. They may know things seem out of control, but be unable to describe the impact their situation is having on them emotionally. They may hope that there is help available without understanding what kinds of resources there are or how these resources could help them specifically.

Preparing for the First Interview

If you did not perform the telephone intake, you will want to look at the intake material that is available before the person arrives. As you do this, ask yourself what more you should know about the person's difficulties. What details need to be clarified? Where are there particular gaps in the information that need to be filled in?

If the person was in your agency before, read past records to fill out the picture of this person. Look at past medical difficulties, medications the person might be on, or medications that were prescribed previously while the person was in treatment.

As you begin to form a picture in your mind of this person, remember that the information you have was collected by others who saw this person under other circumstances. This may not be the whole picture, and it may not be an entirely accurate picture. Rely heavily, therefore, on your own insight and your own competence to form an accurate assessment of the client at present.

Let's look at the specific case of a woman seen by various case managers, physicians, therapists, and psychologists, who conducted some tests. The woman had a problem with anger and had been asked to leave the home of several family members where she had been staying. At last she was residing in a group home. A psychologist was asked to do an evaluation for possible neurological problems (problems in the brain or nervous system that would cause anger and loss of control). In reading the chart with the many records in it, the psychologist came across an early note by the case manager: "Client created a scene at the local Giant food store last week over the fact that another customer took the last head of lettuce as she was reaching for it. Crisis Intervention was called. Client was taken to her home." About a year later, in another note, a therapist noted, "Client made several scenes in the past at the Giant where she shops. Apparently she gets upset over the fact that there is not more produce in the store. Management has called Crisis Intervention." Still later a new case manager wrote, "Client apparently creates violent scenes at the food stores when she feels there is not enough produce. Crisis was called on several occasions. Will advise client to stay away from food stores." Finally, several years later, a doctor prescribing medication noted, "Client was barred from shopping at any local food stores several years ago because of violent outbursts of rage over a lack of store items she meant to purchase. These outbursts resulted in contacts with Crisis Intervention and indicate a difficulty with anger control and poor communication skills."

Between the individual notes in the chart, there was no other reference to outbursts at the food store. It appeared, from looking at the chart carefully, that each person who mentioned the incident was summarizing the previous note and magnifying it in the process. Think how differently you would approach this client if the first note you read was the doctor's note as opposed to the original case manager's note.

In this example, you do not know if the client changed her story each time she met someone new or if the workers read the charts and records and misinterpreted the information. There are other reasons the notes you read may not be accurate. The person who wrote a note might have been hurried in her assessment. She might have felt hostility or prejudice toward the client for some reason. The note may have been made by someone who was inexperienced in interviewing. For all these reasons, you will have to rely on your own insight and competence in doing your assessment.

If you see inconsistencies in the previous history, make a note of them for further exploration.

Your Office

Most case managers have an office or place where they see the people to whom they are giving service. Sometimes case managers share an interview room. Look at your office or interview room. Be sure it is a place in which you would feel comfortable while confiding in another person. Is it warm and comfortable or utilitarian? Are there comfortable chairs? Is it free of harsh lighting? Are the walls attractive?

It is probably best not to have personal pictures sitting about because you cannot be sure how your clients will view these or what meaning they may find in them. A picture of a happy, smiling 3-year-old may be upsetting to a person whose children were just removed from the home or to someone who is struggling to find shelter for her own 3-year-old. It can create a barrier, and you are seeking to diminish barriers.

In your office, there needs to be a comfortable place for the person to sit facing you. You want her to be able to talk to you in a normal tone of voice, but you do not want her to feel crowded by your presence. In one agency the psychologist had a very large desk which he placed between himself and the client. His chair was large and very upholstered while the client chair on the other side of his expansive desk was lower and less comfortable. This professional was sending an inappropriate message to others about his importance. There should not be large pieces of furniture between you and the other person and the chairs in the room should in no way denote a disparity in status.

In addition to the type of furniture you have in your office you should also consider how it is placed. Most people coming to see you are not violent. However, as a general rule, you should always be positioned closest to the door. Your chair should be where you could exit the room easily in an emergency.

Meeting the Client

From the very beginning, you want people to know that you respect them, that you wish to be helpful, and that you will be relating to them as a professional, and not as a social acquaintance. The interview begins with your first introduction.

1. Begin by going to the waiting room to meet your clients. Do not make them find their way through the halls to your office.
2. Greet adults, particularly older adults by their title. Hello Mrs. James. Good morning Mr. Thomas. It is rude to call an older person you are meeting for the first time just Mary or Jorge.
3. Introduce yourself as Ms., Mrs., or Mr. _____. Or say, "Hello, my name is Jim Pelham." Do not say, "Hi. I'm Jim" or "Hello, my name is Jim." You did not spend all those years in school to earn this degree in order to be simply "Jim."
4. Make a mental note of your first impressions. How do the individuals respond to your greeting? What do they say first? How do they look?
5. How do people react to your office? Do they seem comfortable? Do they appear to feel awkward? Do they readily sit down or wait to be invited to do so?

6. If people start talking, show interest in what they are saying. Often the first things people tell you will hold the most significance.
7. If people ask about your credentials, tell them about these matter-of-factly. It is part of informed consent for people to know who is seeing them. There is no need for you to sound defensive. Do not go into personal details about yourself, however. If a person insists on knowing if you have ever had children, if you are married, or if you are old enough to know how to help her, point out respectfully that the purpose of her visit is to understand the issues and problems she is experiencing.
8. Describe the agency and explain its purpose to clients who are unclear about it. Some individuals come to a case management unit and expect to see a doctor or a psychologist. Give information about the types of professionals that staff your agency and what they do.
9. Make certain that you or someone else has described payment arrangements to the client.
10. Make sure that you or someone else has explained confidentiality and the limitations of confidentiality to the individual. Be sure the client is given information on the Health Insurance Portability and Accountability Act (HIPAA). People need to know that their diagnosis may go to their insurance company. It is not necessary to go into every exception regarding confidentiality verbally, but let them know that under circumstances where they might be in an emergency, information may be shared.

Taking Notes

Sometimes, note-taking during an interview can be viewed by others as intrusive and negative. Some people feel uncomfortable having what they have said documented, fearing this might come back later to cause them a problem. It is all right to take notes during your interview, but you need to explain what you will be dong and why. For one thing, you are collecting very basic information, and you want to ensure that it is accurate. Explain to clients that notes help you make certain that you have accurate information. Taking notes also helps to ensure that important information is not lost.

During other contacts with the client, jot down significant phrases or information. You can reconstruct your contact in short notes after the client has gone.

Collecting the Information

Allow people to tell their story in their own way, using their own words and expressions. Help them to begin talking about why they are here by asking an open question such as "Tell me a little bit about what brought you here today" or "Can you tell me something about what brought you here?" While the person is talking, develop an empathic understanding of their situation and their perspective. Do not recoil in horror or gasp or squeal with delight. Do not tell the individual how you would have felt under the circumstances. This is not about you. While the person answers your open questions, reflect back the content and feelings you hear.

If clients come from a different race, religion, or culture or have different values from yours, be aware of that without judging them. As we have seen, some case managers are tempted to judge people using themselves as the standard. In other words, these case managers see themselves as the standard against which everything should be compared. By doing so, they miss the unique circumstances and characteristics that make the individual a separate person. This diminishes the case manager's ability to be truly helpful.

Asking for More Clarification

During the interview, ask for clarification. Use open questions primarily ("Can you tell me a little bit more about your father?" or "Could you describe your relationship with her before you were divorced?"). It is all right to ask an occasional closed question if you still need further clarification ("Bill is your boss?" or "You lived there how many years?"). Avoid closed questions for the most part so that your interview does not take on the tone of a grilling.

Avoid "why" questions as much as possible. Even when you have taken special care to ask them respectfully, a person may experience them as prying. Sometimes a "why" question actually asks clients to give an understanding of what motivated their actions or the actions of others. You may be asking for a level of insight people have not yet developed. In that case, they can only feel incompetent and uncomfortable.

Sometimes individuals know why they behaved in a certain way or why something happened, but the reason is upsetting to them or they are having trouble recognizing and talking about it. A "why" question can make them think that you will probe for answers and push them to talk about things they are not ready to discuss.

Finally, people may tell you a great deal more than they had intended to tell the first time. They may go home upset with themselves and embarrassed by the amount of self-revelation in which they engaged. It may be so uncomfortable to them that they do not return to continue with your services. In that case, it is possible that you have intruded into the client's personal information too quickly. As the person conducting the interview and, therefore, the person with the most power, you have an obligation to protect the client from this kind of intrusion. A good way to protect people is not to go too far beyond what they appear comfortable talking about.

Intrusions and discomfort of this sort can be avoided if you recognize from the very beginning that *the facts and circumstances of the client's life and problem belong exclusively to the client.* Those receiving services are under no obligation to tell you more than they feel comfortable revealing. This means that you focus on the information individuals can give you freely without probing and without discussing feelings and motivations.

What Information to Collect

The most important piece of information is to understand why the person is here now as opposed to last week or last month. Some of this information may be on the phone inquiry form, but your task in this first interview is to develop that information more completely. The reason the person has come to the agency now is often referred to as the "presenting problem."

In addition to the presenting problem, you want to understand the extent to which this problem has interfered with the person's ability to function socially, occupationally, and personally. Is this person able to work? Are the client's most important relationships feeling any strain? Is the person taking care of personal hygiene and other needs? You might ask questions like the following:

- "Can you tell me something about how things are going at work?"
- "Could you describe how things are at home?"
- "Can you give me some idea of how this has affected your daily routine?"

Individuals who are alone, or who perceive they are isolated, are at greater risk for stress and suicide than those who have a good support system in place. Does the person have a support system or seem to be all alone? To find out, you might ask such questions as these:

- "Tell me a little bit about your family."
- "Tell me something about your friends."
- "Can you describe what you do in your spare time?"

Client Expectations

No service or treatment plan is entirely useful unless the individual has participated in developing the plan. During this first interview, ask people what it is they would like from your agency. You might ask them questions such as these:

- "Can you describe how you think we might be able to help you?"
- "Tell me a little bit about the services you had in mind."
- "Can you share with me some of your ideas about services you would like from us?"

Often people do not know what services are available or what services they need. Together explore what your agency has to offer, and describe various alternatives for clients to consider. By the end of this first interview, you and the individual need to have developed a tentative plan for services.

Social Histories and Forms

In Chapter 16, we look in more detail at how information is collected. Most agencies have a standard format that gives you the foundation for creating a social history. Many agencies have a form that covers the essential information you need to develop a useful treatment or service plan and to give ongoing support and service to a client.

Obligations and Responsibilities

When two people get together to work on a problem each brings important resources to the solution. Help your client understand that there are expectations that she will contribute to the success of the relationship. For instance, she will need to keep appointments or call ahead to cancel. She will need to bring in things like insurance cards or other papers necessary for processing her case. Her ideas are important and she will be asked to contribute these to the overall process. If plans the two of you devised are not working she will be asked to tell you that and work with you to develop a new plan.

First Interview Tasks

Here are some tasks to complete during the interview:

1. *Do They Have Any Questions?* Ask people if they have any questions, and answer these questions thoughtfully. This is part of giving people information they need to give informed consent.
2. *Do You and the Client Have a Mutual Understanding of the Problem?* Work with people to define their problem in language they can understand. This is very important because it gives you and your clients a mutually clear definition of the presenting problem. The most important thing is to learn the client's understanding of his problem. It is not how you see it, at this point, but rather how does the client see and understand his situation.
3. *What Are the Client's Expectations?* Talk to clients about what to expect as a result of coming to the agency. Ascertain what their goals are for themselves, the sort of service they are looking for, and the expected outcome. You might say, "Tell me something about where you would like to be a month (or 4 months, or whatever) from now."

Digital Download Download from CengageBrain.com

Wrapping Up

At the end of the interview make sure people know what will happen next.

1. Give clients some information about what will happen next. If the case is to be presented to a panel or treatment team, tell people that, and tell them how long it will be before they will have information about a formulated plan for them. If they must go on a waiting list, tell them that, and give them information about where they can get services more quickly. Let them know what will happen after this first interview with you. Never let people leave wondering what will happen next.
2. If people are to return to you in a set amount of time to discuss the implementation of the plan that has been developed with them (or for some other reason), be sure to give them a card stating the time and the date of the next appointment. If someone must bring them to their next appointment (such as a parent, guardian, group home worker), be sure that person is aware of the time and date as well.
3. Finish with a good **transition summary**. Your summary should include

   - A definition of how you both define the client's problems.
   - A summary of what you and the person have concluded he or she needs.
   - Any responsibilities the client might have to accomplish before returning.
   - What will be the next steps in the process.

The Client Leaves

Rise at the end of the interview to indicate that the session is complete. It is always a good idea to walk the person back out to the waiting room.

Be aware of something social service workers refer to as the *door-knob syndrome*, wherein a person begins to tell you something of great significance just as he is leaving. He may have saved this information for last deliberately because it is painful and he did not want to discuss it in depth. In any case, let clients who bring up significant issues at the very end know the session will not continue and that they should bring the subject up with the therapist or with you the next time they see you. Do this in a warm and interested manner. Do not appear to scold.

Do not allow people to leave your office if you believe they are a danger to themselves or to others. If what they choose to bring up at the end indicates to you that this is a possibility, you will need to explore that further or see that someone else is available to do so.

After the individual has left, do not go to the receptionist to place the client's name in the appointment book and discuss the client with the receptionist; and do not use the person's name where other clients in the waiting room can hear. Do not discuss your session in the hall or in another case manager's office where other people can overhear your comments.

Summary

You now have a considerable amount of information about the person you have just interviewed. Your initial phone contact is documented on the new referral or inquiry form. The initial assessment will be documented according to a format used by your agency.

In Chapter 16, we turn to how these social histories and forms are handled by agencies. We look at two general ways agencies arrange initial information about their clients. One is the social history and the other is the assessment form. Our next step is to arrange the information we have assembled in at least one of these formats.

Video Examples

To view the videos that accompany this book, go to CengageBrain.com.

- By watching "The First Interview," you can see Keyanna conducting the first interview with Michelle.

CHAPTER 16

Social Histories and Assessment Forms

Introduction

Every agency has a different way of taking and recording relevant client information during an initial intake process. Some use **assessment forms**, which typically are specific to a particular high-risk population. Forms such as these can be found at the end of each chapter on a high-risk population in *Fundamentals for Practice with High Risk Populations* (Summers, 2002). In that textbook, you will find all the information you need to begin to develop a client from a population of interest to you. After using those forms, you will be able to move easily to other similar forms for specific populations in agencies where you choose to work. In this book, a generic form is provided in the Appendix C to help you become accustomed to assessment forms.

It is important to note that more and more agencies are moving to electronic records. Hard copies of forms kept in charts are becoming a thing of the past as records are maintained online in secure databases. Here we will look at working with hard copies as you begin taking and organizing client information. Assessment forms for six specific populations can be found on CengageBrain where you can also practice keeping records and taking information electronically. The populations are: children and their families, women's issues, substance abuse, mental health, intellectual disabilities, and aging.

In some situations, you will be asked to take a **social history**, either as a supplement to assessment forms used by your agency or in place of these forms. Just as the assessment forms in different agencies may differ, the format for a social history will vary from agency to agency. Nevertheless, having written social histories in the classroom, you will be able to more easily adapt to whatever format is used in your agency.

What Is a Social History?

A social history provides the following information:

1. A description and history of the **presenting problem** (the problem that brought the individual into the agency)
2. **Background** information about the person's life, including background related to the presenting problem
3. The worker's **impressions and recommendations**

Taken together, these three sections of the social history give a picture in summary form of where people were when they came to the agency seeking assistance. In this way, the social history functions as the baseline or foundation for decisions about services and for measuring clients' progress. Using the social history as a tool we devise with the client a plan to help the person address the issues that brought her into the agency.

This document, containing all the relevant information about the client, forms the basis for your initial assessment. That assessment is written at the end of the social history in the final section titled impressions and recommendations. Here you give your impressions of the issues facing the client and make your recommendations for services. This section, like the rest of the social history, is a collaborative effort between you and the client.

Layout of the Social History

Social histories always use subheadings set out to the left of the text. This is done to help people find relevant information quickly without reading an entire text to find buried pieces of information. In most agencies, the outline of subheadings is the same for all clients so that workers are familiar with the outline and know exactly where to look for information. Figure 16.1 is a typical outline for a social history.

FIGURE 16.1 Typical outline for a social history

Description and History of the Problem

Presenting Problem (includes the reason for referral, what the client is requesting, and how the client sees the problem)

Background Information about the Person's Life

Family of Origin	Medical History
Birth and Childhood	Behavioral Health
Marriages and Significant Relationships	Legal History
Current Living Arrangements	Social and Recreational Activities
Education	Religious Activities
Military Service	Successes, Strengths, and Resources
Employment History	

Impressions and Recommendations

It must be stressed that the outline shown in Figure 16.1 is one of any number of formats. For instance, if you worked in an agency that dealt with criminal offenders, the legal history section might be called "Criminal Justice Background," and that section might be closer to the top of the outline. If you worked in an agency that served the needs of victims of domestic violence, the agency might place the marriages and significant relationships section right after the presenting problem to augment the information on the presenting problem. If you were working with children, you might include a school adjustment section and drop the education section. In an organization devoted to helping people with their addictions, you would probably have several sections related to the course of the addiction and attempts to overcome the addiction in the past. Both the criminal justice history and medical history sections would become more prominent in that outline.

In the chart or record, a social history may appear on different-colored paper so it is easily identifiable in the folder. The history will have identifying information on each page so it is not separated from the record in which it belongs.

In the text that follows, we look at each section of the typical social history more closely. Examples are provided to demonstrate how to write these summaries.

How to Ask What You Need to Know

If you follow the outline in Figure 16.1, you will have assembled a considerable amount of personal information. Clients who are giving a social history probably are not very familiar with you or the agency, and talking openly about each of these aspects of their lives may be difficult. For that reason, begin by explaining that you will be taking a social history so that you and the client can determine what the issues are and the best plan for addressing these. During the interview use plenty of open questions.

Certainly you will not use all open questions. Asking, for example, how many brothers and sisters a person has is a closed, but useful question. You might, however, follow that with an open question, "Tell me a little bit about them." In another example, you might ask the closed question, "Were you ever in the military?" followed by, "What were the dates of your military service?" Then you could follow up with an open question such as, "Can you describe your military service?"

Open questions soften the interview, making it less prying so that people can choose the significant details to reveal without feeling grilled. Asked with respect and a genuine interest in the individuals, the questions during these initial interviews can be helpful to people in sorting out the factors in their lives that are relevant and significant. In some cases, this might be for the first time.

In each section that follows, open questions are given that you might use to solicit information in that section. These are only examples; you should become proficient in asking open questions on your own.

1. Description and History of the Presenting Problem

The first section is the description and history of the presenting problem. In a brief summary, the background to the presenting problem is documented. This is done in no more than two or three paragraphs. Usually only one paragraph of summary is needed.

Agencies may differ in what they call this section: background information, presenting problem, or history of presenting problem. Some break this section into two parts: presenting problem and background to presenting problem. For our purposes, we call the first section "Presenting Problem" and will include both the problem and the background to that problem in this section.

Writing the First Sentence. The first sentence of the presenting problem is the first sentence as well of your impressions and recommendations. Therefore, you want to put as much information into that first sentence as possible so that people get a mental picture of the person quickly. Here are some examples:

- Alex is a 42-year-old unemployed male, divorced father of two teenage girls and currently referred for issues regarding use of alcohol and job-related difficulties.
- Mattie is a 69-year-old widow and British citizen visiting her oldest daughter in the United States who was brought in by her daughter because of acute memory loss and confusion.
- Milton is a 36-year-old man, recently returned to college for an advanced degree in chemistry, who is suffering from depression he states began when his wife left with the couple's 3-year-old son and the family dog.
- Camille is a 16-year-old high school sophomore who weighs 47 pounds, referred by her family physician and brought in by her parents concerned about her eating habits.

Following are two examples of a presenting problem section in a social history. The first is written about Kate, a 47-year-old woman who contacted the agency requesting help for a long-standing depression. The second example is written about Carlos, a 38-year-old Mexican man who recently entered the country and is having problems adjusting to the new culture.

Presenting Problem

Kate Kate is a 47-year-old married woman with one daughter away in college who called requesting help for a depression she states has lasted almost 2 years starting with the death of her mother. Kate describes the depression as beginning after a serious episode with the flu and the death of her mother approximately 2 years ago. At the time her mother was ill, Kate did not follow doctor's recommendations that she take off work and stay home. She was very involved in caring for her mother, who subsequently died. Kate states she was not aware of being depressed until after the funeral, but grew depressed during the 7-month period in which she and her husband cleaned out her mother's house and settled her mother's affairs. She describes this work as "heart wrenching" and involving several legal difficulties. Currently Kate describes her depression as characterized by hypersomnia, an inability to go to work several days a month, and a loss of interest in social activities and friends. She states she is here in part because her husband insisted she get help.

Carlos Carlos, a 38-year-old unmarried male Mexican citizen, contacted the agency at the suggestion of his boss, Ronaldo Rodriquez, who felt Carlos was having emotional difficulties adjusting to living in the United States. Carlos states he came to the United States from Mexico because his only family is living here. His brother and parents came to this country in 1982. His brother received a good education, went on to medical school, and currently practices medicine in Maryland. Carlos was left behind with an aunt when the family emigrated "because I was hard to handle." Last year the aunt died and Carlos came to the United States to join his family, "who are all I have." Carlos has been here for 8 months and is not sure if he wants to stay or return to Mexico. He describes feeling "out of place" in American culture, particularly compared to his brother and his brother's lifestyle. He states he has no commitment to Mexico but believes he would be more comfortable in familiar surroundings. In addition, he states his parents are "putting pressure on me" to remain in this country and get a better job. He describes them as critical of the few friends he has made and his lax attendance at church.

In each of these examples, we see the information the worker assembled as being most relevant to the immediate difficulty. We have a picture now of why each of these people called the agency and what might have precipitated the request for help.

Questions you might use include:

- "Tell me a little bit about what brought you here today."
- "Tell me a little bit about what happened."
- "Can you describe this problem a little for me?"
- "Could you give me some idea of what's been going on lately?"

Client's Appraisal

Always ask the clients what it is they are seeking. A person new to the system may not know exactly what services are available or have only a partial understanding of what a service actually is or can accomplish. When you ask for the individual's expectations, you may have to describe and explain what is available and how the service works, not only at your agency but also in other places in the community if these are relevant.

Ask people for their assessment of their problem. Valuable information may be found in listening to how people view what is going on in their lives.

Questions you might use include:

- "Tell me a little bit about what you see as the main problem."
- "Could you tell me something about what you think is most important here?"
- "Could you give me some thoughts on how you see the problem?"
- "Do you have any thoughts about the service you would like to have from us?"
- "Give me some ideas you have for how you feel we could best help."

2. Background Information about the Person's Life

As noted earlier, this part of the social history has a number of sections on various aspects of the client's personal background. We look at each of these.

Family of Origin. Here you document the relevant information about the family of origin, the family into which the person was born. Information on the parents, their occupation, siblings, outstanding characteristics, or information on the family would be placed here.

Questions you might use include:

- "Can you tell me something about your brother?"
- "Could you describe your parents for me?"
- "Tell me a little about what your family was like."
- "Can you describe what your home was like?"

Birth and Childhood. In this section, you want to note if the pregnancy and birth of the person were in any way complicated. Ask about important features of the person's childhood and what the client remembers. You want to elicit comments from the individual that give a flavor of the client's perceptions during this period of life. Was it happy or fraught with conflict? As a child, did the person feel appreciated or ignored? Was the person asked to shoulder very adult burdens or allowed to remain a child? Do not write, "the client reports an unhappy childhood" or "the client says her childhood was a happy one." You need supportive details. Why was her childhood happy?

Questions you might use include:

- "Can you tell me something about your childhood?"
- "Tell me a little bit about what growing up was like."
- "Describe a little more about your happy childhood."

Marriages and Significant Relationships. In this section, you document the marriages and other significant relationships of the client. Be sure to include here all significant relationships, whether the couple actually went through a marriage ceremony or not. If the person lived for several months or years with someone, note that here. Some people will have more than one marriage or relationship. A sentence or two on each is important, such as when it took place, how long it lasted, and why the marriage or relationship ended. Information about the current relationship of the person to an ex-spouse is also relevant. In addition, always mention children, their ages, whether they reside with their parents, and where they are now. If the children are no longer in the home, document the degree of contact the parents and children have.

Questions you might use include:

- "Tell me a little bit more about her [him]."
- "Can you describe something about what that marriage [relationship] was like for you?"
- "Tell me about your children."

Current Living Arrangements. In this section, give a brief description of the home, how the client feels about the home, and who lives there. Does the person feel the home is adequate or comfortable?

Questions you might use include:

- "Tell me a little bit about your home."
- "Can you describe your life when you are home?"

Education. Document here the person's highest level of education. Note, too, any difficulties or successes the client experienced while in school.

Questions you might use include:

- "Can you tell me more about school?"
- "Can you explain a little bit about your problems in school?"
- "Tell me a little bit about college."

Military Service. If the individual served in the military, give the details of that service here in a summary. Always mention the type of discharge from the military and the status as a veteran. If the person was never in the military, simply state "No military service."

Questions you might use include:

- "Can you tell me about your military service?"
- "Could you describe a few of the things you did in the service?"

Employment History. Here you will document the type of employment the individual has held. Note breaks in employment and give the reasons for the period of unemployment. Also indicate how the person views the work she has done. If the person never held a job, write "No employment history."

Questions you might use include:

- "Tell me a little bit about working there."
- "Can you give me some examples of the work you did?"
- "Talk a little more about the work you did."

Medical History. Medical history will be extremely important to medical personnel who may be called to assess and give service to your client. A psychiatrist or a nurse may spot a possible underlying medical problem based on the information you assemble in this section. Ask about childhood illnesses and any other illnesses, allergies, or surgeries.

Questions you might use include:

- "Tell me a little bit about your health."
- "Could you describe this surgery a bit more for me?"
- "Can you tell me a little bit about the polio?"
- "Describe those allergies."

Behavioral Health. This is a general term used to cover common mental health problems and problems with substance abuse. You would use this section if these were not the presenting problem but you are seeking to know if these problems were present in the client's background at some point.

Questions you might use include:

- "Could you tell me about any issues you've had with drugs or alcohol?"
- "Can you describe any mental health problems you might have had in the past?"
- "Tell me a little bit about any previous treatment you might have had for mental health issues."
- "Describe any problems you might have had with drugs or alcohol."

Legal History. In many cases, people will have no legal history. Write "No legal history" if this is the case. However, involvement in a lawsuit or criminal case, as either the defendant or plaintiff, is usually an important source of stress. Petty criminal activity that is current gives insight into how people view authority and their place in society. Previous criminal activity may show how much a person has been able to turn her life around or may indicate an unfortunate pattern.

Questions you might use include:

- "Can you explain a little about the lawsuit?"
- "Tell me about those early encounters with the law."
- "Could you tell me a little bit about what brought you into contact with law enforcement?"

Social and Recreational Interests. Because we are always interested in the strengths of the people to whom we give service, we want to note what it is that interests them and the social and recreational activities in which they participate. If a person has no activities to report, be sure to note this as well. It gives important clues to the client's social involvement or withdrawal. Ask if this lack of social involvement has always been present or whether it is more recent in the individual's life. Recent lack of interest in activities that used to be important can be a sign of depression. Note what interests the person most and what activities he pursues.

Questions you might use include:

- "Tell me a little bit about what you do in your spare time."
- "Fill me in on what you do for fun."
- "Can you tell me a little bit about what you like to do most?"

Religious Activities. Some people are very involved in their church, synagogue, or mosque. Others are not so involved but hold firm spiritual beliefs that they find very sustaining. Still others have neither religious involvement nor any interest in spiritual matters. Ask clients about their religious affiliation and activity or involvement. For people who do not have anything to report, explore with them any spiritual beliefs they might have that give them strength and comfort. This sort of strength and comfort is often enormously helpful to people as they recover or cope with illness and difficult problems. This is a sensitive area, however, so move on if you feel your clients are reluctant to discuss their beliefs. For some people, this line of questioning may be construed as your attempt to push a specific religion. Be sure to note, matter-of-factly, a client's discomfort with this topic.

Questions you might use include:

- "Could you tell me about your synagogue?"
- "Tell me something about some of the beliefs you feel are most helpful to you."

Client Successes, Strengths, and Resources. This is a section you may not find on most social history outlines. As much as we want to see clients as whole people, several factors—agencies, their policies, the pressure of time—often prevent us from exploring anything other than problems with clients. This focus on the negative aspects of people's lives often causes workers to create a skewed picture of the people they want to help and to register barely disguised surprise over the successes clients have had that come out during their social histories.

Asking people what they are most proud of, or what things they consider accomplishments, makes it clear that you expect a client to be a whole person, not a collection of problems. It is good practice for you too, as you will get into the habit of asking about and documenting the positive aspects along with the difficulties.

Questions you might use include:

- "Tell me a little bit about the things that make you proud."
- "Can you tell a little about the things you consider successes for you?"

In addition, note here the strengths and resources your client brings to the situation. Personal skills, financial assets, education, and social supports are all important to note in this section.

3. Impressions and Recommendations

This last section in the social history contains your own impressions and recommendations. Much of the information included here about your client is referred to as a mental status exam.

Start with the first sentence of your presenting problem and go on to give a brief one- or two-sentence summary of what you have already written. Include the way the client appeared during the interview, and any problems you see with memory or reality, anxiety or depression.

Then, give your recommendations for services that might be considered when creating the treatment or service plan for the person noting the person's input into these recommendations. Further details on writing impressions and recommendations are found on page 312.

Capturing the Details

Sometimes when case managers are taking social histories for the first time, they are inclined to write the barest number of details. They might write something like this: "Alice worked at Kmart for 4 years." In addition, it would be useful to know when

FIGURE 16.2 Capturing the details

Fair	Much Better
Madelaine's health is good. She states she had one surgery in 2007, but since then she has been fine.	Madelaine's health is currently good. She had an appendectomy in 2007, but since that time she has been fine.
Marie is the mother of two children.	Marie is the mother of two teenage children. Oliver, age 12, lives at home, and Michael, age 17, is currently in a boot camp in Mount Allen. She states she and Oliver get along well and that he is doing well in school. She sees Michael every other weekend and is hopeful that the boot camp experience will help him in the long run.
Bill is currently active socially.	Bill states he has a number of friends and sings in his church choir. Last summer he joined a local baseball team and intends to this summer as well. He likes sports and goes to games with friends.
Carl describes his childhood home as happy, but today does not know where two of his siblings are.	Carl describes his childhood as happy. He talks about a close relationship with his parents before their death and a number of activities, such as scouting and wrestling, that he participated in with his family's support. Carl has two sisters whose whereabouts are unknown. He states that at the time of his parents' death they were already married and living in other states. He went to live with the family of a friend to finish high school and then went into the Army. During that time they lost touch with each other.

© Cengage Learning®

she worked at Kmart, what she did there, and why she is no longer there. Figure 16.2 provides some other examples.

Your history should be a concise summary of the main points of a person's life, but too much brevity can leave a number of unanswered questions that, if answered, would shed considerable light on the person's life and problems now. Figure 16.3 shows a completed social history on Kate, the 47-year-old woman described earlier who contacted the agency for help with a long-standing depression. Examine the figure carefully so that you can see how a completed history is constructed.

Who Took the Social History

Your name goes in two places on the social history. First, your name should appear in a heading at the top of the history. Second, your signature should go at the end of the history. Your name should not go in the top right hand corner as if this is a school assignment.

FIGURE 16.3 Example of a completed social history

NAME: Kathryn (Kate) Carter   Agency # 04587
SOCIAL HISTORY   Date: *7/10/2014*

Prepared by: Winston Cramer

Presenting Problem

Kate, a 47-year-old married woman with one daughter away in college, is requesting help for a depression she states has lasted almost 2 years starting with the death of her mother. At the time her mother was ill, Kate did not follow doctor's recommendations that she take off work and stay home. She was very involved in caring for her mother, who subsequently died. Kate states she was not aware of being depressed until after the funeral, but grew depressed during the 7-month period she and her husband cleaned out her mother's house and settled her mother's affairs. She describes this work as "heart wrenching" and involving several legal difficulties.

    Currently Kate describes her depression as characterized by hypersomnia, an inability to go to work several days a month, and a loss of interest in social activities and friends. She states she is here in part because her husband insisted she get help.

    Kate believes she needs medication "to jolt me out of this." She blames herself for letting it go so long but says she felt it would lift on its own. She also states she didn't want to disturb her husband with the problem. She is asking for a session with a "doctor" and a prescription. She seems uncertain that she needs therapy, stating "I don't think there is anything wrong in my life, really."

Family of Origin

Kate's father died when she was 6 years old, and Kate describes feeling responsible for her mother most of her life. She describes her childhood as a happy one. A number of aunts and uncles took an interest in her, and she grew up with a number of cousins close by. She described happy family gatherings for holidays.

    She depicts her mother as living from the Social Security that came after her father died and being unable to sustain a consistent work history. She states her mother sought her advice often, and Kate feels that she made many of the important decisions for the family. At present only one aunt remains and is in a nursing home, and the cousins have moved out of the state. Kate has some contact with them at Christmas.

Birth and Childhood

Pregnancy and birth were uneventful. In addition to what is noted above, Kate and her mother never had enough money. "That's why I think my aunts and uncles took an interest in me." She spent weeks away from home in the summer at the homes of her cousins and often went to camp with them. Her mother would come for family picnics and was always warmly received.

Marriages and Significant Relationships

Kate has been married to her husband for 25 years. They have one daughter who is 20 years old and currently a student at the University of Minnesota. She is studying engineering. Kate remembers the pregnancy as easy, but she suffered severe and incapacitating depression immediately following. She was unable to return to work at the end of her maternity leave, thus losing her job.

(continued)

FIGURE 16.3 (continued)

She describes her husband as "steady" and reports that he is an accountant with a local accounting firm. Kate wanted more children, but he discouraged her, fearing she would again suffer postpartum depression. As a couple they are fond of going to symphony concerts and plays. Her husband is a model railroader, and Kate helps with the activities of the club from time to time.

Her daughter is "quiet like her father." She did very well in school and got a scholarship to the university. Kate worries about her in that she has had few friends and no boyfriends. She is concerned that she "may be prone to depression the way I am."

Current Living Arrangements

Kate and her husband live in a three-bedroom home on a half-acre in Meadowview. The couple has been married 25 years. Except when their daughter is home from college, the couple lives alone. Kate has a cat that she is very fond of. She describes her home as "in need of work" and says that she would like to do more to fix it up but is not able to find the energy. The couple has lived in this house since they were married.

Education

Kate has an associate's degree in early childhood education. She recalls that she did well in school and it was suggested she go on in college and become a teacher, but her mother discouraged her due to the financial situation of the family. She has rarely used this education. Recently Kate started taking courses at the local college to improve her computer skills but is not taking any this current semester.

Military Service

There was no military service.

Employment History

Right after Kate got her degree, she worked for 3 years in a day-care center, but she was attracted to a job as an office manager in a small insurance firm. She has done office management for the last 27 years except for a time when she was off for a prolonged postpartum depression. This occurred at the time of the birth of her only child, a daughter. Following this episode she obtained an excellent position in a large law firm and has, until now, moved up steadily with promotions and pay raises.

Medical History

Aside from the postpartum depression and the bad flu she suffered 2 years ago, Kate reports her health is good. She had few childhood illnesses and has a medical check-up about every 3 years. She denies using drugs, and states that she never smoked and that she will have a glass of wine once or twice a year when she and her husband go out to dinner. Recently, along with the depression, she has noticed that she has more headaches. She has not seen a doctor about these.

Legal History

There is no legal history to report.

FIGURE 16.3 (continued)

Social and Recreational Interests

Kate is very interested in sewing and has made all the curtains and draperies for her home. "I would make my own clothes, but I haven't the time with working." She is also an avid reader of mystery novels and does some "modest gardening" on the weekends. She talked at length about her cat and the pleasure she gets from being with her pet. Since her depression, she notices a gradual withdrawal from activities she used to enjoy.

Religious Activities

Kate states she is a Methodist and she and her husband regularly attend the church just six blocks from their home. She used to sing in the choir but dropped out after her mother died and has not been able to "find the strength to add that to my list of things to do." She and her husband used to be active in Sunday school, but she doesn't want to go any more, and he has dropped out too.

Client Successes, Strengths, and Resources

Client was unclear what she could list as successes. She denies that raising her daughter was an accomplishment or that helping her mother to the extent she did was important. "I was just doing what anybody would have done." She smiled and shook her head, unable to think of anything she would list as a success. She does not see her degree as "anything special" because she never went on for further education or used the degree.

Client feels she can be open with her best friend, Sue who works at the law firm as well. She reports no financial problems or marital problems.

Impressions and Recommendations

Kate, a 47-year-old married woman with one daughter away in college, is requesting help for a depression she states has lasted almost 2 years starting with the death of her mother. She states her husband is insisting she go for help.

She is a competent, but modest, person who is coherent and oriented × 3. Her appearance was very neat, her affect was flat, and she expressed concern that she give accurate and useful information to the worker. She reports a loss of interest in activities that formerly gave her pleasure. Impaired social functioning and hypersomnia have increased recently, causing her to seek help. Husband supports this. During the interview client had difficulty discussing herself and her accomplishments. She tends to minimize her successes and focus on things she could have done better.

She is requesting medication and, after a discussion with CM, agreed to three sessions of counseling to address general feelings of depression and an appointment for a psychiatric evaluation. Recommend medication and psychiatric evaluation with Dr. Crumlich and three initial sessions with Bay View Counseling Services.

Generally agencies have you sign your social histories at the end after a phrase such as "taken by," "submitted by," "prepared by," or "filed by." If you have credentials such as BSW or MSW, these follow your signature.

In the example shown in Figure 16.3, the name of the worker who took the history is also typed in at the top of the history so that anyone in the agency can quickly see who did this history. It is not appropriate for you to place your name anywhere but in the two designated places. Headings for social histories often look something like this:

NAME: Kathryn (Kate) Carter Agency # 04587
SOCIAL HISTORY Date: 7/10/2014

Prepared by Winston Carter

Social Histories in Other Settings

Limited Time for Intake

As a case manager having ongoing contact with your client, it may be possible to assemble an entire social history like the one discussed in this chapter. Certainly developing this complete picture of the person at the time of intake is best practice. Today, however, many agencies do not have the opportunity to spend the time it takes to assemble such a history. Often the person is in and out of service, sometimes in a matter of days, at the direction of the funding source. Many managed care organizations and insurance companies have severely limited the amount of service a person may receive. This is particularly true for adult services in areas such as mental health or drug and alcohol agencies.

Brief Intakes

In some agencies, the emphasis during intake is on the presenting problem, the background of that problem, and your impressions and recommendations. In all cases, you still would be expected to discuss the services you provide with the client and to seek the client's input; and you would be expected to document this discussion. Unfortunately, you may not have the time to do this thoroughly or to note many of the other aspects of the person's life in the history. This happens when people are given only a few days of service by their insurance company or other funding source.

When the individual will not be with you very long and you must do only a brief history, it is important to focus on the most immediate problem. Therefore, carefully document that problem, and talk about the most important points in the background to that problem. For example, when Harry came for services, he was recovering from a stroke in which he lost the use of his left hand and foot. The intake focused on the level at which Harry functioned before the stroke, the stroke itself, and the goals Harry wanted to pursue in recovering from the stroke. His acrimonious relationship with his brother in another state was not covered in detail, although the worker noted it in the intake material and, in her assessment, indicated that this relationship may have contributed to the recent stroke due to the client's anger and agitation over things that had been said in the week before the stroke. Left out of the history were extensive details about previous health problems unrelated to the present stroke, social interests, client accomplishments, military and legal histories, and details about his childhood.

In the current environment, brief intakes may be required more than we would like because of the limited time we have with a client. When the intake must be brief, the goal is to put sufficient focus on the immediate problem and the background to the problem to provide treatment for the client without delay.

Writing Brief Social Histories

A brief social history has three parts:

1. Presenting problem
2. Background
3. Impressions and recommendations

Presenting Problem

In the presenting problem section, describe why the person came into your agency. What precipitated his admission? Why is he here? Here is an example explaining Fred's presenting problem:

> Fred is a 26-year-old unmarried male who returned Sunday evening from a 2-week trip to the Philippines and Japan. On admission he was accompanied by his mother and was complaining of hearing voices, some confusion, and a lack of coordination. He states he had these same feelings briefly in September, but they cleared within several days. During that episode, he describes rigidly maintaining his routines as a means of dealing with these symptoms. Client further reports feeling as if he is standing outside himself watching his condition. He recognizes that he "is not right."

Background

In the background section, describe the client's background briefly along with any additional information regarding the presenting problem. Here is the background information about Fred and his presenting problem.

> Fred was attending the wedding of a high school friend in Manila when he became ill. He was traveling with friends of his and the groom. He states that he left New York at 2:18 P.M. on a Sunday afternoon and arrived in Manila 19 hours later at 9:00 P.M. the same day. He reported that the celebration went on for several days, during which time he drank a considerable amount of alcohol and slept very little. Toward the end of the journey he began to experience confusion and could hear his friend's voices, but not what the voices were saying. He came here immediately upon return home.
>
> Fred is employed as a computer analyst for ProWeber, where he has worked for the past 4 years. He currently shares his apartment with another male whom he describes as a childhood friend. According to client, they rarely see each other and travel in different social circles.

Fred is the oldest of two boys. His parents separated while he was still in high school and were subsequently divorced. The divorce is described as amicable. Father lives in Alabama but returns for holidays and graduations. When asked about his father's side of the family, client replied, "That whole side of the family is crazy." He did not elaborate further. When discussing his mother's side of the family, he reports that his aunts and uncles are supportive and have always shown considerable interest in him and his other cousins.

Fred graduated from high school and completed one year of college, but dropped out after he experienced intense stress. He claims his grades in both high school and college were "good."

Currently Fred is not dating anyone and socializes with a group of people with whom he went to high school. They have returned to the area after completing college. He expresses an interest in music and cars. He denies drug use, but says that he and his friends drink on the weekends.

How to Write Impressions and Recommendations

Impressions and recommendations come at the end of all social histories, regardless of the length or method in which the history was taken. After taking the social history, use this section to express your impressions of the client. In addition, include your recommendations for what needs to happen to support the client in the present situation. Later, in Chapter 18, we discuss how to write better impressions.

Impressions and recommendations have three parts. If you think of three very short paragraphs you will be able to address the three parts which are:

1. A summary of the presenting problem, beginning with the first sentence of the presenting problem
2. Your impressions
3. Your recommendations

Writing the First Sentence. Begin with the same sentence you used to open your social history:

> Lisa is a 42-year-old married woman, the mother of two girls ages 10 and 12, who is complaining of depression and lack of energy.

You can use this comprehensive sentence to open the social history and again to open your impressions and recommendations.

Next Two or Three Sentences. The next sentences can further describe the client's situation:

> She states that her depression began 4 months ago and has persisted during a time of marital strain. Husband is threatening divorce and appears to be seeing another woman.

Next State Your Impressions. Give your impressions of the client in the following areas where these areas seem to apply:

- Functioning (is the person able to function at work, school, home)
- Affect (how does the person seem?)
- Vegetative functions (Appetite, sleep, elimination)
- Insight (Does the person appear to understand the problem, the origin of the problem, his or her contribution to the problem?)
- Motivation (for change or help)

Here are the case manager's impressions of Lisa:

> Lisa reports that she has trouble waking up, is unable to work at home or at her job effectively, and has neglected parenting responsibilities. She appears to have impaired occupational and social functioning accompanied by insomnia. Her affect is blunted, and at times she was tearful. She has good insight about her current marital situation and the role it may play in her depression. In addition, she expresses a desire to obtain help and "change some things in my life."

You will have a much better idea about how to address the areas listed above after you have read Chapter 18.

End with Recommendations. End with a recommendation for what you believe should happen next. Do not use the word "I." Recommendations for Lisa are as follows:

> Lisa is requesting a therapist, with possible marital therapy in the future. Recommend psychiatric and medication evaluation of her depression and referral for six sessions of individual counseling, initially to address depression and marital issues.

When writing your recommendations, state the number of sessions, the number of days of hospitalization or group therapy sessions, and the reason for this recommendation. For example, "Six therapy sessions with Carlo Baldini to address anger management."

The points in the impressions and recommendations will have been elaborated upon in the social history itself, and the history will include details on each of the points not included in the impressions and recommendations. This section allows others to get a brief summary of the client's situation and what your impressions were at the time you took the intake. For a physician in a hurry, or for another case manager dealing with a crisis in this person's life, reading your impressions and recommendations can be an invaluable place to start.

Here is an example of impressions and recommendations for 56-year-old Landon:

> **Part 1.** Landon is a 56-year-old married man complaining of anxiety and depression, recovering from a recent stroke which left his left side paralyzed and forced him to give up an administrative position at the local community college. Landon reports a desire to "become active again but feels incapacitated by anxiety."

Part 2. Landon states he feels incapacitated by anxiety and irritability and worries his physical care and irritation are straining his marriage. He appeared anxious and irritable with the C.M. [case manager] during the interview. Due to the stroke, functioning is impaired in all areas. He appears motivated to get assistance, but skeptical that anything can change. He has good insight into how difficult his current situation is for his wife.

Part 3. Recommend review of medications by psychiatrist to see what psychotropic medications can be taken with those he is currently taking for the stroke. Recommend 4 sessions of individual counseling to help Landon better develop insight and coping skills and 6 sessions of marital counseling to strengthen his marriage. Further recommend planning meeting with Landon, his family, and the home health organization to see how services can relieve some marital strain. Will check on support groups for individuals or couples in similar situation.

COMMON ERRORS WHEN WRITING A SOCIAL HISTORY

- People don't "admit" things unless we are sticking them with a hot poker. They tell us things. "Adele admitted her address is 2346 Lincoln Way" sounds adversarial. Instead try, "Adele gave her address as 2346 Lincoln Way."

- Don't state things as a fact if they are things you don't know personally or have not observed personally. Use "according to the client" or "the client stated." For example, instead of "Alice's mother is a vicious gossip in the community," write "according to Alice, her mother is a vicious gossip in the community."

- Don't recommend something without saying what it is for. For example, "recommend psychiatric evaluation for depression" not "recommend psychiatric evaluation." Not "recommend 12 counseling sessions at Susquehanna Counseling," but "recommend 12 counseling sessions at Susquehanna Counseling to address loss of job and depression."

- Leaving gaps in the history. "She lives with four other people," but we haven't a clue who they are; or "they were married 10 years and divorced three and have a 20-year-old son." What is the explanation for that?

Using an Assessment Form

Because of the limited time in some agencies and the need to be sure that specific questions are covered in the initial interview, many agencies provide an intake or assessment form. The form contains certain information the funding source requires, and use of the form makes that information easily accessible. This assessment form may be very similar to, but a bit longer than, the phone inquiry form you used when learning to do phone inquiries. These forms usually have names such as "Initial Assessment Form" or "Intake Evaluation Form." Other intake or assessment forms may be lengthy and contain numerous questions to determine the client's needs and capabilities.

Remember, when you are using a form, that it is important to stop and ask open questions as you go through the form. This increases rapport and reduces the sense on the client's part of being grilled by you. Long forms that spell out what questions need to be answered create the potential for you to begin to sound like an interrogator. You may be asking, "Name? Address? Phone number?" and then move on to more involved questions such as "When were you married? How many children? Names and ages of your children?" Where is the warmth and empathy in such questioning?

The assessment form does contain many of the questions you need to ask to create a more complete picture of the person. If people seem unwilling to answer any of the questions, move on to others in a matter-of-fact way. Do not try to persuade them that they should answer something they feel uncomfortable discussing. Chances are this information will readily come out as you establish rapport.

In addition, most forms have a place for interviewer comments. Be sure to take advantage of these spaces to elaborate on what the person has told you. This is where your notes give a more individualized picture of the client.

The assessment form is simply an outline of what is important. If your intention is to fill in all the blanks on the form and close the interview, you have not really conducted an adequate interview. Put these spaces to good use by filling them with relevant comments or summaries. Obtain that information with good open questions that help you fill out the picture.

In one agency, a worker took the information required on the form. He also carefully inquired and documented additional information in the interviewer comment spaces. After one interview with a client, his supervisor came into his office and said, "We aren't here to make friends. That form should only take 15 or 20 minutes to complete, max!" This supervisor is an example of a person who is poorly trained in establishing rapport and does not understand the importance of the first contact for the future success of the client.

It is expected that you will talk with the client, that you and the client will discuss the situation, and that the client will volunteer additional information. Most forms soliciting information from people are focused on the clients' problems and deficits. In order to round out a complete picture of the person, you need to look for strengths. Take the time to do so—and remember, people only share such information when they feel comfortable with you. If you sound like a machine or rush through a series of questions, the person will not connect with you at all.

A blank intake assessment form is provided in the Appendix C at the end of this book. In addition, the companion book, *Fundamentals for Practice with High Risk Populations* (Summers, 2002) or on-line at CengageBrain, provides assessment forms for six high-risk populations. These forms are tailored to each specific population and contain detailed inquiries into common issues for each population. You may use those forms as you create and follow a client and these forms can be found on-line at CengageBrain.

Taking Social Histories on a Computer

In some agencies, case managers are asked to take the information from the clients and place it in electronic forms on a computer. This is seen as more efficient than taking a social history and then having it typed up. However, case managers have raised two valid concerns about this form of history-taking: First, can the case manager look at and engage the person if the case manager is typing on a computer? Second, does the electronic form have enough space to allow for the real details of the person's life and current problem? These are legitimate concerns, and I hope you will work for an agency that has addressed these possible obstacles. Using a flat screen computer allows the case manager to engage the client, who sits facing the case manager at the side of the desk. You do not want your computer to obstruct eye contact with the person.

The need to know the particulars of a person's situation is important because the goal is always individualized planning. If the electronic form you use does not allow for individualized descriptions of the person's problem, the plan is jeopardized. Even when using electronic forms, there should be plenty of room to give particular information about your client and to spell out in detail what the person sees as the major problem and wants to accomplish by coming to your agency.

Taking Social Histories in the Home

There are reasons why a person cannot get into the agency to give a social history. Perhaps the person is elderly and too infirm to come into the agency or a child whose mother has other children at home and is unable to find someone to look after the other children. Some agencies have a policy of talking the first time with a person in her own home. Taking a social history in the home has definite advantages and drawbacks.

Among the advantages is the glimpse case managers receive of the setting in which the person functions. The setting in which the person lives may be supportive of his situation or may present obstacles. Visiting him there gives you first hand information about this. Further, your client may feel more relaxed in her own home whereas she might have felt somewhat intimidated by an office in a downtown building.

On the other hand, problems can arise during home visits when confidentiality is compromised because other people are present or nearby during the interview. In addition, there can be things that distract both you and the client such as pets, children running in and out, others calling on the phone or coming to the door. Some case managers have experienced situations where well-meaning neighbors have felt they should sit in during the interview to explain Mary's situation because Mary is too infirm, or confused, or unable to speak for herself.

When deciding to take a social history in someone's home be very aware of privacy issues and confidentiality. Be sensitive to the fact that the client may not want to talk freely because her son is in the next room or her bother-in-law and his friends are on the porch and the front door is open. She may not feel comfortable saying to the other person that she wishes to speak to you privately.

The Next Step

Having completed the initial intake with the person, it is time to begin to put together a chart for the client if your agency keeps hard copies of your forms and contacts. Charts or files are kept in a particular order. The order of the contents is precise, making it easier to find information (see "Arrangement of the Client's Chart" in the Appendix C). Sometimes information is color-coded. For example, financial information may be on yellow paper and assessments on blue.

As for the social history in the chart, check these points before filing it:

1. Keep the headings with their content. Many put the heading for something on the bottom of one page and the content for that heading on the following page. They should stay together.
2. Bold all your subheadings so they are easily found.
3. All social histories or assessment forms should be stapled together.
4. All social histories must be typed so they can be read easily and quickly.

It is important to maintain your charts in the order specified by your agency and to return them promptly to the records section of your agency or file them properly.

In addition, it is important to make sure that the charts are locked up safely when you leave for the night and that no one has access to them but those who should. Finally, never remove a chart from the agency. The possibility that the information could be lost, stolen, or read by others is too grave.

Summary

When people come in for their first interviews, we are there to collect important information that will illuminate the problems these people are experiencing. Our task is to assess and document the details of clients' problems as well as the background to the problems. When people must be served quickly, we are asked to focus more closely on the reasons the individuals are seeking help and the histories of their immediate problems. In all cases, however, your information will serve at some point as the foundation for the development of services and treatment. Accurately understanding people and assessing their moods and motivation gives others valuable information with which to work.

If you are asked to use a form to collect information, be sure to ask questions and discuss the clients' situations with them. Note the information you obtain on the form so that the form will better illustrate each particular person's needs and concerns.

Finally, it is important to keep in mind that you may be the first person from your agency with whom the clients meet. Their contact with you needs to be positive for them to move on and begin to heal and resolve issues. As a representative of the agency, you are responsible for setting a tone that is warm, accepting, and safe so that people can talk freely about what has brought them to seek help. In this particular step you are also responsible for formulating with the client an assessment of the problem and suggestions for addressing that problem.

Exercises

These exercises can also be filled out online at CengageBrain.com.

Exercises I: Practice with Social Histories

1. On a single sheet of paper, write a note on how Kate appears to you from the social history (see Figure 16.3 in this chapter). What are your impressions of her? What do you think she would be likely to do or not do? How likely is she to commit suicide following this visit? How readily can she stand up for herself? What strengths and supports does she have? What contradictions do you see in looking at her life?

2. Write up a social history of Carlos using the presenting problem paragraph provided in this chapter to begin the history. Invent other information as needed. This will give you practice in organizing information.

3. Take a social history from a friend or classmate. Be sure to explain that the person does not have to answer any question that is uncomfortable and that the person can make up information to fill in the gaps. The important thing is for you to practice taking a social history and then organizing the information in a useful format such as the one discussed in this chapter.

Exercises II: Assessment of a Middle-Aged Adult

Instructions: Using one of the blank assessment or evaluation forms found in the back of each chapter in the companion textbook, *Fundamentals for Practice with High Risk Populations* (Summers, 2002) or on-line at CengageBrain, develop further one of the clients for whom you did a phone intake. Choose the assessment form that fits your client's problems. Develop details about the client's life and gather information relevant to the reason the client called the agency based on information that you found in *Fundamentals for Practice with High Risk Populations*.

In developing your client further, piece together the circumstances you believe might be reasonable for a person who has this particular problem. Assign the client a socioeconomic situation, amount of schooling, and other particulars. Develop as well your client's problem by again consulting the current *DSM*, if relevant, and books and articles on this specific problem. Look at the chapter in the companion textbook, *Fundamentals for Practice with High Risk Populations*, related to the assessment form you chose to use for your client. Think about how your clients might be feeling as they come in for the first time and what issues or problems clients might be facing for which they are looking for help.

Or

Develop a believable social history on the client you have chosen, consulting the same material so that you are familiar with the issues and common problems faced by clients in this population.

Or

Use the generic assessment form in the Appendix C at the back of this book, again consulting relevant material and the current *DSM*, to create a believable client.

Exercises III: Assessment of a Child

Instructions: Using the assessment form for children found in the back of Chapter 3 on children in the companion textbook, *Fundamentals for Practice with High Risk Populations* (Summers, 2002) or on-line at CengageBrain, develop further the child for whom you did a phone intake. Develop details about the child's life and gather information relevant to the reason the parent or guardian called the agency.

In developing your client further, piece together the circumstances you believe might be reasonable for a child who has this particular problem. Assign the child's family a socioeconomic situation, amount of schooling, and the other particulars. Develop as well the child's problem by again consulting the current *DSM*, if relevant, and books and articles on this specific problem. Look at the chapter on children in the companion textbook, *Fundamentals for Practice with High Risk Populations*, and think about how you, the parent (or guardian), and the child might be feeling as the child comes in for the first time and what issues or problems the family might be facing for which they are looking for help.

To complete this assignment, you may use the chapter on children's mental health (Chapter 3) or the chapter on intellectual disabilities (Chapter 7) found in the companion textbook, *Fundamentals for Practice with High Risk Populations*.

Or

Develop a believable social history on a child, consulting the same material so that you are familiar with the issues and common problems faced by children and their families.

Or

Use the generic assessment form in the back of this book, again consulting relevant material and the current *DSM*, to create a believable client.

Exercises IV: Assessment of an Infirm, Older Person

Instructions: Using the assessment form at the end of Chapter 8 on older people found in the companion textbook, *Fundamentals for Practice with High Risk Populations* (Summers, 2002) or on-line at CengageBrain, further develop your client who is an older person, a person over 80 years of age. In this case, a child or close friend or neighbor may be present when you do the assessment and, depending on the condition of the older person, may give you most of the information. In addition, you may need to go to the person's home to meet with the person because the client is too infirm to come to the office.

In developing your client further, piece together the circumstances you believe might be reasonable for an older person who has this particular problem. Assign a socioeconomic situation, amount of schooling, and the other particulars, such as marriages, number of children, former occupations, and interests. In developing your client's problem, read the current *DSM*, if relevant, and look at books and articles on the specific problem your older person appears to have. Look at the chapter in the companion textbook, *Fundamentals for Practice with High Risk Populations*, for information on how your client and the concerned caller might be feeling and what issues or problems the caller might be facing when meeting with you for the first time.

Or

Develop a believable social history on the client you have chosen, consulting the same material so that you are familiar with the issues and common problems faced by clients in this population.

Or

Use the generic assessment form in the Appendix C at the back of this book, again consulting relevant material and the current *DSM*, to create a believable client.

Exercises V: Creating a File

Instructions: At this point, you need to create a file on the people you are seeing as clients. Create a separate file folder for each client you intend to follow throughout the remainder of the course. Place the client's name on the tab of the file folder—last name and then first name, so that the cases can be filed alphabetically by their last names. Place the "New Referral or Inquiry" form on top, followed by the verification letter you sent. Under that, place the long assessment form you used or the social history, clipping the pages of the form or history together.

CHAPTER 17

Using the DSM*

Introduction

The ***Diagnostic and Statistical Manual of Mental Disorders*** (***DSM***; APA, 2013) is a collection of diagnoses of mental disorders accompanied by the typical behaviors and symptoms you might see in a particular diagnosis. The idea behind this manual is to provide a common set of criteria for each mental disorder so that practitioners will be more likely to give the same diagnosis to people with similar symptoms and behaviors regardless of where they are being treated or who is seeing them. Thus the manual provides a common language that everyone in the helping professions can use in diagnosing individuals, discussing their symptoms and issues, and planning their care. Students may want to use the *Quick Reference to the Diagnostic Criteria from DSM 5*, a smaller book with the basic information sufficient for students to work with this material.

    Many students wonder why they need to learn about the *Diagnostic and Statistical Manual of Mental Disorders* (*DSM*; APA, 2013) when it appears to be a tool used exclusively by mental health practitioners. Actually the *DSM* is a valuable tool you will use in many different settings. Although the majority of people receiving services in the broad human service system do not have mental disorders, the *DSM* sometimes helps to define what the client is experiencing and what that person needs. For instance, people who come to agencies as victims of abuse or assault often suffer from post-traumatic stress disorder. Workers in agencies dealing with the problems of growing older will encounter people who have dementia or cognitive symptoms that resulted from a stroke or other long-term, debilitating illness. Those who work with children in

---

* Adapted from *Using DSM-IV: A Clinician's Guide to Psychiatric Diagnosis* by Anthony L. LaBruzza in collaboration with José Méndez-Villarubia, 1997, 1994 Jason Aronson, Inc., an imprint of Rowman & Littlefield Publishers, Inc.

a variety of settings will encounter children who have learning difficulties or behavior issues. Familiarity with the language and process of the *DSM* enables you to participate in planning for the client more competently.

Is *DSM* Only a Mental Health Tool?

Today, with deinstitutionalization of the mentally ill, those with mental disorders come for services at many social service agencies, and more often than in the past, we see people who have more than one problem. You might be working at a shelter for victims of domestic violence and do an intake for a woman who also suffers from bipolar disorder. You might find that a person has both a mental disorder and an addiction to heroine. People who seek our help no longer fit into neat boxes with no overlapping problems. For that reason, it is important to be familiar with this system.

The *DSM* is the language of insurance companies and other funding sources with regard to behavioral treatments such as drug and alcohol treatment or treatment for those with mental health problems, addiction issues, or intellectual disabilities. In addition, the *DSM* contains information about situations and problems that may not constitute a mental disorder but may be the focus of attention in a clinical setting. Many of these disorders come to the attention of social service agencies not equipped to treat them. You will need good information to make sound referrals.

Your ability to understand the *DSM* and your acquaintance with the various classifications of mental disorders will enable you to be more conversant with others in the field and to recognize a mental disorder when you encounter one.

Cautions

Having spelled out why the *DSM* is important in human service practice, it is equally important to understand that most people who come for services in social service agencies are not suffering from a mental disorder. The *DSM* cannot be used to help you understand every client. If you try to give a psychiatric label to everyone you see, you will unnecessarily burden individuals who are well but are grappling with life transitions and disruptions such as unemployment or grief. Labeling people can have lifelong consequences for them. In addition, these labels can make a person seem much sicker than they really are. Diagnosing calls for caution.

Further, the *DSM* comes from the medical model. That is, the model suggests that individuals are labeled with an illness and are then treated as sick. This is a view of the client that can cause you to lose sight of the fact that the person has strengths and successes. Although a diagnosis is useful to clinicians in providing treatment, to case managers it can have the subtle effect of diminishing the client as a whole person.

Moreover, people have a right to know what their diagnosis is. This is part of informed consent. They also have the right to know who else will see the diagnosis: The insurance company? The companies and businesses where clients work? Knowing who will see the diagnosis is also part of informed consent.

Finally, be cautious about diagnoses that do not seem to take into account what might be normal for people of a particular culture. There is always the danger that we judge behavior and problems according to what is normal in the dominant culture or in our own culture and forget to consider the culture from which the person came.

The agency where you work will have policies and guidelines about using the *DSM*. Many agencies do not rely on the manual at all. Agencies that do rely on the manual generally are required to give diagnoses in order to be reimbursed for services. When you must rely on the *DSM*, be very careful not to categorize people or to allow their diagnoses to color your complete understanding of them as individuals.

Who Makes the Diagnosis?

You are not studying the *DSM* to make final diagnoses. The responsibility for overseeing how people are diagnosed generally lies with a physician or a senior staff person, usually with a PhD. Nevertheless, the *DSM* contains a language that is universally understood. Your experience with this language and with the mental disorders in the manual will facilitate your communication and reports to those responsible for giving the diagnoses. It could happen, on a rare occasion, that a harried emergency room physician with a waiting room filled with medical emergencies would turn to the emergency worker from a social service agency and ask that the worker give a provisional diagnosis to facilitate admission to the hospital (where the diagnosis will be reevaluated in less pressing circumstances). Further, it is becoming common practice for insurance companies and other payers to require a diagnosis at the completion of intake. Case managers responsible for intakes may need to give a provisional diagnosis at the time of the intake so the agency can be reimbursed. These diagnoses may be changed later by senior professionals, but case managers need to be familiar with the common diagnoses seen in their agencies. For this reason, the exercises at the end of this chapter are for you to become acquainted with the *DSM*. The discussions you and your classmates have will help you to explore the manual and learn more about it.

It is important for you to keep in mind that additional clinical information is *always* needed to help round out the picture and make the best diagnosis and treatment plan. Much of that additional information in many settings will come from your social histories and notes.

Portions of this chapter are based on the work of Anthony L. LaBruzza (1994), whose book *Using DSM-IV: A Clinician's Guide to Psychiatric Diagnosis*, gives excellent background on how *DSM*s have evolved over the years until the *DSM-IV*.

Background Information

Until the 1600s, physicians used a patient's horoscope to diagnose mental disorders. Medieval physicians looked at the four humors to account for differences in human personality and temperament. The humor that predominated accounted for the patient's disposition—with blood accounting for a happy temperament; choler contributing to a fiery, competitive temperament; phlegm resulting in a cold, delicate disposition; and bile causing melancholy.

Psychiatry Attempts to Classify Mental Disorders

In colonial times, most individuals with mental illness were managed at home by their families. Many were abused and exploited or were confined to workhouses and almshouses in which varying theories about the reasons for their illnesses caused harsh treatment in most cases. Between 1800 and 1860, a number of people became concerned with placing those with mental illness in "asylums" in which a more humane approach and more respect for the patient would be the rule. Such treatment was referred to as "moral treatment" (LaBruzza, 1994). Dorothea Dix was active in this movement; and when her attempts to start a federal asylum program failed, she became instrumental in founding state hospitals in Pennsylvania and New Jersey, which bear her mark to this day.

Mental illness was little understood; and in the census of 1840, people were classified as either sane or "idiocy/insanity." The shift from the asylum to treatment, research, and education occurred in the late 1800s and early 1900s. At that time, research was beginning to provide a clearer picture of the anatomy of the brain, and the diagnostic system became more refined. By 1880, there were seven categories of mental disorder.

Diagnosis continued to be the focus of research. Wilhelm Greisinger (1817–1868) in Germany looked at the mental disorders as diseases of the brain, an organic view. Another German, Emil Kraepelin (1855–1926), looked at syndromes or collections of symptoms and made statistical records of the symptoms patients exhibited, the course of their diseases, and the outcomes. His goal was to be able to accurately predict the outcome of a disorder for a patient based on certain combinations of symptoms. He used a behavioral and descriptive approach that made it easier for others to use his concepts.

Others also contributed their views of the brain and nervous system in creating a diagnostic classification system. Most influential in the United States was a Swiss-born psychiatrist, Adolph Meyer (1866–1950). Mental disorder, in his view, was a response to psychosocial stressors. This view was widely accepted because individuals drafted into the military during both world wars appeared to break down under the stress of combat. Had his view continued to be influential, mental illness would have been seen today as an adaptive response. Instead, mental disorders gradually came to be seen as discrete psychiatric diseases.

In the 1920s, the American Psychiatric Association (APA) decided to find a way to standardize the medical terminology psychiatrists used. A national conference in 1928 looked at how diseases were named. The classification system that emerged focused only on the most severe forms of mental disorders, those that would most likely cause the patient to be institutionalized. The classification became broader when World War II veterans returned with less severe disorders. In the 1940s, there were 10 types of psychoses, 9 neuroses, and 7 disorders related to behavior, intelligence, and character. In 1952, the APA published the first *Diagnostic and Statistical Manual*.

In an attempt to keep up with international changes in the way mental disorders were classified, in 1965 the APA revised the original manual and brought out the second edition, *DSM-II*. This manual seemed to return to the Kraepelian descriptive model for diagnosis. Those who did the revisions eliminated terms that implied a particular theory of etiology (or cause) for the disorder. This successfully did away with Meyer's idea of seeing mental disorders as a response to stress. Nevertheless, psychoanalytic terminology remained because psychoanalysis was still quite popular and influential among psychiatrists.

The 1950s, 1960s, and 1970s

At this point, the manual was still quite unreliable. Psychiatrists would give different diagnoses to the same symptoms, making replication of research impossible. Anthony LaBruzza (1994) stated, "[T]he possibility that two psychiatrists would agree on the same diagnosis in the 1950s and 1960s was nearly random." In the 1960s, psychiatry was out of favor with the public as famous court cases pitted psychiatrists against each other in what appeared to be a nebulous theoretical system, and motion pictures, such as *One Flew Over the Cuckoo's Nest*, introduced moviegoers to the possibility that institutions were punitive and that the staff in such places were not much healthier than the patients. This was a time when all authority was challenged, and a number of books challenged the authority of psychiatry, particularly Thomas Szasz's book, *The Myth of Mental Illness*. Many saw psychiatry and psychiatric diagnoses as stigmatizing and as wielding undue social control.

In addition, insurance companies began to cut back on the amount of psychiatric care for which they were willing to pay, in part because the diagnosis of mental illness was unreliable and there seemed to be no consensus on the best treatments. No studies had been conducted to determine which illness responded to which treatment.

Psychiatry Becomes More Medical

The third edition of the manual, *DSM-III*, came out in 1980. Every edition of the manual since *DSM-III* has been an expansion or refinement of that document. This manual relied on a more medical, research-oriented model of disease, and it also relied more heavily on the Kraepelian use of descriptions. In addition, it was no longer slanted toward psychoanalytic descriptions or causes; in fact, causes were, for the most part, left to research to determine. Responding to the concerns voiced about psychiatry, the third edition of the manual contained 14 discrete and specific mental disorders with very explicit descriptions. These descriptions had operational criteria that allowed them to be measured statistically. All references to unconscious motives were removed, and the clinician based the diagnosis strictly on what could be seen.

The changes in the third edition of the manual could be summarized as follows:

1. There was every attempt to use clear English, and not mental health scientific jargon.
2. Disorders were labeled, and not people.
3. *Patient* was dropped in favor of words like *person* or *individual*.

4. The manual was tested for reliability for the first time by clinicians using it in the field.
5. A **multiaxial** system was adopted to give a fuller diagnostic picture of the person.
6. Decision trees were included to help the physician rule out similar disorders and narrow the diagnostic choice to one.
7. The words *disease* and *illness* were dropped in favor of the word *disorder*.
8. All the pet theories about causes of disorders were eliminated.
9. Each disorder had a working definition that contained operational criteria (criteria that could be observed and measured).

After publication of *DSM-III*, psychiatrists were far more likely to make the same diagnosis for the same set of symptoms. This enabled research to be done more effectively, particularly field trials of medications that treated specific psychiatric symptoms. In other words, it became more likely that practitioners would all agree on the diagnosis for certain clusters of symptoms, regardless of where they were practicing. If everyone was seeing the same thing when they looked at a cluster of symptoms, then it was possible to treat that cluster of symptoms in various ways to determine the best approach to alleviating the symptoms. Now clinicians could communicate reliably in a common language about diagnoses. This common language facilitated good research. Pharmaceutical companies supported this research for products they developed for these specific disorders.

From DSM-III to DSM-IV

With all the field testing that took place as a result of *DSM-III*, revisions were inevitable. Thus, in 1987, *DSM-III-R* (or revised) came out; this edition included 27 new categories and revisions to some older diagnoses. The number of categories went from 265 to 292. An appendix contained further categories requiring additional research.

The *DSM-III-R* made another important shift, moving from the *monothetic diagnosis* to the *polythetic diagnosis*. The old *DSMs* used the monothetic diagnosis. They gave a series of symptoms that constituted a disorder, and unless all of them were present you could not use the diagnosis. This meant that a diagnosis was only as useful as the least useful item in the series of symptoms. In a polythetic approach, the series of symptoms is given, and the patient must have several, but not all, of them. This has improved the reliability of diagnoses.

Another important shift was the move to give a patient more than one diagnosis if the patient met the criteria for more than one. Previously the clinician had to choose the diagnosis that was most obvious or urgent. Other diagnoses that coexisted with the first diagnosis or were, perhaps, part of a larger clinical problem were not mentioned. This narrowed the clinical picture of the patient. Now a fuller clinical picture was possible. The *DSM-III-R* also lined up with the new version of **International Classification of Diseases (ICD-10)**, which made it easier for American clinicians to talk to clinicians internationally.

The *DSM-IV* contained as few changes as possible, and those changes were based on good research with empirical results. To establish the empirical basis for changes, the work committees (those committees working on various classifications) systematically reviewed the literature for different diagnostic categories, reanalyzed previous data, and conducted field trials to make certain the diagnoses were reliable in many different settings and in many different types of clinical work. The *DSM-IV* also did away with all sexist language.

The *DSM-IV-TR*

The *DSM-IV-TR* (or text revision) refined the diagnostic categories still further and brought new information to the diagnostic process.

The following are some of the features you would find in this manual:

1. Every disorder has a name, numerical code, the criteria needed to give the diagnosis, the subtypes of the disorder, the **specifiers (modifiers)**, recording procedures, and examples that illustrate the disorder.
2. Associated features and associated disorders may include such items as clinical features that may be present but are not always seen in the disorder; disorders that precede, often co-occur, or generally follow the disorder in question; typical laboratory findings; physical signs and symptoms; and typical medical conditions.
3. The typical age at onset and any cultural and gender-related information.
4. The prevalence of the disorder, the incidence, and the risk.
5. A description of the typical clinical course of the disorder.
6. Any complications that might be applicable to the disorder.
7. Typical predisposing factors discovered through research.
8. Family patterns if there are genetic or suspected genetic components to the disease.
9. Differential diagnoses or disorders that share similar symptoms and information on how to distinguish among similar disorders (LaBruzza, 1994, pp. 57–58).

The Transition from DSM-IV-TR to DSM 5.

A new manual, *DSM 5,* was developed and came out in 2013, but it will take some time for the *DSM 5* to be used universally. Insurance companies, social service agencies, medical records, and physicians' offices will have to find ways to accommodate the new system. For that reason, you are likely to see *DSM-IV-TR* diagnoses in client charts and on other forms for some time.

Figure 17.1 shows you how the diagnosis was made using *DSM-IV-TR*. The individual received a diagnosis along five separate dimensions, referred to as *axes*. Each axis gives different information about the person. This was thought to provide a more accurate clinical picture than would be possible with a single axis. The diagnosis was called a multiaxial diagnosis. You will find that the diagnoses entered on axis I and II will have both names and numbers, the numbers being useful for insurance and billing purposes. Each axis served a different purpose.

Figure 17.1 shows what information is coded on each axis. The clinical disorders were all listed on Axis I or II. Every disorder has a name, numerical code, the criteria needed to give the diagnosis, the subtypes of the disorder, the specifiers (modifiers), recording procedures, and examples that illustrate the disorder.

As you practice you are very likely to find individuals whose diagnosis is written according to the *DSM-IV-TR* because the new *DSM 5* has only recently been published. Therefore, specific information on *DSM-IV-TR* can be found in the appendix and you can use this to better understand diagnoses you come across that are written according to this former system.

DSM 5, the Current Diagnostic Manual

When you begin practice, the *DSM 5* will be the diagnostic manual that is in current use. Initially this material may seem daunting. Keep in mind, however, that this is a manual written for clinicians and not for case managers. Whereas previously a clinician had to spell out a diagnosis on five separate axes, related to five different aspects of a person's clinical picture, now a clinician only needs to write out the diagnosis or list diagnoses if there is more than one. All you need to do as a case manager is

FIGURE 17.1 Dimensions used in multiaxial diagnosis

Axis I: All clinical syndromes listed in the *DSM-IV* are coded on this axis *except* personality disorders and mental retardation. Axis I includes developmental disorders and other conditions that might be a focus of clinical attention.

- V71.09   No diagnosis on Axis I
- 799.9   Diagnosis deferred on Axis I (meaning too little time or information to establish a diagnosis)

Axis II: Coded on this axis are personality disorders, mental retardation, significant maladaptive personality traits, and habitual defense mechanisms.

- V71.09   No diagnosis on Axis II
- 799.9   Diagnosis deferred on Axis II

Axis III: This axis is used for all general medical conditions that are relevant to planning and understanding the patient's diagnosis. *International Classification of Diseases (ICD-10)* codes can be used here.

- None (meaning no medical conditions)
- Deferred

Axis IV: Psychosocial and environmental problems that affect the prognosis, management, or treatment of the case are coded here.

Axis V: This axis is for the rating on the Global Assessment of Functioning (GAF) scale, which is usually a single number between 1 and 100, indicating the current level of functioning the patient possesses.

Digital Download   Download from CengageBrain.com

become somewhat familiar with this book in order to be able to understand discussions about the conditions and treatments of people seeking help and to understand how categories of disorders are grouped by chapter.

The chapters are arranged so that similar disorders are in adjacent chapters. Each chapter also presents the course and development of a particular disorder so that you can see how it differs as people age. Where it seemed appropriate the *DSM 5* also looks at gender and cultural issues that could affect a diagnosis. To help you better understand and work with the exercises at the end of this chapter, it would be a good idea to purchase the *Quick Reference to the Diagnostic Criteria from DSM 5*. The *DSM 5* has some new features worth noting:

1. The *DSM 5* looks at genetics and neuroimaging. The APA contends, "Given the explosion in neuroscience, neuropsychology, and brain imaging over the past 20 years, it was critical to convey the current state-of-the-art in the diagnosis of specific types of disorders." (*DSM 5*, 2013).
2. In addition to the manual and numerous print materials available about *DSM 5*, you can find what the APA calls "on-line supplemental information" to help clinicians make an accurate diagnosis.
3. The *World Health Organization Disability Assessment Schedule 2.0 (WHODAS.2.0)* is found on page 747 of the manual and can be self-administered by the client. This is used to give a better picture of how well the person is functioning with a severity scale for scoring the person's capacity related to the various tasks and abilities. It can be used to track improvement or deterioration over time.
4. *DSM 5* contains (on pages 743–44) a symptom severity chart for psychosis allowing a person to define better the degree of psychosis and how severe the symptoms are for the person.
5. The *ICD* numerical codes are included to bring diagnoses into line internationally. We will look at this further.

Before Making a Diagnosis

Before you can make a diagnosis you need to have an idea where in the *DSM* your client falls. If a person tells you that he or she is having trouble sleeping you would look in the chapter titled "Sleep-Wake Disorders." If a person presents hearing voices or hallucinating you would look at the chapter titled "Schizophrenia Spectrum and other Psychotic Disorders." Each chapter contains disorders that are to some extent similar. For instance, if a young person tells you she is not eating because she believes she is fat and she is really painfully thin, you would look in the chapter on "Feeding and Eating Disorders." You would also go to that chapter if a young mother tells you that her child is eating dirt and leaves. Both of these problems have to do with eating. Making yourself familiar with the 21 categories of disorders in the *DSM* will help you to better pin down what disorder your client is presenting. The chapters and the disorders contained within them are listed in your *Quick Reference to the Diagnostic Criteria from DSM 5*.

Each chapter, as noted, contains a group of somewhat similar or related disorders. Eating disorders are together, psychotic disorders are together, learning and developmental disorders are together, and so forth. Once you have turned to the appropriate chapter you would be looking for a diagnosis that looks very much like what your client is presenting.

Each disorder has a list of criteria to use in making the diagnosis. Often this list is followed by specifiers. In other words, you could be asked to specify if the onset of the disorder is recent or in the past. You might be asked to specify if the disorder is in partial remission or full remission. In another instance you might be asked to specify if the disorder is mild, moderate, severe, or extreme.

Making the Code Using *DSM 5*

In *DSM 5* there are no more axes, greatly simplifying the coding process. A person coming in with a clear mental disorder would simply have that written out. If there is more than one disorder, you would list these.

Writing a diagnosis, often referred to as coding, is done in the following order:

- The number of the disorder
- The name of the disorder
- The severity of the disorders (when required)
- Any specifiers that apply (when appropriate or required)

The Number of the Disorder. All the disorders in the *DSM* have a numerical code. This is the ICD-9 code. Next to the number is a second number preceded by a letter. This second number is the number that will be used by the ICD-10. In order to have a common language about these disorders internationally, the United States will adopt these second numbers on October 1, 2015, and the first numbers will no longer be used. It appears that from this point on the *DSM* will coordinate the *DSM* numbering system with the *ICD* numbering system.

The Name of the Disorder. Every disorder has a name. For example, if a person had a diagnosis of schizophrenia it would be written as

    295.90, schizophrenia

Or when using the ICD-10 code it will be written as

    F20.9, schizophrenia

Other examples

WRITTEN WITH *DSM 5* ICD-9 CODE Before October 1, 2015	WRITTEN WITH ICD-10 CODE After October 1, 2015
300.3, Hoarding Disorder	F42, Hoarding Disorder
780.52, Insomnia Disorder	G47.00, Insomnia Disorder
308.3, Acute Stress Disorder	F43.0, Acute Stress Disorder

The Severity of the Disorder. By noting the severity of a disorder you are indicating the degree to which the disorder interferes with the person's ability to function. Not all diagnoses ask you to specify severity. However, if you are asked to do so, you would write that next to the disorder.

Take for example 312.31, gambling disorder. The *DSM* asks you to specify if this disorder is mild, moderate, or severe. For 300.82, somatic symptom disorder, you are asked to specify if the disorder is mild, moderate, or severe. When looking at the eating disorders you are asked to specify if the disorder is mild, moderate, or severe or is it extreme.

These terms are defined so that you know what constitutes a mild case or what constitutes a severe case for that particular disorder. With anorexia it is Body Mass Index (BMI). The lower the BMI, the more severe the disorder is for that person. With Bulimia these terms are defined by the number of "compensatory episodes" a person has in a week. For example, a person engages in binge eating and then uses a compensatory method to prevent gaining weight such as self-induced vomiting. How many times in a week the person engages in this compensatory behavior tells you how severe the condition is for this person. If the person only uses self-induced vomiting one to three times a week, it is considered mild. However, 14 or more episodes would be considered extreme.

Specifiers That Apply. Many of the diagnoses will have other specifications you are asked to note. With enuresis (urinating into clothes or bedding), you are asked to specify whether this is "nocturnal only," "diurnal only," or "nocturnal and diurnal."

A disorder may ask for several different specifiers to distinguish clearly what the diagnosis involves. For 300.4, persistent depressive disorder, the *DSM* asks for a list of specifiers. You are asked to specify if the person experiences additional distress or features and these are listed for you. They include things like "with anxious distress" or "with melancholic features." Next you are asked to specify if the disorder is in partial or full remission. You are then asked to specify early onset (before age 21) or late onset (21 or older). Finally, you are asked to specify for the most recent 2 years of persistent depressive disorder whether it was pure dysthymic syndrome, which is defined, or if it involved other forms of depression such as major depressive episode or intermittent major depressive episodes, and so forth. Think of specifiers as clarifying and individualizing the person's disorder.

Multiple Diagnoses

It is possible for people to have more than one diagnosis and you would want to note all of those that apply. If a person comes to your office seeking help with a disorder and the professionals in your agency diagnose the person with more than one disorder, the disorder that brought the person in or the one that is most prominent and most in need of relief would be listed first. All others would be listed below in order of importance.

Other Conditions That May Be a Focus of Clinical Attention

There are many times people seek help for their problems, but they do not have a *DSM 5* disorder. Not everyone will have one, but there are many personal problems that bring people into agencies for help. Usually agencies simply describe the problem in the notes but you may work in a place that is required to note some diagnosis in order for your agency to be reimbursed. In the back of the *DSM* is a section titled "Other Conditions That May Be a Focus of Clinical Attention." Here you will find an array of problems people can have without having a psychiatric disorder. Usually their distress is a normal reaction to what is going on in their lives.

In this section you will find a number of relationship problems, abuse and neglect issues, domestic violence, and work and income troubles. All of these have a number and many of the numbers begin with the letter V causing them to be referred to by clinicians as V codes.

Sometimes a person does have a disorder as well as one of these conditions. For example, a man might come in and he is diagnosed with 300.2, Generalized Anxiety Disorder. In the course of the intake interview the case manager learns that the person is having considerable difficulty with a hostile neighbor. The neighbor has threatened the man repeatedly and threatened the man's pets. The police have made only half-hearted attempts to stop this behavior. When the man comes to see you he is torn between bringing charges, which could further inflame the situation or trying to live with this frightening neighbor next door. After listing 300.2, generalized anxiety disorder, you would list V60.89, discord with neighbor. Your social history will describe the situation in more detail, but listing this makes it clear that the man is experiencing stress related to his neighbor. Combined with the first diagnosis we get a clearer picture and some sense of why this man is anxious.

When the Diagnosis Does Not Quite Fit

People may describe their symptoms and those symptoms sound a lot like depression but the person's symptoms do not meet any of the criteria for the disorders listed in the chapter on "Depressive Disorders." You look there and you see that the person does not have disruptive mood dysregulation disorder because there are no "severe recurrent temper outbursts. . . ." The person does not really have "(F)ive or more" of the symptoms for major depressive disorder and the picture does not look like dysthymia either. It can't be premenstrual dysphoric disorder because the person is a man and he is not using substances or medications, which rules out substance/medication-induced depressive disorder. Further, it does not appear that the person has a medical condition that would give him a diagnosis of depressive disorder due to another medical condition.

In each chapter you will find at the end a section that gives you other choices. This section is titled "Other Specified Disorders." For our person we would look at that section titled "Other Specified Depressive Disorders" and see if any of those options fit better.

If the other specified disorders do not fit what you are seeing in the client then turn to the end of the chapter. Each chapter ends with a section titled "Unspecified Disorders." In our example we would turn to 311, Unspecified Depressive Disorder, as the diagnosis. The *DSM* states, "(T)his category applies to presentations in which symptoms characteristic of a depressive disorder that cause clinically significant distress or impairment in social, occupational, or other important areas of functioning predominate but do not meet the full criteria of any of the disorders in the depressive disorder diagnostic class." This statement, as it applies to each category of disorders, can be found at the end of every chapter. It is used when we know the symptoms belong in a certain category of disorders, but the person does not entirely meet the criteria for any of them.

When There Is No Number

Sometimes you will find a diagnosis in the *DSM 5* that has no number. For example, your client has a major depressive disorder but when you turn to page 160, Major Depressive Disorder has no number with which to code it. In cases like this one the *DSM 5* is asking you for more specific information. If you turn to page xvii you can see the information that is required. First, you are asked if this is a single episode or is it recurrent. Is this the first time this has happened or has this person been depressed on other occasions? For our purposes we will say she has experienced depression several other times in the past. We would go to recurrent episode and see that we need to indicate what this episode is like. Is it mild, moderate, or severe or does it have psychotic features? Is this episode in partial remission, in full remission, or is this recurrent episode unspecified? The choice you make here has a number and that number gives the next person looking at the diagnosis more information about the current depression than that the person is simply depressed.

Summary

While *DSM 5* has simplified diagnosing mental disorders, the *DSM* is a complex manual. It takes practice and good clinical skills to use the manual effectively. Nevertheless, entry-level individuals are being asked to understand the categories of disorders and discuss diagnoses with clinicians. In this course, we begin to look at how you would use the manual in your work as a case manager.

As you practice, you will begin to understand more clearly how disorders are defined and treatments are assigned. As you work with the *DSM* over a period of time, diagnoses will become more familiar to you and easier to use. You will also be able to note such diagnoses more quickly with practice.

Exercises

These exercises can also be filled out online at CengageBrain.com.

Exercises: Using the DSM 5

Instructions: Working together in small groups, see how many of the following exercises you can complete. These are designed to familiarize you with where different material is located in the *DSM 5*. When it comes to actual diagnoses, there will often be discussions about what diagnosis to use. In other words, don't expect that there is only one right answer. Discuss the cases with your colleagues and try to seek the best answer instead. After discussion, in which you will no doubt cover many of the issues raised in a real work situation, assign the diagnosis you feel is most appropriate.

1. It seems to you that Jim is having trouble in school. The teacher reports that Jim can read and speak well but has trouble writing out his thoughts in a coherent and organized manner. What chapter would you turn to? _____. You suspect the diagnosis is _____. What might you want to have done in order to confirm that diagnosis? _____

2. A man comes in and indicates he is suffering from depression. He tells you he has felt this way for several months. He appears to have a flat affect, some tearfulness, and tells you he is not sleeping well. In the course of the interview, you learn that he is an intermittent cocaine user. He tells you he tried unsuccessfully to quit but wasn't successful. What chapter or chapters would you turn to? _____. Do you think this man has two disorders or one? How do you code these two disorders?

3. A woman has breast cancer and reports being depressed since the diagnosis was first given. She has stopped going to choir practice, let her garden go to weeds, and is no longer interested in her book club. Her family reports that she seems to be getting worse. What chapter do you turn to? _____. What diagnosis do you use? _____.

4. Marie has been concerned for months that she is very ill with a serious illness. When you ask her about her medical symptoms that lead her to think this, she is very vague. Sometimes she has a headache. On occasion she has a stomach ache. Marie worries about her health constantly. She has a family doctor but claims he doesn't listen. She sees a gynecologist and has recently started seeing a doctor of internal medicine at the medical school. In addition, she regularly consults with a "healer"

she found in a rural county nearby. He prescribes various herbal concoctions for her to take. Sometimes Marie thinks she has cancer, but other times she is sure she has low thyroid and on occasion she has told you she believes she has a brain tumor. Today Marie tells you she has purchased yet another blood pressure machine and a glucose meter "so I can have a set on each floor." Marie's worries about her health interfere with her ability to do her job effectively in the billing office of a local hospital. She is asking for help "to sort out what is wrong with me" even though you are not a medical facility. What chapter do you turn to? _____. What do you think the diagnosis might be? _____

5. Catherine reports binge eating and self-induced vomiting in order to avoid gaining weight. "I just eat," she tells you. "I feel better when I am eating. The more I eat the better I feel until I can't eat any more and then I feel terrible and guilty, I guess, so I throw up." She tells you she often eats a dozen donuts in an afternoon and downs these with chocolate milk. "Or I will just buy a tub of Kentucky Fried Chicken and eat it all afternoon with a couple of cokes, you know, and then by evening I feel terrible and I get rid of it all." When you ask she admits that in the last few weeks she has been doing this more often. "oh, say about 6 times a week. That's why I decided to come in here. I mean it has gotten to about once a day. I just feel better when I eat, but I can see this is not so good." What chapter do you turn to? _____. What diagnosis do you think this might be? Write out the full diagnosis including severity.

6. Milton and his wife are separating and he is worried she will ultimately sue for divorce. He is here today to see what options he has for marriage counseling and issues regarding his children going through this. What chapter do you turn to? _____. What code would you use to describe his concerns? _____.

7. Among other things Henry says at night he gets "terrible sensations in my legs." He goes on, "I have to move my legs to make it stop but then it starts again in a few minutes. It is getting to the point where I have this about every night. Can't sleep!" It appears that his lack of sleep is interfering with his work and because he is often up pacing during the night in an attempt to relieve these sensations he is disturbing other family members. "My wife says I should 'just roll over and go to sleep,' but there is no sleeping!" He has had a physical examination and was referred to your agency. You are doing an intake. What chapter would you turn to? _____. What diagnosis would you choose? _____.

8. Ray has been using amphetamines for several years now, but the actual usage has increased in the last 9 months. He claims he wants to control the use. "I don't really want to quit, but I would like to use less," he says with some uncertainty. He goes on to say the reason he can't cut down is that he craves the stimulant and soon he finds a way to get more. "I get it that this is no good. I just don't really want to give it up altogether but I gotta do something." What chapter do you turn to? _____. What diagnosis do you choose? _____. Which criteria apply here? Give the numbers for each of the criteria that apply to Ray's case. _____.

9. Scott is 16 and was referred by the school. His mother who comes with him describes him as irritable and angry. Scott sits sullenly and listens to her. The school report states that he refuses to follow directions in the classroom, talks back to the teachers and most recently to the principal, and often annoys his classmates so that they can't really pay attention to the class assignments. Scott tells you sarcastically that the teacher is "dumb" and if she could do something interesting he'd listen. He admits that he won't follow her directions. "They're stupid." Last week another boy in the class offered to help him with his homework. Later it turned out the other boy blamed Scott for stealing his homework and turning it in as his own. Scott tells you, "the guy's a nerd." What chapter do you turn to? _____. What diagnosis do you think this might be? _____.

10. You are working in a drug rehab unit of the local hospital. Pearl has come in to detox after prolonged use of fentanyl and the unit physician has ordered an opioid antagonist, naltrexone. At present Pearl is complaining of muscle aches and is vomiting. She lies on her side groaning about feeling nauseated. She seems a little confused and weak. What chapter do you turn to? _____. What diagnosis do you choose? _____.

11. Angela has brought her 84-year-old mother in for evaluation. Her mother sits quietly but appears to be paying no attention to the conversation going on between you and Angela. Angela reports that her mother began to "fade away" beginning last month. Angela tells you, "she became forgetful and sometimes confused and she had been fine a few weeks before. She and I go to lunch about once a month and this time I went to pick her up and she was still in her nightgown and didn't remember that I was coming. She didn't even seem to know what time it was. I got her to go with me but I could tell at lunch she was not following the conversation. When she was still like this last week I decided we should come in here." What chapter do you turn to? _____. What diagnosis do you choose? _____. Consult the Case Manager's Toolbox in the Appendix to help you make this diagnosis.

CHAPTER 18

The Mental Status Examination

Introduction

The *mental status examination* (**MSE**) is based on your observations of the client. It is not related to the facts of the client's situation, but to the way the person acts, how the person talks, and how the person looks while in your presence. A mental status examination can be an abbreviated assessment done because someone appears to be in obvious need of hospitalization, or it can be an elongated process that takes place over several interviews. The MSE always has the same content, and you write your observations in roughly the same order each time.

Although a formal MSE would be done by a physician or psychologist, you can do an informal MSE in which you systematically look at the person's thinking process, feeling state, and **behavior**. You will want to understand the way the person functions emotionally and cognitively.

Much of the examination is done by *observing* how people present themselves at the interview and the manner in which they spontaneously give information about themselves and their situations. The examination is not done separately but is an integral part of the assessment interview. Questions that relate to mental status are framed as part of the overall assessment and not as a separate pursuit. There will be times when you or a clinician might ask for psychological testing to confirm your evaluation of the person, but during your own MSE of the person, this is not done.

Some of the terms you learn in this chapter are not necessarily words you will use in describing your clients and their **appearance** or behavior. This chapter is meant to familiarize you with the way some professional practitioners describe their clients and patients. If you know these terms, you will be able to follow the notes and discussions better.

Observing the Client

What to Observe

Your mental status examination of the individual involves observations of the following:

General appearance
Behavior
Thought process and content
Affect
Impulse control
Insight

Cognitive functioning
Intelligence
Reality testing
Suicidal or homicidal ideation
Judgment

A good case manager is a good observer. You pick up many details about the person, all of which are relevant to understanding the client's mental status. In a sense, you watch for the most obvious and the most subtle visual and verbal clues as to who your client is. Use what you see and hear to give you direction in regard to what questions to ask.

How to Observe

Throughout the interview note how the person communicates verbally and nonverbally and how the person behaves. In addition, you look at the content of the communication. You are looking at both *what* the person tells you and *how* the person tells it.

As people talk about why they came to your agency for services and about the main problems they are confronting, you will make some judgments about how they functioned in the past and how well they are functioning currently. You will note how they tell their stories. Is the person cooperative and friendly? Does he appear to be relieved and eager to talk to you, or is he mute, guarded, and uncooperative? Is she weepy and hesitant as she speaks, or is she forthright and stern? Does the person twist a tissue in her hands or rock back and forth in her chair, or does she use appropriate gestures? Does he relax during the interview or remain guarded and irritable?

At times you may need to assess clients' mental status through the observations of others who are close to them. Your clients may not always be able to tell you much about past events or functioning, and you will need to turn to others for that information. If there is no reliable source, you may not be able to perform a complete MSE that has a clear degree of certainty.

Documenting Your Observations

To back up your observations, use both descriptions of the individual's behavior during the interview and direct quotes made by the person in the interview. In this way, you carefully document your observations and your resulting conclusions.

When you describe the person, be sure that your values and prejudices do not appear in your notes. Use adjectives that describe the individual, but are objective. All editorial comments and value judgments should be omitted. Figure 18.1 defines some general terms that are commonly used when documenting observations of clients.

FIGURE 18.1 General terms used in documentation

Primary language: When you see this on a form, give the person's native language, and if it is not English, tell how well the person functions with English.

Presenting problem: In one or two sentences, tell why the person is coming to see you *now*. Use the person's own way of telling about it.

Past psychiatric history: Use incomplete sentences. Give dates, approximately how long, and summarize if there is much detail.

Functional ability: Note particularly if the person is able to display and carry out age- and stage-appropriate skills and tasks. Also note any recent change.

Moods/emotions: What does the person or the person's family say? How does the person seem to you?

Physiologic: What does the person or the family say about the person's appetite, sleep, and sexual activity?

Thinking: What is the person saying about how she is thinking? Are you able to follow her thinking? Does the story make sense? Are there delusions?

Perception: Are there any hallucinations?

Orientation/cognition/memory: Does the person think he can find his way? Does he know where he is? Does he remember well? Does he know what day this is?

Mental status examination: This is a word picture that tells what the person looks like *now*, not all the time.

Digital Download | Download from CengageBrain.com

Mental Status Examination Outline

Anthony LaBruzza (1994), in his book *Using DSM-IV,* provides a good outline for the mental status report that you will complete after the interview. He stated that his outline is not meant to be followed precisely, but it does give the major points and a framework to determine what is important. The outline shown in Figure 18.2 provides the major categories you must cover in a mental status report.

FIGURE 18.2 Outline for the mental status examination

I. General Description
   A. Appearance
      1. Dress and grooming
      2. Physical characteristics
      3. Posture and gait
   B. Attitude and interpersonal style
   C. Behavior and psychomotor activity
   D. Speech and language
      1. Rate
      2. Clarity, pitch, volume, tone, quality, and resonance
      3. Abnormalities

(continued)

FIGURE 18.2 (*continued*)

II. Emotions
   A. Mood
   B. Affect
   C. Neurovegetative signs of depression
III. Cognitive Functioning
   A. Orientation and level of consciousness
   B. Attention and concentration
   C. Memory
      1. Immediate registration, retention, and recall (a minute or less)
      2. Recent memory (a minute to days or weeks)
      3. Remote memory (weeks to years)
         a. Memory for recent past
         b. Memory for distant past
      4. Client's subjective report of memory difficulties
   D. Ability to abstract and generalize
   E. Information and intelligence
      1. Fund of knowledge
      2. Estimate of intelligence
IV. Thought and Perception
   A. Disordered perceptions
      1. Illusion
      2. Hallucinations
      3. Depersonalization and derealization
   B. Thought content
      1. Distortions
      2. Delusions
      3. Ideas of reference
      4. Magical thinking
   C. Thought processes
      1. Flow of ideas
      2. Quality of associations
   D. Preoccupations
      1. Somatic
      2. Obsessions and compulsions
      3. Phobias
V. Suicidality, Homicidality, and Impulse Control
VI. Insight and Judgment
VII. Reliability
VIII. The Environment

Source: Adapted from *Using DSM-IV: A Clinician's Guide to Psychiatric Diagnosis* by Anthony L. LaBruzza in collaboration with José Méndez-Villarubia, 1997, 1994 Jason Aronson, Inc., an imprint of Rowman & Littlefield Publishers, Inc.

Digital Download Download from CengageBrain.com

This section discusses the outline for the mental status examination and report in detail, defining terms to use and identifying items on which to focus for each category you will cover in mental status examinations and reports. Pay particular attention to the terms that have **Always** in boldface in the descriptor, as these are important items to which you must *always* give attention.

I. General Description

A. Appearance

1. Dress and Grooming. You may find the person's appearance to be average, meticulous, slightly unkempt, or disheveled. The person may have body odor, no makeup, makeup that is skillfully applied, or garish makeup.

- *Meticulous:* The appearance is too perfect, unusually so.
- *Skillfully applied:* The person is made up to look like a model.
- *Garish:* The person looks outlandish.
- *Self-neglect:* **Always** indicate when you think this is present. It involves such things as having body odor or looking disheveled and unkempt. Dress would be dirty, stained, or rumpled. This can be a sign of a mental illness such as depression or schizophrenia.
- *Dress:* You may find it casual, business, fashionable, unconventional, immaculate, neat, stained, dirty, or rumpled.
- *Immaculate:* This means the person is too neat.
- *Unconventional:* Use this term to refer to clothes that are inappropriate to the setting.
- *Fashionable:* This is fine unless the person looks like something out of *Vogue* in an office in a small town or average city.

2. Physical Characteristics. Note those features that are outstanding. Look at body build, important physical features, and disabilities. Note voice quality. Is it strong, weak, hoarse, or halting?

3. Posture and Gait. Note gait and any need for devices such as a cane or crutches. Look at coordination and gestures. For instance, does a right-handed person make most of her gestures with her left hand? Something like this could be a clue to neurological difficulties. Does the person limp or appear to slump? Does the person seem unsteady or shuffle?

B. Attitude and Interpersonal Style

Look at the attitude the person has with you. You may find it cooperative, attentive, frank, playful, ingratiating, evasive, guarded, hostile, belligerent, contemptuous, seductive, demanding, sullen, passive, manipulative, complaining, suspicious, guarded, withdrawn, or obsequious.

- *Hostility:* **Always** note when the person is hostile.
- *Uncooperative:* **Always** note when the person does not or cannot cooperate.
- *Inappropriate boundaries:* **Always** note if the client is too friendly, touches you, or attempts to draw you out personally.
- *Seductive:* Too close a relationship too soon; might call you by your first name or touch you

- *Playful:* Jokes, uses puns, self-deprecating humor
- *Ingratiating:* Goes along with whatever you think; wants to please
- *Evasive:* Talks, but gives nothing
- *Guarded:* Is more reserved than evasive; contributes the bare minimum, often with suspicion
- *Sullen:* Angry and somewhat uncommunicative
- *Passive:* Barely cooperates, needs to be led; generally without overt hostility
- *Manipulative:* Asks for special favors, uses guilt, solicits pity, or threatens
- *Contemptuous:* Superior, sneering, or cynical
- *Demanding:* Sense of entitlement
- *Withdrawn:* Volunteers little, appears sad

Watch your own emotional reactions to the people. Your reactions will give you important clues. Also pay attention to the person's *facial expression*. You may find it pleasant, happy, sad, perplexed, angry, tense, mobile, bland, or flat.

- *Bland:* Intense material, but looks casual
- *Flat:* No facial expression
- *Mobile:* Rapid changes in facial expression and mood

C. Behavior and Psychomotor Activity

Look at the quality and quantity of the person's motor activity. You may find the individual is seated quietly, hyperactive, agitated, combative, clumsy, limp, rigid, or has retarded motor function. You may find the person has mannerisms, tics, twitches, or stereotypes.

- *Seated quietly:* Uses normal gestures, but does not move around much
- *Hyperactive:* Is busy with hands and possibly feet
- *Agitated:* Cannot sit still (could be secondary to antipsychotic medication)
- *Combative:* Looks ready to hit, threatening
- *Awkward:* Unable to manage activity like sitting in the chair or writing; drops things (may be part of the illness or reaction to medication)
- *Rigid:* Sits like a tin soldier
- *Mannerisms:* These are unconscious repetitive actions
- *Posturing:* The person assumes certain postures and holds them inappropriately
- *Tics and twitches:* Less voluntary body movements
- *Stereotypes:* Four mannerisms strung together
- *Motor hyperactivity:* **Always** report this when you see a lot of hyperactivity, restlessness, and agitation. It may indicate a manic state, reaction to medication, or anxiety.
- *Motor retardation:* **Always** report this when you see the person moves slowly, in a constricted manner and with minimal motor responses. Speech and thought are slowed, often depressed. Depression can give the appearance of cognitive impairment. It could be the person is over medicated.
- *Mannerisms and posturing:* **Always** indicate mannerisms you see and any posturing.
- *Tension:* **Always** note tension, particularly if the person seems tense and the interview does nothing to relax the person.

- *Severe akathisia:* **Always** note severe restlessness. Sometimes it may be part of an illness, and sometimes it may be due to medication. If the physician believes it is due to an illness and increases the medication, the person may grow much worse. Therefore, try to establish when it started, how long it has gone on, and whether it has grown worse recently.

Always note the following when present: pacing, fidgeting, nail biting, trembling or tremulousness (a common side effect of lithium carbonate and tricyclic antidepressants), and abnormal movements such as rocking, bouncing, or grimacing (particularly strange facial movements).

- *Tardive dyskinesia:* **Always** note this condition if you see it or suspect this is what you are seeing. It occurs among psychiatric patients who have been on antipsychotic medications over a long period of time. The term literally means "late appearing abnormal movements" and seems to involve the muscles of the face, mouth, and tongue. Sometimes the trunk and limbs are also affected.

These movements can be slow and irregular (*athetosis*) or quick and jerky (*choreic*). All the movements are brief, involuntary, and purposeless. A person may twist the tongue and lips, make odd faces, bounce or tap the feet, or actually writhe and squirm in the seat.

- *Catatonic behavior:* **Always** note this behavior. It is generally a sign of severe depression or schizophrenia, catatonic type. It usually appears as a rigidity of posture wherein attempts to reposition the person are rigidly resisted. The person may voluntarily pose in bizarre and inappropriate ways. In waxy flexibility, the limbs of the person will remain in the position in which they are placed.

There is also a *catatonic excitement* wherein the patient engages in almost continual, purposeless activity that is nearly impossible to interrupt. Sometimes the patient engages in echolalia (repetition of everything that is heard) or mimics and imitates others during this episode.

D. Speech and Language

Speech is important because it is the primary means of communicating. Important to note are such things as rate, clarity, pitch, volume, quality, quantity, impediments, use of words, the ability to get to the point, and articulation.

You may find speech to be a normal rate, slow, hesitant, rapid, pressured, monotonous, emotional, loud, whispered, mumbled, precise, slurred, accented, stuttering, stilted, rambling.

- *Pressured:* Often rapid but constantly talking; cannot be interrupted (often a sign of a manic episode). Person appears to have racing thoughts.
- *Monotonous:* No variation in tone
- *Emotional:* Very expressive
- *Accented:* Note a native accent and also if the patient seems to accent certain words or syllables inappropriately

- *Impoverished:* May say very little either because of depression or because he is being interviewed in a language other than his native one; may also indicate a lack of facility with language.
- *Neologisms:* **Always** note when the person makes up entirely new words with idiosyncratic meanings. (This can occur due to **aphasia** or brain injury due to accident or stroke.)
- *Loose associations:* The person moves from topic to topic and the topics are only minimally related.
- *Flight of ideas:* The person moves from topic to topic or from idea to idea and these ideas are not related. The person may move rapidly through several different unrelated topics.

You should be able to identify any *neurological language disturbances*. Strokes, head trauma, and brain tumors can cause patients to lose their facility with language. Try to determine if the client has always had a language difficulty. Patients with schizophrenia may use loose associations as they talk. Those in a manic state may be prone to flight of ideas.

- *Aphasia:* Loss of ability to understand and produce language; damage usually to left hemisphere of the brain (left-handed people often have this in the right hemisphere)

The type and extent of aphasia depends on location and extent of brain injury.

- *Global aphasia:* Can neither speak nor understand, read, write, repeat words, or name objects
- *Broca's aphasia:* Can understand written and spoken language, but has trouble expressing own thoughts verbally
- *Wernicke's aphasia:* Inability to understand language and uses fluent, bizarre, nonsensical speech (The person may also act strangely and appear euphoric, paranoid, or agitated. It is easy to think this is a psychotic thought disorder, but in schizophrenia the person is generally able to write and speak in her language, repeat words, and name objects.)
- *Dysarthria:* Difficulty articulating due to problems with the mechanisms that produce speech. This sometimes produces distorted or unintelligible speech. The person usually can read and write normally. Ask the patient to repeat "No ifs, ands, or buts" to hear dysarthria better.
- *Perseveration:* Defined as the persistence "in repeating a verbal or motor response to a prior stimulus even when confronted with a new stimulus" (LaBruzza, 1994, p. 113). The client may give the same answer to different questions, stay on the same subject, or repeatedly return to the same subject.
- *Stereotypy:* "Constant repetition of speech or actions" (LaBruzza, 1994, p. 113). The patient may pull a shoe on and off, twist and untwist the hair, or repeat the same phrase or word over and over. These behaviors appear to be ritualistic and are common in childhood autism.

Give verbatim examples of what the individual has said to support your assessment of speech.

II. Emotions

A. Mood

This is the way a person is feeling at any given time. You may find it euthymic, depressed, sad, hopeless, empty, guilty, irritable, angry, enraged, terrified, expansive, euphoric, elated, sullen, dejected, or anxious. Ask yourself, what seems to be the dominant mood of the person?

- *Euthymic:* Normal mood
- *Expansive:* Feels very good and is getting better
- *Euphoric:* Out-of-sight happy
- *Anxious:* Worried and distressed

B. Affect

Affect refers to the underlying flow of moods. This would be the outward expression of the emotional state. You can see it in the way people use and position their bodies and in their tone and manner of speaking. You may find it broad, appropriate, constricted, blunted, flat, labile, or anhedonic.

- *Broad:* Normal range of moods
- *Appropriate:* Appropriate to the situation
- *Constricted:* Restricted range of emotional expression
- *Blunted:* Even more restricted
- *Flat:* No change of mood, unemotional
- *Labile:* Rapid change in mood (crying, then laughing)
- *Anhedonic:* Incapable of any pleasurable response, depressed
- *Blunted affect:* **Always** note a blunted affect where you find no change in mood throughout the interview and no change in facial expression. It generally indicates depression.
- *Emotional withdrawal:* **Always** note if the person seems emotionally withdrawn to you. The person would be inexpressive and probably have a blunt affect.
- *Excitement:* **Always** note if the person seems inappropriately excited to you. It means the person is overly enthused or terrified about the given situation.
- *Full range of affect:* This refers to an appropriate affective response to the entire interview. **Always** note inappropriate affect (such as giggling when there is nothing funny happening), as this can be a sign of schizophrenia.

C. Neurovegetative Signs of Depression

In major depression, body functioning often becomes irregular. Always inquire about sleep and appetite, and report a loss or gain of more than 5% of body weight. Listen for symptoms such as changes in energy levels, interest, enjoyment of everyday activities, or sexual functioning; constipation; and weight changes (LaBruzza, 1994, p. 115).

- *Initial insomnia:* Trouble falling asleep
- *Middle insomnia:* Middle-*of*-the-night wakening

- *Terminal insomnia:* Early morning *wakening*. Depressed individuals will often wake several hours earlier than usual and feel most depressed in the morning.
- *Hypersomnia:* Some depressed *individuals*, especially those with bipolar disorders, tend to sleep a great deal.

III. Cognitive Functioning

A number of medical and neurological problems, as well as substance abuse, affect one's cognitive functioning. The concern is that many people who have a disease of the brain may appear with what seems to be emotional and behavioral changes. In taking the history from the person, note previous levels of functioning and any previous emotional problems. If these are appearing in middle or late life, it is quite possible the person has a neurological problem.

A. Orientation and Level of Consciousness

Nearly all of the people who come to you will be alert and aware of their environment and their body. Occasionally, however, you may see individuals who are inattentive, drowsy, or who have a clouded consciousness. If these symptoms are present, use the proper term to indicate the person's level of awareness and briefly describe how the person exhibits this level. Medication can contribute to these stages as well.

- *Lethargy:* The person has trouble remaining alert and appears to want to drift off to sleep, but can be aroused. The person has trouble concentrating on the interview and seems unable to maintain a coherent train of thought.
- *Obtundation:* The person is difficult to arouse and needs constant stimulation to stay awake. The person may seem confused and unable to participate in the interview.
- *Stupor:* The person is semicomatose, and it takes vigorous stimulation to arouse her; she cannot arouse herself. There is no normal interaction during the interview as a result.
- *Coma:* This is the most severe consciousness problem wherein the person cannot be aroused and does not respond to any stimulation.
- *Oriented x3:* Means the person is oriented as to who he is, where he is, and when it is. Even when a person is having difficulty with consciousness, he may be oriented. If orientation problems occur as a result of lack of consciousness, it typically happens that the sense of time is affected first, followed by the sense of place, and finally by the sense of person. To be fully oriented requires an intact **memory**; thus, disorientation means there are memory deficits.
- *Ask for current date:* Reasonably accurate dates are acceptable.
- *Ask where the person is:* You can also ask for a home address, the present city or state, or for directions from here to the person's home or another familiar place. Sometimes people confused about place will behave as if they are at home or in another very familiar setting while in your office.
- *Ask who the person is:* Ask for personal identifying information (age, birth date, name). Ask if the person recognizes or knows other people who might be present. Does she know her relationships to these other people?

B. Attention and Concentration

Always note inability to pay attention and if the person appears easily distracted.

- *Attention:* Can the person remain focused on the interview?

If you feel a need to test this in the person, you can use digit repetition. Say five numbers, and then ask the person to recite them back to you.
Concentration is needed to learn new tasks and for academic success.

- *Concentration:* The person can concentrate on one thing for an extended period of time.

You can test the person's concentration by asking the person to perform a complex mental task. (Serial 7s is one way of testing; in this method, you ask the person to add in increments of 7 or subtract from 100 by 7s. Be sure your instructions are on the client's level of education, and do not use this exercise if severe academic problems are present. Be careful not to humiliate people!)

C. Memory

Memory involves the ability to learn new material, to retain and store information, to acknowledge and register any sensory input, and to retrieve or recall stored material. When there are problems, they usually have to do with three areas:

1. Registration
2. Retention
3. Retrieval

Destruction of significant parts of the brain causes problems with memory. All memory deficits should be noted. The physician or clinician will want to do further tests. If you suspect something, ask others who know the individual about their perceptions of the patient's memory functioning.

- *Short-term memory:* Refers to immediate recall limited to about seven items and generally lasts for about one minute. Some problems may be due to inattention, so evaluate attention before memory.
- *Long-term memory:* Rehearsal allows material in short-term memory to convert to long-term memory. Anxiety about the interview or the person's situation or even depression can interfere with this.
- *Amnesia:* Inability to remember
- *Anterograde amnesia:* Cannot learn new material
- *Retrograde amnesia:* Cannot recall recent past events
- *Head injuries:* Most common deficits are inability to recall names, recent events, and spoken messages, and forgetfulness or forgetting to do something important. The person may have trouble telling you what she is experiencing with her memory. Memory loss may be permanent if there was severe or repeated head injury.

- *Transient global amnesia:* Lasts minutes to several hours and is usually seen in older people. The person experiences sudden confusion, loss of memory, and disorientation and cannot recall what happened during the time period in question. Retrograde amnesia will be present. Person will be distraught, asking for reassurance as to where he is and what he is doing. This is caused by an insufficient amount of blood to the brain.

Memory Testing. First, ask the person if she has been having any problems with memory. A family member may be able to shed some light on memory issues if any exist. During the interview, note memory lapses and difficulty recalling what the interviewer has just said. If you notice memory loss, note it so that further testing can be done. All the memory tests described in the following would be done only if you had considerable questions about a person's memory:

- *To test immediate recall:* Use a random list of digits, saying them in a normal tone of voice, about one digit per second. Ask the person to repeat them. Start with two digits and keep adding until the person fails. Give the person two times to try this. If the person fails at five digits or less, there is reason for concern about sustained effort, attention span, and immediate memory. Anxiety and depression are the most common reasons people fail this test (LaBruzza, 1994, p. 125). Strokes and other brain injuries can also affect recall.
- *To test recent memory:* Ask the person to recall events that have happened in the last few hours or days before she came to see you. You might ask what she had for lunch or where she parked the car. It is helpful if you can validate the answers with someone close to the client who knows. Another way to test is to ask about something that may have happened or been discussed earlier in the interview.

    You may get several different versions. In cases of assault or trauma and where the victim feels comfortable with you, the different versions may indicate the she is able to recall more of the details each time she goes over what happened. With some people, you might give three or four unrelated words and ask them to recall these words after a short interval. Begin by saying the words in a normal tone of voice and ask them to repeat the words back to you. Note how many times a client must do this before learning the words. About 3 to 5 minutes later, ask clients to recall the words. With a normal memory, a person should be able to recall them (LaBruzza, 1994, p. 126).
- *To test remote memory:* You can ask people about personal events in their lives and commonly known public events that happened in years past, such as major news stories. Use material that should be known by a person who is reasonably well informed. If the person does not appear to be able to do this test because of a lack of education, a difference in culture, or an intellectual disability, decide carefully what you will ask the person (LaBruzza, 1994, p. 128).

Additional information on memory and aging and how to assess memory can be found in Chapter 8 of *Fundamentals for Practice with High Risk Populations* (Summers, 2002).

D. Ability to Abstract and Generalize

Proverbs. Cultural background and intelligence can influence how well a person thinks abstractly or how well the person can deal with similarities. Proverbs are generally used to see how well a person thinks abstractly. You need a general fund of information to be able to use proverbs in this way. Tell the person you are going to say a proverb and you would like the person to tell you in his own words what he thinks the proverb means. Then judge how concrete or abstract the reply is. Repeat the person's response verbatim in your report.

Individuals who are psychotic or on the verge of psychosis will often indicate this in their response to a proverb. Use proverbs that are free of gender and racial bias. The following are some proverbs you can use (LaBruzza, 1994, p. 129):

- A stitch in time saves nine.
- A rolling stone gathers no moss.
- Don't judge a book by its cover.
- Two wrongs don't make a right.

"A rolling stone gathers no moss" could be explained by a person who thinks concretely as, "If you roll a stone down the hill, it can't collect moss." A more abstract response might be, "If you keep moving, life remains interesting and challenging." Looking at "Don't judge a book by its cover" a person who thinks concretely might say, "well, the cover on the book doesn't always tell you what is inside the book." A person who is more abstract might say, "first impressions aren't always correct."

Similarities and Differences. Ask the person to tell you how two objects or two events are different or alike. This will require the individual to think somewhat abstractly about categories and relationships. Name two items and ask the person how these differ and how they are similar. The following are some combinations you might use (LaBruzza, 1994, p. 129):

- Apples and oranges
- Trees and flowers
- Houses and cars
- Dogs and cats

E. Information and Intelligence

To get an idea of the person's overall intelligence, ask questions that tap the person's fund of general information. It should be information known by the general public. Again, you must be sensitive to the person's cultural background, level of education, and intelligence. The following are examples of some questions you might ask (LaBruzza, 1994, p. 130):

- Who were the last four presidents?
- Who is the governor of the state?
- How many weeks are there in a year?
- What is the capital of the state (or the country, or France)?
- Who was Mark Twain?

IV. Thought and Perception

When a person's perceptions are disordered, it offers important clues to what the diagnosis might be. Here you want to know how people actually perceive themselves, the world around them, and others in their world. What does the person think, and what thoughts and concepts are most on his mind? Perception is the way in which we form an awareness of our environment. People who have difficulties with perceptions often perceive their world inaccurately (LaBruzza, 1994, p. 131).

A. Disordered Perceptions

Following are some terms that describe various disordered perceptions:

- *Illusions:* The person either misperceives or misinterprets a sensory stimulus. A tree branch brushing the side of the house in the wind sounds like people entering the house, or a dishwasher running sounds like people talking in another room.
- *Hallucinations:* In the absence of external stimuli, the person perceives something. The most common hallucination is hearing voices. Voices generally increase when the person is around white noise. White noise is even background noise, such as the dishwasher running, a roomful of people chattering, or rain drumming on the roof. If you can, find out who is talking, what they are saying, and how the person feels about it. Is there a command for the person to do something? If so, include the command in your report. Some commands are dangerous to the person or to others. **Always** note hallucinatory behavior.
- *Depersonalization:* **Always** note depersonalization wherein the person feels estranged or detached from herself.
- *Derealization:* **Always** note derealization wherein the person feels detached from what is going on around her. Be sure to note this. A person who dissociates cannot always be sure that what is happening is real (LaBruzza, 1994, p. 132).

B. Thought Content

The following terms are used to describe thought content:

- *Distortions:* A person distorts a part of reality. A woman with anorexia believes she is fat when she is thin. A person who is well believes his cough indicates tuberculosis. A person whose neighbor does not think to wave assumes the neighbor is angry.
- *Delusions:* An inappropriate idea from which a person cannot be dissuaded using the normal means of argument or evidence. Sometimes it is culturally inappropriate as well. Evidence to the contrary has no effect. For example, a client might insist that she has a case in court that will eventually yield her a great sum of money. No amount of persuasion or documentation can dissuade her from that belief and convince her that this isn't so. **Always** report the content of a delusion. Note if the delusion is incongruent with the client's mood. Delusions indicate psychosis. **Always** note if delusions are present.

People with *paranoid delusions* believe they are being singled out for harassment or are being controlled by forces outside of themselves. They may have an entire system of interconnected ideas developed that support their delusions. Common to schizophrenia are:

- *Thought withdrawal:* Belief that one's thoughts are being taken out of one's mind by an outside force
- *Thought insertion:* Belief that thoughts are being placed into one's mind by an outside force
- *Thought broadcast:* Belief that thoughts are being taken and broadcast so that others know what one is thinking
- *Suspiciousness:* **Always** describe this and the object of the suspicion.
- *Grandiose delusions:* The false belief that one is extremely important or a false belief that one is imbued with special powers. **Always** describe ideas of grandeur and any grandiose behavior.
- *Somatic delusions:* False beliefs about one's physical health
- *Delusional guilt:* Falsely believing that one is the reason or cause for terrible things that have happened or will happen
- *Nihilistic delusions:* A false belief in the meaninglessness of life and all events and circumstances, in nothingness; hopelessness; belief in the end of the world
- *Ideas of inference:* Refers to the ideas the person holds about what others do to affect him
- *Ideas of reference:* Refers to beliefs that people are talking and thinking about him. Messages on TV and radio are meant specifically for this person.
- *Magical thinking:* Means belief in astrology or a superstition, or the person thinks he has magical powers in his words, thoughts, or actions. This thinking is found in children who have not developed **reality testing**. It is part of human development and is not pathological until it becomes extreme, as in obsessive-compulsive disorder or a delusion. **Always** check religious beliefs or cultural background to see how this thinking fits with what is going on in the person's life and these aspects of the person's life (LaBruzza, 1994).
- *Thought content:* **Always** specify unusual or important thought content such as: (a) what the person is suspicious about, (b) what the person feels guilty about, and (c) what the person is preoccupied about.
- *Bizarre behavior:* **Always** note any that you witness or any that is reported to you by reliable others.

C. Thought Processes

You may find the form of the person's thoughts to be spontaneous, logical, goal directed, coherent, impoverished, blocking, nonspontaneous, incoherent, perseverative, circumstantial, tangential, or illogical. You may find the thoughts to have loose associations or flight of ideas. You may find that the content contains neologisms or is distractible.

- *Flow of ideas:* Refers to the quality of the associations the person makes between ideas or between points in the person's discussion. Note the stream of the client's thoughts, the rate of thinking, the coherence, the continuity, and whether the thought process is goal directed (LaBruzza, 1994, p. 134).
- *Spontaneous:* Means you do not have to keep asking questions. The person readily volunteers information.
- *Goal directed:* The person answers the main questions about why she came and what she needs, and does not stray to other related topics.
- *Impoverished:* The person uses words but is very skimpy with them. There are too few ideas, and thinking is slow. Often attributable to depression or schizophrenia.
- *Racing thoughts:* The person thinks rapidly. Speech appears pressured. Often attributable to manic or hypomanic state.
- *Blocking:* The person stops, pauses, and starts somewhere else. There is an interruption to the normal flow of speech. The person may appear to forget where she was in the conversation when she resumes talking.
- *Circumstantial:* The person appears to throw in too many irrelevant details. The person has too many ideas associated with one another and too many digressions. Often thought to be a defense against dealing with troubling issues or feelings (LaBruzza, 1994).
- *Perseverative:* The person goes over and over the same point or idea.
- *Flight of ideas:* The person goes from one thought to another in logical sequence but is headed far from the original topic.
- *Loose associations:* The person's points do not hang together logically. Ideas shift in an apparently unrelated way. Characteristic of schizophrenia.
- *Illogical:* What the person is saying does not make sense.
- *Incoherent:* There is no meaning; the speech is disorganized; the person may be schizophrenic.
- *Neologism:* The person makes up new words.
- *Distractible:* Person cannot stay focused; may indicate mania.
- *Clang association:* "The sound of a word, rather than its meaning, triggers a new train of thought" (LaBruzza, 1994, p. 136).
- *Tangentiality:* Means "veering off" on somewhat related, but irrelevant, topics. May show a difficulty with goal-directed thinking. Common in mania and hypomania (LaBruzza, 1994, p. 136).
- *Overvalued ideas:* The idea might be possible, but it is used or seen by the person to explain more than it could possibly explain.
- *Conceptual disorganization:* **Always** note conceptual disorganization. This refers to an inability to conceptualize the problem clearly and may involve a number of the terms previously noted, such as loose associations, flight of ideas, tangential thinking, or incoherent content.

A note about the word *confused*: Students often say a client is confused when they are describing a person who is having trouble deciding what to do or a person who is feeling very ambivalent about a particular decision. In the mental status examination, however, the word *confused* means that the client was not oriented to time and place and person. In other words, confused clients do not know where they are, cannot understand what is being said to them, or do not recognize familiar people.

When a client is *ambivalent*, use that word; and when a client cannot make up her mind, say just that. Reserve the word *confused* for describing true cognitive confusion.

D. Preoccupations

These are thoughts and issues that appear to be the primary focus of the patient's thinking. There is an "obsessive quality" to the preoccupation (LaBruzza, 1994, p. 136).

- *Somatic preoccupation:* Focus on bodily functions, physical health. There is a hypochondriachal quality to the preoccupation. List and describe somatic concerns. Do not assume these are not real problems without proper medical documentation.
- *Obsessions:* Persistent thoughts that are intrusive and unwanted and that appear to haunt the person (LaBruzza, 1994, p. 136). The person may hold an idea that is not true in that intensity.
- *Compulsions:* Actions that are often the "counterpart of the obsession." These are "persistent, intrusive and unwanted urges" to take some action. If one does not complete the action, there is intense anxiety. These actions can be repetitive and ritualistic, such as checking the stove, counting steps, and straightening picture frames (LaBruzza, 1994, p. 136).
- *Phobias:* These are "irrational, intense, persistent fears" of such items as dogs, heights, elevators, insects, leaving home, closed spaces, and flying (LaBruzza, 1994, p. 136). The person will go to great lengths to avoid the situation or object of the phobia.
- When writing out notes we need to be clear about the difference between *perseveration* and *obsession*. Note that *perseveration* is the term we use when a person focuses on one topic and comes back to it over and over, unable to change to another topic or stay on another topic relevant to the interview. Sometimes students describe this as an obsession. *Obsession*, as noted above, consists of thoughts that intrude on the person beyond their control and are experienced as intrusive and unwanted.

V. Suicidality, Homocidality, and Impulse Control

Suicidality and Homocidality. You have a clinical and a legal responsibility to assess whether the person is a danger to herself or to others. A person who is dangerously impulsive or holds thoughts of suicide or homicide needs special observation. When conducting an interview, look for these thoughts. These are called suicidal ideation or homicidal ideation. Discover if there are plans to carry out these ideas or if the person seems to have an undeniable intention. Thoughts present some danger, plans make the situation more dangerous, and the clear intention to go forward with the plans creates an extremely dangerous situation.

The mildest form would be thoughts, followed by actually making plans. If the individual tells you she has purchased a gun and learned the intended victim's work schedule, you may assume there is a plan. If the individual tells you he has saved 3 months' worth of medication for an overdose, this would appear to be a plan. If the individual has a plan and expresses to you her obvious intention to carry out this plan, the situation becomes more serious (LaBruzza, 1994).

Always include the plan, the means, and the timetable in your report. If the person is homicidal, include toward whom the thoughts are directed as well.

Impulse Control. When you assess **impulse control**, you want to know how the person deals with aggressive urges, sexual urges, and strong desires to carry forward any plan not particularly well considered. Look at ways the client has handled stressful situations in the past. Is there a history of acting on impulse without much thought as to the consequences? Does the person seem unable to tolerate stress? How did this person handle stress in the past? Is there a history of uncontrolled aggressive behavior, either sexual behavior or hostile behavior? Can this client tolerate frustration?

The three behaviors to look for in childhood are setting fires, cruelty to animals, and bedwetting. When these behaviors all occur in childhood, they appear to be significantly associated with cruel adult behavior. In adulthood, you might find behaviors such as punching holes in the wall, smashing furniture, slitting one's own wrists, drinking excessively, or turning to drugs or a drug overdose.

VI. Insight and Judgment

Insight. The person understands that she is "suffering from an illness" or has an emotional or personal problem. Make a note if the person completely denies any problems or denies having any part in a problem that obviously affects the client's relationships with significant others. For instance, some people have extremely acrimonious relationships with their relatives but seem unaware that their hostile responses actually elicit more hostility from these relatives. As another example, a person may come in seeking help for a substance abuse problem and genuinely believe that there is no problem and that his behavior is appropriate. In addition, he may indicate that those who are concerned are actually inappropriately concerned. People hold **insight** in "varying degrees" and may show only partial understanding of their difficulties (LaBruzza, 1994, p. 137).

Judgment. The person can critically evaluate her situation and make good decisions about a course of action. Look for risky behavior in the past that could have been potentially harmful (practicing unsafe sex, binge-drinking and reckless driving, etc.). "Assess whether the person is able to understand the potential consequences in their behavior" and can plan preventive measures. You might ask what the individual would do if he spotted a fire in a movie theater or what he would do if he found a stamped, sealed, and addressed envelope (LaBruzza, 1994, p. 138).

VII. Reliability (Accuracy of the Client's Report)

You need to state briefly your impression of person's reliability and accuracy in giving you the details of their situations. If a person is psychotic, the material presented is likely to be extremely unreliable. A person who is suffering from dementia or delirium may be having considerable difficulty remembering what has happened or what is happening now. Some people deliberately tell falsehoods to qualify for disability or to give false impressions of themselves to the worker (LaBruzza, 1994, p. 139). Others, with memory problems, sound entirely plausible but have learned to make things up to fill in the gaps in their memories, referred to as confabulating.

VIII. The Environment

Sometimes you will be asked to go to someone's home to do an assessment or to do an interview. People's surroundings often hold clues to the way they are currently structuring their lives.

- *Inappropriate surroundings:* Means the person has arranged furnishings inappropriately. It may be that furniture blocks doors and windows or that the windows are covered oddly, perhaps with tin foil or some other material. There might be strange wires leading nowhere, odd decorations, or strings of odd things hung across a room. You might find household objects being used inappropriately. Sometimes a person believes the people on TV can see into his home or the people on the radio can hear what he says. In such cases, you may find these devices covered or blocked in some way.

    Be very careful in making these judgments. At one time a worker decided that windows partially covered with foil-wrapped insulation were a sign of something amiss in the person he was seeing. It turned out the man was doing that to save energy. In another home, a worker decided the blanket over the TV indicated the woman was paranoid. In fact, it turned out the TV was very new and she was very proud of it; she had put a blanket over it the day before when her young nieces came to visit so the TV would not get scratched. It helps to inquire matter-of-factly about what you see.

- *Waste and Trash:* Look at the way the person keeps his home. Are there unusual collections of junk or trash? Are there piles and piles of paper and magazines everywhere, or collections of string, bags, and other objects? Sometimes the home is very cluttered or dirty, with unwashed bed linens and with unwashed dishes stacked all over the kitchen. This tells you something about the person's capacity to attend to the routine details of living, or it may indicate a debilitating mental illness, such as hoarding. You might find urine or feces on the floor or walls. You might find the pets in the home neglected, their food bowls full of rotten food. Sometimes the house might be overrun with strays the person has taken in, but is unable to adequately care for.

Note: It is not unusual for a middle-class worker to assume that the manifestations of poverty are the signs of a person who is mentally ill. You must be very careful not to ascribe mental problems to someone who is actually too poor to live by middle-class standards.

Always report environmental cues that seem to support the rest of your mental status examination.

Clinical Definitions. There are two words we tend to use in everyday conversation that mean one thing when we do but mean something entirely different when we use these words clinically.

- *Confused:* As noted above, students often use the word "confused" to mean that a person is not clear what decision to make or what would work in the person's best interest. They describe their client as being confused when it would be better to say the person was uncertain, ambivalent, or unsure. Confused is a term we use clinically to describe someone who is not oriented as to person, place, and time.
- *Obsessed:* Students often use "obsessed" to indicate that the person is focused on a topic to the exclusion of most other things. If the person does keep returning to one topic and it is hard to steer the discussion to other topics or issues we would say the person is perseverating, not obsessed.

Summary

Every contact you have with clients must be documented, and every documentation of a client contact must contain your impressions of how the person seemed at the time of that contact. In social histories and evaluations, these impressions are lengthy and contain material observed by you during your contact with the client.

The mental status examination is a snapshot of where the individual is at a particular time. You are not trying to document how the client will always seem. Tomorrow the person may be different. Your notes can be used to follow peoples' progress or regression. Using the material in this chapter will help you to write full, accurate impressions of clients that will help clinicians understand what has been going on with them between sessions and determine the best course to follow with each individual.

Exercises

These exercises can also be filled out online at CengageBrain.com.

Exercises: Using the MSE Vocabulary

Instructions: See if you can fill in the blanks for each of these questions. This exercise is simply to acquaint you with words you might encounter in the course of your work.

1. The chart on Mr. Kling reads that he has conceptual disorganization. What was the psychologist referring to? _____.

2. In a staff meeting, the psychologist and the case manager discuss the fact that Mrs. Purdy seems to have racing thoughts. They are talking about what? _____.

3. Mr. Garrison is withdrawing from alcohol and he sees green bugs on the wall and is afraid of them. We call this a _____

4. The neurologist's report comes for Mr. Engler. The diagnosis is Broca's aphasia. What does that mean? _____.

5. When the psychiatrist saw Mrs. Nguyen, he said her mood was euthymic. He meant _____.

6. Mr. Kissel's speech is described as impoverished. That means his speech is _____.

7. Another term for severe restlessness is _____.

8. Another way to describe a broad range of moods is _____.

9. When Dr. McCoy said Mr. Perkins used neologisms, Dr. McCoy was referring to _____.

10. Mrs. Dell has been on antipsychotic mediation for some time, and now she shows signs of tardive dyskenesia. That means she _____.

11. When Mrs. Jones was described by the psychologist as seductive, he meant that Mrs. Jones was _____.

12. The chart says that during the interview with Mr. Landon at the prison, where he was being held after being arrested for drug possession, Mr. Landon's speech was guarded. That means his speech was _____.

13. Bill has been on medication for schizophrenia for over 30 years. Today you notice he shuffles as he walks and grimaces with his mouth and tongue. You suspect he has developed _____.

14. When the ambulance crew called in about 93-year-old Mr. Keller, they said they were bringing him into the hospital, but he was oriented ×3. They meant _____.

15. Mrs. Harris complained of trouble falling asleep. In the chart, this was written as _____.

16. When Alice stops taking her medication she tends to believe she can fly. We call this a _____.

17. Discussing Mr. Rodriquez's delusions with you, the psychiatrist remarks that Mr. Rodriquez has thought withdrawal. The psychiatrist means _____.

18. The psychologist is describing his interview with Mrs. Carter. "She tends to perseverate," he noted. He means that Mrs. Carter _____.

19. Mr. Trong has been depressed ever since he came to this country from Vietnam. Lately the depression has worsened. Today his therapist calls from the International Center asking if you can arrange hospitalization for Mr. Trong because he is catatonic. Mr. Trong is _____.

20. Miss Aller is homeless and mentally ill. The mission calls to say that Miss Aller believes that her thoughts are being taken out of her head and broadcast so that others know what she is thinking. Another way to write this is _____.

21. When Mr. Cruz talks about the accident, he tells you he feels detached from himself. Another word for that feeling is _____.

22. When the ambulance arrived the student had had so much to drink that they noted he was experiencing obtundation. They meant _____.

23. The chart tells you that after the accident Mr. Cruz had retrograde amnesia for a while. That means he _____.

24. The record that comes from the hospital on Ms. James states that she suffered from terminal insomnia during her stay there. That means that she _____.

25. After the stroke, Mr. Torres was described as having Wernicke's aphasia. That means _____.

26. During the interview, Miss Bell constantly pulled her gloves on and off, on and off, in a ritualistic fashion. One word for this kind of behavior is _____.

27. Miss Bell was asked to interpret several proverbs to test her _____.

28. Mr. Lincoln was seen with his wife for marital difficulties the couple was having. He said that he personally had no problems at home and that if his wife said there were problems at home, these were her problems and she was bringing them on herself. What might we say about Mr. Lincoln's insight? _____.

29. The family doctor of Mrs. Fong calls to arrange for an appointment. She believes Mrs. Fong is depressed and "is showing strong neurovegetative signs of depression." What does the family doctor mean by that? _____.

30. When you interview Mr. Marks, you notice that the *sound* of a word seems to trigger a new line of thought for him. You call this _____.

CHAPTER 19

Receiving and Releasing Information

Introduction

You have done an assessment of the individual who is seeking treatment or services through your case management unit. Some people come in having been seen or treated in other places or at other times. Before you can adequately plan, it may be a good idea to obtain records or summaries of past treatment or services the person might have received. This information gives you a clearer picture of the client and will help you to plan for that person. Knowing what has been done in the past will prevent your trying programs and services with which the person has had little success.

Sending for Information

During the assessment, have the person sign the appropriate release forms. Then, immediately after the interview, mail or fax those forms to the other agency so the information can be obtained in a timely manner.

If You Release Information

Sometimes people will bring you a form and ask you to release their records to another agency. It is important to protect clients when releasing information. Even though they tell you it is all right to do so, people may not fully realize the consequences of giving out information. It is your responsibility to see that people understand what could happen when information is released.

For example, a mother of a teenage boy wanted to have his records sent to a summer camp he would be attending. The camp was for students who tested well but seemed bored with school and whose grades did not match their potential. The records related to his depression were not entirely relevant to the camp activities and the application simply asked for any history of depression or other mental health issues. The case manager was able to talk to the mother about what to release and a short statement was finally constructed that indicated the boy currently being successfully maintained on his particular medication. The details were not shared with the summer camp.

The facts of a person's situation belong to the person. These facts are not yours, even though they are in your possession. Releasing them without written consent is unethical—and also illegal, unless covered by the exceptions we looked at in Chapter 2 on ethics.

Before continuing your work in this chapter, review the discussion in the section entitled "Confidentiality" in Chapter 2. In addition, review the HIPAA regulations discussed in the same chapter under the section "Health Insurance Portability and Accountability Act" so that you are entirely familiar with those guidelines. Reviewing these materials will prevent you from erroneously releasing information and placing your agency in legal jeopardy.

You may find, when you are practicing case management, that some places still use a blanket release form. The client signs the blanket form approving **release of information** to a number of places listed on the form for an unspecified length of time. These forms, while saving the staff time, are no longer legal.

Directions for Using Release Forms

Included in Appendix C of this book are two release forms for your use. The first is a standard release form, and the second is a release form specifying that information regarding the client's HIV/AIDS status may be released. These forms will allow you to obtain the information you need to plan for the client.

General Release Form

Here is information you will need to correctly fill out the general form:

1. RE: refers to the client. Neatly and clearly write in the client's name on that line.
2. DOB: refers to the person's date of birth. Place that date on this line.
3. "I hereby authorize..." is the client speaking. It is the client who is authorizing that her information can be either released or received by your case management unit. If the client wishes to have you release information for another agency, check the first box. The purpose of the release is generally "developing a service plan" or "goal planning." You might also write for the purpose of "developing a treatment plan."
4. If the person wants you to receive material from another agency where the individual received services in the past, check the second box, and then provide name of the agency to which the form will be sent and from which you will **receive the information.**

5. The phrase "for the purpose of" refers to why the material is being sought. Generally, filling in the blank with "goal planning" or "treatment planning" or "developing a service plan" will be sufficient. Sometimes a doctor with whom you are working is seeking the information for "ascertaining past medications for the purpose of prescribing." Do not go on a fishing expedition, asking for material that you think would be interesting to read, but which bears little impact on what you are planning for the person.
6. Next write in the date on which the permission to release information will expire. This date should be 90 days from the date below next to the signatures.
7. Information you are seeking is always best obtained in summaries. You do not want everything because you cannot store it all or read all of it. Only seek what is relevant. For instance, do not ask for school records if the person is an adult with no school problems. Do not seek all the hospital records if the discharge summary will suffice.

    On the line marked "other," it is perfectly all right to write in "medical records pertaining to . . . ." This way you will not get all medical record information, including that which you do not need for planning, but only the information pertaining to what is relevant (e.g., the most recent thyroid surgery or the two hip replacements or the emergency room records from April 16, 2015).
8. The authorized signature is the signature of an adult or the signature of the parent or guardian if the client is a child. You countersign the form, as well, and date it.
9. Your signature goes on the line next to the authorized signature as the person who will receive the information from the other agency. Date your signature.
10. If your client is not able to write but is able to indicate that he wants his records sent to you, you will need to get two witnesses to sign. By signing, they are saying that they witnessed the person indicating the desire to have the records sent.

Again, as the staff person, you must sign the form and date your signature.

HIV/AIDS-Related Release Form

The HIV/AIDS–related release form is used only when:

- The information to be released contains references to the person's HIV/AIDS status (including testing for HIV that was negative).
- The client has given permission for this information to be released.

The two forms go together if there is any information related to the client's HIV/AIDS status. You may staple them together so that they do not get separated. The expiration date on the general release form will also apply to the attached HIV/AIDS release.

Most states have laws providing that HIV/AIDS information is to be released only to individuals who "have a need to know." These laws take the position that only those directly involved in the medical and physical care of a person with this disease "need to know." Therapy, therapeutic programs, and other services generally can give quality services without knowing the client's status. Therefore, it is important to use universal precautions with every individual in your program who might need first aid at some point.

If your client does not want HIV/AIDS material released, you must send records with all references to the client's status deleted. Blacking out portions of the record is insufficient, as such copies can give the person's status away; therefore, they are not adequate in preserving the client's confidentiality. Instead records must go out with no mention and no obvious deletions of HIV/AIDS status.

Examples of the Release Forms

General Release Form

A sample release form signed by a woman who has been in a drug and alcohol treatment program in another state is shown in Figure 19.1. The case manager needs to understand what took place there in order to plan adequately with this woman.

FIGURE 19.1 General release of information form

Wildwood Case Management Unit

Request/Release of Information

RE: *Marcella Renova*   DOB: *3/19/79*

To Whom It May Concern:

[ ]   I hereby authorize the Wildwood Case Management Unit to release information about services rendered to the above-named, for the purpose of:

[√]   I hereby authorize the Wildwood Case Management Unit to receive information about services rendered to the above-named from:

Spring Garden Substance Abuse Clinic, 123 First Street, Anywhere, PA

for the purpose of *service planning*

Such information may be transmitted under the conditions stated below, and/or as required by Federal or State statute or order of the court. This release will be effective for a period of ninety (90) days from the date signed below and will expire on *7/25/16*.

Information to be released/received may include

( ) Medical records   ( ) Social/developmental
(x) Discharge summary   (x) Psychiatric evaluation
( ) Psychological evaluation   ( ) Educational record
( ) Vocational evaluation/summary ( ) Substance abuse treatment history
( ) Treatment summary   ( ) Social/developmental history
( ) Personal information including Social Security no(s) address(es) and telephone no(s)
( ) Other _____

To the agency or professional person receiving this release:

THIS INFORMATION HAS BEEN DISCLOSED TO YOU FROM RECORDS WHOSE CONFIDENTIALITY IS PROTECTED BY STATE LAW. STATE REGULATIONS PROHIBIT YOU FROM MAKING ANY FURTHER DISCLOSURE OF THIS INFORMATION WITHOUT PRIOR WRITTEN CONSENT OF THE PERSON TO WHOM IT PERTAINS.

FIGURE 19.1 (*continued*)

> THIS CONSENT TO RELEASE OF INFORMATION CAN BE REVOKED AT THE WRITTEN REQUEST OF THE PERSON WHO GAVE THE CONSENT.
>
> I have read this form carefully and I understand what it means.
>
> *Marcella Renova*   4/25/16   *Kathryn Parsons*   4/25/16
> Authorized signature   Date   Staff person signature   Date
>
> I have read this carefully and I understand what it means. As I am not physically able to give my written consent, I am giving my verbal consent to release these records.
>
> _____
> Witness signature   Date   Staff person signature   Date
>
> _____
> Witness signature   Date

© Cengage Learning®

HIV/AIDS-Related Release Form

For our purposes, we will assume that the woman on our general release form, Marcella Renova, has contracted HIV/AIDS. We will need to attach the HIV/AIDS–related form. Figure 19.2 provides a sample of this form.

While these forms may not be precisely the same as the forms you will find in the agencies where you practice, they will give you valuable practice in sending for and releasing information and understanding the kind of information needed before information can be released.

When the Client Wants You to Release Information

You will receive, from time to time, requests that you or your agency release information about people you have served to another person or agency. Be sure that the release form is in order and follows the guidelines that you have learned in this chapter and Chapter 2 on ethics. If there is HIV/AIDS material in the file, be sure that the proper forms have been received. If there are no HIV-related release forms, you cannot send the material until all references to the client's HIV status have been eliminated.

When the Material Is Received

When you have received the material from the other agency, you have the elements to construct a fairly accurate clinical picture of the person who has come to you for help. You have the person's initial description of the problem, followed by a more

FIGURE 19.2 HIV/AIDS–related release of information form

Wildwood Case Management Unit

Release of HIV/AIDS–Related Information

RE: *Marcella Renova*   DOB: *3/19/79*

To Whom It May Concern:

[√]   I hereby authorize the Wildwood Case Management Unit to receive information about services rendered to the above-named,

[ ]   I hereby authorize the Wildwood Case Management Unit to release information about services rendered to the above-named, for the purpose of:

service planning.

[√]   Such information to be released includes information regarding my HIV/AIDS status and/or treatment.

To the agency or professional person receiving this release:

THIS INFORMATION HAS BEEN DISCLOSED TO YOU FROM RECORDS WHOSE CONFIDENTIALITY IS PROTECTED BY STATE LAW. STATE REGULATIONS PROHIBIT YOU FROM MAKING ANY FURTHER DISCLOSURE OF THIS INFORMATION WITHOUT PRIOR WRITTEN CONSENT OF THE PERSON TO WHOM IT PERTAINS.

THIS CONSENT TO RELEASE OF INFORMATION CAN BE REVOKED AT THE WRITTEN REQUEST OF THE PERSON WHO GAVE THE CONSENT.

I have read this form carefully and I understand what it means.

Marcella Renova	*4/25/14*	*Kathryn Parsons*	*4/25/14*
Authorized signature	Date	Staff person signature	Date

I have read this carefully and I understand what it means. As I am not physically able to give my written consent, I am giving my verbal consent to release these records.

Witness signature	Date	Staff person signature	Date

Witness signature	Date

---

detailed explanation and a thorough social and medical history. You have sought and noted the individual's expectations for services and done some preliminary planning with the person. You have at least a general idea of how the *DSM* categories might or might not apply to this individual. You have assessed the person's mental status and received information to support or enhance your own conclusions. It is time to structure a service plan with the person.

Other Issues Related to Releasing Information

Generally the client requests that his or her records be sent to another agency or individual practitioner. Sometimes, however, the client is not the person doing the requesting. There are several situations in which you might receive a request for records from someone other than the client.

If a person dies the family may seek the records. People have the same privacy rights after death that they had with your agency before they died. Your state has laws regarding who is eligible to receive the records of a deceased person. Generally the executor of the estate or the spouse of the deceased has the right to request and receive records after a person has died. There could be times when several family members are vying for the records, but this is not an issue for case management but rather an issue for the courts. When the courts decide who may have the records the, agency would comply. Some people die of a disease they wish to keep secret from the family. One such case involved a client who was diagnosed with HIV/AIDS. He informed case management that he did not want family members to have access to his records in the event of his death. The case management unit honored that request which he had put in writing. Some agencies routinely get in writing the name of the person who may have the records should the person die.

Occasionally records are requested by someone other than the client in court proceedings. As we saw in Chapter 2, when a person sues the agency the records can be introduced into the court proceedings. Sometimes a record is requested by an attorney who is not representing the client. For example, a lawyer representing your client's spouse may request your client's records. You would not give records to anyone without a signed release from the client. If the court subpoenas the records in a court proceeding, the agency administration will deal with how to respond to the subpoena.

Summary

At many different times in your career you will be involved in collecting and organizing important information about clients for the purpose of understanding their history and for planning current services. When you request information for your use or the use of your agency, be sure to collect only those summaries and evaluations that are relevant. Do not send for large quantities of material that will be bulky to store and difficult to maintain in the record.

When you are asked to supply information about a person, take care to operate within the guidelines and the law regarding confidentiality. Be particularly careful about what is released when the individual is a child, has a mental disability, or is infirm. We are the ones who are charged with maintaining the individuals' privacy.

Social service agencies frequently make requests, with clients' permission, for information from other agencies. It will be your role to choose carefully what to request and what to release.

Exercises

These exercises can also be filled out online at CengageBrain.com.

Exercises I: Send for Information Related to a Middle-Aged Adult

Instructions: Using the blank form in Appendix C entitled "Request/Release of Information," send for information relevant to planning for one of the cases you are developing. Choose the information for which you are sending carefully, and be able to explain how this information will assist you in planning for your client.

Exercises II: Send for Information Related to a Child

Instructions: Using the blank form in Appendix C entitled "Request/Release of Information," send for information relevant to planning for the child's case you are developing. Choose the information for which you are sending carefully, and be able to explain how this information will assist you in planning for your client. Be sure you obtain all the proper signatures.

Exercises III: Send for Information Related to a Frail, Older Person

Instructions: Using the blank form in Appendix C entitled "Request/Release of Information," send for information relevant to planning for an older person's case you are developing. Choose the information for which you are sending carefully, and be able to explain how this information will assist you in planning for your client. Will you need additional signatures?

Exercises IV: Maintaining Your Charts

Instructions: Place your completed release forms into the appropriate clients' charts. They should be placed under the long assessment forms in the charts. If there is more than one form, staple them together. For questions about how the chart should be organized see Appendix C, "Arrangement of the Client's Chart."

CHAPTER 20

Developing a Service Plan at the Case Management Unit

Introduction

Using the information you received from other agencies (see Chapter 19) and the information you took to develop the social history at the time of your interview with the client, the next step is to develop a plan. Depending on the agency in which you work and the population your agency serves these plans may have different names. They can be called *service plans*, *treatment plans*, *goal plans*, or *care plans*. For our purposes we will use the term service plan.

A service plan contains broad, general **goals** for the case management unit to follow with regard to a particular person. Case managers then refer people to the agency where the actual service will be provided to help the person reach the goals. These agencies are referred to as *provider agencies*. Generally, the case management unit develops an outline for the **provider agency** to follow. The provider agency takes these broad, general goals and turns them into very specific and measurable goals with objectives for the person. These will be in effect and constitute the guidelines while that person is receiving service from the provider agency.

In agencies that provide both case management and services or treatment this step may be skipped as the agency moves directly to developing very specific goals and objectives for the client.

This chapter describes the step-by-step process for developing a general service plan before referral to a provider agency. Remember that you will not be giving these services. Instead, you will be determining what the person **needs**, based on your evaluation or intake assessment, and based on your discussions with the client about services the client feels are important. The services and agencies you choose for people will be those that will help them attain goals that you and they see as important for them.

There are several types of goals for clients. You might develop treatment goals to treat conditions such as mental illness, emotional problems, or drug or alcohol abuse. When an individual has treatment goals, the case manager generally refers the person to mental health programs, drug or alcohol rehabilitation, or physical health interventions.

Not all of your clients will need treatment goals. Some people will have problems involving poverty, home arrangements, and other material needs. For these people you will develop a service plan containing general living goals that will improve their situation, perhaps by providing appropriate in-home care, better transportation, or a more appropriate living arrangement. Individuals housed in institutions often get care plans that address the goals the institution will help the resident meet. These care plans may include things like recreation, nutrition, and physical therapy.

Involving the Client and the Family

Only 10 or 20 years ago, when people requested help from an agency, they were told by that agency what help they would receive. Often the agency prioritized the person's problems for them. People had little or no input into these decisions, and those who objected or had other ideas were often viewed as difficult and uncooperative. Agencies developed services in part based on the skills and expertise of the employees at those agencies or in their community. They might not offer services that were badly needed in their community, and they might offer services that did not meet the needs of any but a select group of people. Clients were expected to fit into the services offered and were often labeled untreatable if they did not.

Today, all that has changed. Several factors have fostered these changes. When the state mental hospitals began to shrink their patient populations, those residents were brought back to their communities and back in contact with their families and other support systems. It soon became obvious that many families were unable to cope with the mental illness of their family member without support, and without that support the former patients were often abandoned by their families. To a lesser extent this abandonment happened to senior citizens when caretakers were struggling to care for an older person with complex problems. In the field of substance abuse, clients' families often gave up and cut all ties after trying to support their family member to abstinence. Agencies gradually began to recognize the value of including families in planning and implementing treatment. In recent years, children have been receiving services in ever-greater numbers. This has involved families in planning and implementing the treatment or service plans for their children.

Today agencies are expected to engage both the client and the client's family (with permission from the client) when this inclusion is appropriate. Individuals and their families assist in developing service or treatment plans. Families receive agency support to remain involved with their family member whenever possible and receive the support they need to remain an intact family. Agencies that work to enable the identified client to live successfully in her community have found that individuals do much better if their families have not cut them off. Children can take better advantage of services if the entire family is considered in the planning. For these reasons, human service professionals are careful to listen to what the individual wants to receive from services and to what the family may need to support the client.

An inappropriate use of family involvement would occur if a worker ignored the wishes of an adult seeking services for emotional or situational problems who did not desire family involvement, or if a worker involved family members after a person asked specifically that his family not be involved in any way. In these situations, you follow the wishes of the person and strictly observe confidentiality guidelines.

Using the Assessment

You will receive an evaluation or assessment for each person in your caseload. This will be either the social history or the assessment form. In our hypothetical case management unit, you will have done the assessment or social history yourself, which is often the case. The evaluation tells you what the person's major problems are and what areas of the person's life need attention. Generally, these problems are the reason the person is seeking help or is being referred to a therapeutic or supportive program. The assessment alerts you to the expectations and desires of the client as well. All of this will help you to determine what goals you and your client should address first. Figure 20.1 lists some common goals to consider for people. You may think of others.

The goals listed in Figure 20.1 are very broad, long-term, and general. Choose those goals that fit the person's most outstanding needs and give them more specificity—that is, a treatment or goal plan shown in Figure 20.2. For example, a person might need "to

FIGURE 20.1 Goals to consider for clients

- To obtain (or maintain) housing
- To make positive changes regarding relationships with family members
- To obtain (or maintain) mental health treatment or rehabilitation services
- To obtain (or maintain) physical health treatment or rehabilitation services
- To recognize precipitants to hospitalization and take appropriate action
- To obtain (or maintain) social support
- To obtain (or maintain) recreation and leisure-time activities
- To obtain (or maintain) employment
- To obtain a new role in life
- To continue living independently
- To obtain (or to maintain) food and clothing
- To acquire (or to improve) daily living skills
- To obtain (or to maintain) drug and alcohol treatment or rehabilitation services
- To make positive changes regarding stressors and coping mechanisms
- To obtain legal assistance on legal issues
- To obtain (or maintain) transportation
- To obtain (or maintain) an education program
- To obtain assistance with financial issues
- To develop realistic goals for the future

Digital Download Download from CengageBrain.com

FIGURE 20.2 Treatment or goal plan

Client _Larry T. McCune_ # _03468_ Next of kin _Lydia McCune (Mother)_
Initial plan [√] Updated plan [ ] Date _3/6/14_ Review date _6/6/14_
Developed with _Larry McCune_
Level of case management _Resource Coordination_ Case manager _Kathy Torres_
Provisional DX: _____

Type	Strength/Need	Goal(s)	Comments	Referral
Income/financial situation	(Strength) Need		Receives income from disability insurance policy	
Housing/living arrangement	(Strength) Need		Owns home	
Vocational	Strength (Need)		Would like to return to job, has job to return to	
Educational	(Strength) Need		Has engineering degree	
Transportation	Strength (Need)	To be able to drive independently	Can be addressed in therapy	Linden Counseling Center
Medical	Strength (Need)	To be free of headaches		Michael C. Fillippo, MD Neurologist for neurological evaluation
Activities of daily living	Strength Need			

370   Section 5   Developing a Plan with the Client

Legal	Strength Need		
Recreation & leisure time	Strength Need		
Mental health	Strength (Need)	To reduce or eliminate symptoms of PTSD	Linden Counseling Center
Substance abuse	Strength Need		
Family relationships	Strength (Need)	To re-establish ties with 9-year-old son	Linden Counseling Center
Social supports	(Strength) Need		Continues to receive support from friends at work and from mother
Other	Strength Need		

Chapter 20   Developing a Service Plan at the Case Management Unit   **371**

make positive changes regarding stressors and coping mechanisms." You would not word it in just this way. You would give it more specificity by writing something like this:

- Goal: "Job with better work hours" or
- Goal: "Resolution of current marital problems"

When the client is referred to an agency for help in meeting the goals, the agency will set up smaller, more specific goals and develop a precise plan for the client to meet those goals.

Creating the Treatment or Service Plan

If your agency holds a meeting in which intakes are discussed, you should have a provisional plan for your client to take to the meeting. After this meeting, your next step is to create the final service plan for the person. In putting together the provisional plan, write out the goals so you have a clear understanding of what you and the person see as important in this case, and then use the "Treatment or Service Plan" form found in the Appendix, following these steps:

1. Place the individual's name and next of kin on the top line of the form.
2. Check the box for an initial plan, as you have never done a plan before on this person.
3. Fill in the date the plan was created. Make the review date 3 months from this date, unless in your judgment a review should take place sooner. For instance, if the person is in need of inpatient drug and alcohol services and those services are only expected to last 10 days, check on that service sooner.

NOTE

Not every person who comes for help has a mental health problem. A person who has run out of fuel oil might have a mental health problem, but if the reason that person is seeking help today is to get a voucher from you for more fuel oil, the mental health problem is not relevant. Be careful not to assume that everyone you see must be mentally or emotionally ill. Look at the conditions in the back of the *DSM 5* for the problems people often encounter in life that might become the focus of clinical attention, without meaning that the person is mentally ill. These are found on page 715 of the *DSM 5* in the chapter "Other Conditions that May be a Focus of Clinical Attention." For many people who depend on the social service setting, even these will be irrelevant.

---

4. Indicate who helped to formulate this plan (the client, daughter, mother, father).
5. Wait until you are in the planning meeting to decide the level of case management, and determine that with others. You may have an idea about the level when you go to this meeting.

6. Print your name as the case manager.
7. The box containing the material on the *DSM 5* should be left for completion in the planning meeting. Again, you can have a rough idea of what you will suggest if a diagnosis is to be given, but wait to make it final until you have heard from others.
8. Sign and date the form at the bottom, placing your name on the line for the case manager. Your instructor will sign and date the supervisor's line.

To fill in the boxes:

1. For each category on the left:
    a. Circle or highlight "Strength" if this is an area of strength for the client—that is, if the client brings resources to this area. You can make a notation or comment by way of explanation in the comment box. ("receives income from Social Security" or "works as librarian")
    b. Circle or highlight "Need" if there is a problem in this area that you intend to address. You can make a notation or comment by way of explanation in the comment box.
    c. Do not circle either "Strength" or "Need" if the area in the box is not applicable to the client's problem.
2. Place relevant comments and explanations in the comment box.
3. Write the goals in the boxes underneath the heading "Goal(s)." You can base the goals on the list shown in Figure 20.1, or you can come up with others.
4. In the boxes in the last column, indicate the name of the agency to which you will refer the person to meet the goals. Try to get as much service from one agency as possible. Sending the person to five or six different agencies can be confusing. For people who have many needs, choose agencies that give more comprehensive services. Providing names of agencies in the community where you plan to practice as the agencies to which your client will be referred for service will help you become familiar with the resources in your area.

Look at the example of a goal plan in this chapter to see how this is done.

How to Identify the Client's Strengths

Clients have strengths that may be useful in working toward the goals. Be sure to use these strengths in planning for your client. When you are looking for the strengths of the client, here are some factors to check (these apply to both children and adults):

1. Look for supports in the community for people and their families. For example, is the person involved in a church, a club, or a recreational program? Are there any other social services involved? Does this individual have a strong circle of friends at work or strong ties to a neighborhood or community?
2. What are the religious or cultural beliefs, practices, or values observed by this person, and how does she use these for support and comfort?

3. When your client interacts with others, such as staff, family, and pets, what interpersonal skills does he appear to have?
4. What special abilities or skills does the person possess? What is his level of education? Does he have any special credentials?
5. If you gave the person a choice as to what she would prefer to do, what would she be most likely to choose? This is particularly important with children.
6. If the individual has contact with his family, what does the family do together? Do they eat together, go to church, meet for special occasions, or watch TV?
7. What hobbies, recreational activities, or talents does the individual pursue? What interests this person?
8. What activities, people, or groups give comfort to your client? Does your client have a pet?
9. With whom is the person most likely to want to spend time? This is important with regard to children.
10. Who outside the person's family has shown an interest in this person?

Individualized Planning

All of this has led to individualized planning for clients. People no longer have to fit into a certain program or go without treatment. Instead, case managers develop plans for people that are specific to that person. No two plans should be exactly alike.

An example of poor planning can be found in the caseload of Alicia, whose clients were all over 70. Alicia was only 24, having recently graduated from school, and to her people over 60 seemed elderly. In planning for her clients, Alicia assumed they were all in the same need of supervision and support. She referred all of them to Homemakers, Inc. for household help. She insisted, when she could, that each of them attend the local senior citizen center. In addition, she tried to line up Meals-on-Wheels for many of her clients.

The clients began to talk among themselves about Alicia's lack of appreciation for what they needed in their lives right then. Marguerita did need Meals-on-Wheels, but only temporarily while her broken arm healed. She found going to the senior center a chore because she wanted to stay home and read and watch her favorite shows on TV with her neighbor. Delbert belonged to a retired businessmen's club and wanted to go there in the afternoon. He found it annoying that Alicia called to check on him when he failed to go to the senior center. Delbert had come in for help with portable oxygen. When Leslie refused to use the Homemaker, Inc. services, Alicia went to Leslie's home to find out why. Alicia appeared astonished that Leslie was able to keep her own home and often entertained family and friends.

Alicia is an example of a case manager who does not view her clients as individual people. She lumped them all together in a category of "elderly people." She devised her service plans based on her assumptions about what elderly people should need.

Understanding Barriers

Barriers prevent workers from fully understanding and helping clients, and they prevent clients from being able to take full advantage of the plans being developed for them. Even when you have identified as many barriers as possible and addressed them, you may refer your client to a place where barriers will appear again. If you are aware that barriers will diminish the effectiveness of your goals, you can modify the plan accordingly. This helps you to plan in a way that prevents problems later and helps the individual to take full advantage of the plan. It is a two-way street. Care in planning is the best way to make sure people can truly take advantage of the service or treatment.

Following are some common barriers:

- *Language:* The individual may not be able to communicate adequately with others because of a difference in primary language. The worker may not be able to communicate with the client because the worker does not speak the client's language.
- *Culture:* The individual may be unable to negotiate an unfamiliar culture. The worker may not understand the person's culture or may be inclined to judge the person's culture by the worker's own cultural standards.
- *Disability:* The individual may not be able to handle all the details of the plan. The worker may overestimate or underestimate the extent of the person's disability.
- *Lack of resources:* The individual may lack the resources to fully participate in the plan, such as lacking transportation or clothing suitable for a job interview. The worker may see the person's poverty as a barrier to the plan or fail to take the lack of resources into account.
- *Mental illness:* The individual may be unable to communicate clearly or follow through with the plan. The worker may be afraid of the illness or may fail to understand how the illness affects the person's capabilities.
- *Intellectual Disability:* The individual may be unable to communicate clearly or follow through with the plan. The worker may see the client as a child or may fail to understand how the disability affects the person's capabilities.

Sample Goal Plan

Figure 20.2 shows the goal plan for a 34-year-old man, Larry McCune, who was in a severe car accident 4 years ago. He has come into the case management unit requesting help with symptoms he originally thought would disappear or that he could handle on his own. The police report states that he was driving his car with his family in the car. He approached a busy intersection and failed to stop for a yellow light that turned red while he was in the intersection. He was hit and two people were killed, one of them his 4-year-old daughter.

Larry is well educated, with several degrees, and works as an engineer in a position he enjoys. Since the accident, however, he has been unable to work consistently and has had numerous arguments with his subordinates. His wife left a year ago, taking the remaining child with her, a 9-year-old boy.

Larry is complaining of severe headaches and is not clear if these are from stress or the accident in which he suffered what was diagnosed as a mild head injury. He is asking for help because of recurring nightmares that involve the accident and an intense fear of riding in cars, which recently has spread to using any form of public transportation. He states he feels detached from other people and lately has thought that perhaps his life is really over.

Note that the case manager has put together the plan (see Figure 20.2) so that Mr. McCune can receive his services almost entirely from one source, in this case the Linden Counseling Center. Personnel at that center will address Mr. McCune's specific needs. A neurological assessment will be done by a neurologist to address the headaches. The neurologist's findings will ultimately be taken into account at the Linden Counseling Center in working with Mr. McCune. Study Larry McCune's plan carefully to see how the case manager has individualized to plan for him.

Summary

For every person we see, we must create a service or treatment plan that specifically addresses that person's needs. In addition, this plan must specifically address the direction the individual states she wants to go. No two plans should be exactly alike because people can have the same symptoms or conditions but for different reasons, and they also can have the same problems but react to them differently. Couple that with other individual concerns that each person brings and the need for an individualized plan becomes obvious.

Case managers, for their part, need to know about a variety of services ranging from formal agencies to social supports and support groups in their communities. The variety allows workers to easily put together unique plans that address each person's individual needs.

When creating plans, always involve the client and, when appropriate, the person's family. Plans without client input are not acceptable. As people discuss their problems and the issues they feel should be addressed, you can formulate general goals for them that will become your plan. Once you have some ideas about goals, ask people if you may share your ideas with them, and always ask for their opinions or additions. Following the guidelines in this chapter makes it more likely that the plans you develop will have a lasting and beneficial effect.

Video Examples

To view the videos that accompany this book, go to CengageBrain.com.

You can watch online as Danica works with Alison to develop some personal goals for Alison to pursue.

Exercises: Broad Goal Planning

These exercises can also be filled out online at CengageBrain.com.

Exercise I: Planning for a Middle-Aged Adult

Instructions: Using one of the blank "Treatment or Service Plan" forms found in the back of your book, develop a plan that addresses the immediate or most important needs of the client you are following. Look at both strengths and weaknesses.

This will be the tentative plan that you will take to the planning meeting. Make certain that you have addressed those areas of your client's life that are most troublesome right now to the client. If one of your clients is facing a poverty situation, do not assign a *DSM* diagnosis. Remember, these are not the goals that will be developed at the provider agency. Your client is not there yet because you have not written a referral for that agency yet. Before you can do that, you must have the broad, general goals that should be pursued at this time with your client.

Exercise II: Planning for a Child

Instructions: Using one of the blank "Treatment or Goal Plan" forms found in the back of your book, develop a plan that addresses the immediate or most important needs of the child you are following. Look at the strengths and weaknesses of both the child and the child's family.

This will be the tentative plan that you will take to the planning meeting. Make certain that you have addressed those areas of the child's life that are most troublesome right now to the child and his family with regard to this child. Remember, these are not the goals that will be developed at the provider agency. Your client is not there yet because you have not written a referral for that agency yet. Before you can do that, you must have the broad, general goals that should be pursued at this time with your client.

Exercise III: Planning for an Infirm, Older Person

Instructions: Using one of the blank "Treatment or Goal Plan" forms found in the back of your book, develop a plan that addresses the immediate or most important needs of the older person you are following. Look at both strengths, including natural supports the person might be able to call on, and weaknesses.

This will be the tentative plan that you will take to the planning meeting. Make certain that you have addressed those areas of your client's life that are most troublesome to the client right now. Remember, these are not the goals that will be developed at the provider agency. Your client is not there yet because you have not written a referral for that agency yet. Before you can do that, you must have the broad, general goals that should be pursued at this time with your client.

Exercise IV: Maintaining Your Charts

Instructions: The "Treatment or Service Plan" form should be placed under your release forms. This is a tentative form and does not need signatures. You will clean up and revise your recommendations in the planning conference, and then put the revised form in the chart in place of this original.

Be sure that every chart you are following has a plan in it. Charts without plans are often the reason funding and accreditation sources withdraw support.

Exercise V: Checking Services

Instructions: If you have a question about a service in your community that you believe might provide the services your hypothetical client needs, call that agency for information. Most agencies have brochures or other informational literature they will send to you.

CHAPTER 21

Preparing for a Service Planning Conference or Disposition Planning Meeting

Introduction

After you have completed your assessment on each new client and done a tentative plan with the person, your agency might hold a meeting in which more specific plans are made for the individual's care or **services**. Some agencies call these meetings service planning or disposition planning meetings. Some call these care planning or simply case planning meetings. In other agencies this is done informally. In small agencies, particularly, individual case managers may make those decisions by themselves, referring people to other services in systems that will have more formal case management and there is no meeting. This chapter is designed to help you attend a planning conference in which you present your client and other professionals help put the finishing touches on the final plan.

In some places, children who come into the system are presented by their case manager to a "children's panel" consisting of child psychologists, child psychiatrists, social workers, pediatricians, and others who serve children. Many other places use panels of professionals for creating plans for clients from different populations; in this situation, the case manager presents the case to representatives of any number of agencies serving or specializing in that population. Together the group decides what combination of services would best suit people in their current situation based on the assessment and gives a **diagnosis**, if appropriate. Much of what is accomplished in these meetings is done to support client success in their current environment.

If a person has both a substance abuse (SA) problem and a mental health problem and the agencies that address these two problems are not combined, representatives from each of the agencies working with the client should meet together to decide what should be done. In the past an individual could be turned down for mental health services because he was drinking and turned down for SA services because he was suicidal. That kind of "turf" exclusion at the expense of the person is no longer tolerated by funding sources that expect people to be served.

In these meetings, decisions regarding the service an individual will receive are made with others who have experience and come, perhaps, from different disciplines. When the meeting is over, a formal plan will be drawn up.

What You Will Need to Bring to the Meeting

You should consider bringing three items to these planning meetings.

1. *Tentative service plan:* You have already developed a tentative service plan with the individual. Bring this tentative plan to the service planning conference.
2. *Human service directory:* Until you are familiar with what services are available in your community it is important to have some sort of directory of the places your client can receive services. As you work within the same social service system, you will come to know, without consulting a directory, which agencies are reliable and which services are used most often by your agency when referring people. If there is a directory, bring that to the meeting so that you can work with your peers to find the best placement for your client. A good place to look is the local phone book, where social services are usually listed together. You might copy these pages and bring them to the meeting.
3. *DSM Handbook:* The *DSM* is a large volume containing considerable information. If you are working with a population that is likely to have a *DSM* diagnosis, you might consider purchasing the *DSM Handbook*, which contains only the most basic information and is easier to carry with you when you go to meetings of this sort. Bring your *DSM Handbook* to assist in making the provisional diagnosis.

Goals for the Meeting

Goal One: Diagnosis

If the person being discussed is seeking mental health, intellectual disabilities or SA services, you will probably need to give a provisional diagnosis at this meeting. Older people who appear to have some type of dementia also receive diagnoses. Children with intellectual disorders, clinical disorders, or learning disorders also require diagnoses from the *DSM*, as would someone suffering a significant emotional response after an assault. Diagnoses in these service delivery systems are usually required for payment purposes and can be changed after the individual is in the system and has been thoroughly observed. Not everyone, however, will have a diagnosis. You may work in an agency where diagnoses are not required.

When a diagnosis is required, planning meetings generally have a professional present who is able to give the diagnosis.

Goal Two: Level of Case Management

In addition to the provisional diagnosis, you may assign the level of case management if your agency uses different levels. For our purposes, we use the four levels discussed in Chapter 2.

1. *Administrative case management:* People receiving services are placed in a pool with other clients who require little service or **follow-up** beyond the original referral. The client tends to function independently.
2. *Resource coordination:* Clients are often in need of services and assistance on issues such as housing, medication, or therapy, but they generally do well with the services offered and do not pose a risk to themselves or others.
3. *Intensive case management:* Clients are at high risk for rehospitalization or for behavior that poses a danger to themselves or other people.
4. *Targeted case management:* Clients of varying needs are given to a case manager who carries a smaller caseload as a result; the client has the same case manager through stable times and times of crisis, giving the person continuity of care.

Goal Three: Services

In every case management situation, the most important part of the intake process is to make decisions with the client about the service the individual will receive. Your assessment prepares you to make those decisions wisely. Remember that you do not give the service. Your task is to look at the material in your assessment interview, such as:

- The strengths and weaknesses of the person
- What the person said she wants or expects
- What services this person could use well or fit into well
- What the major problems or presenting problems were (presenting problems are those that brought the person into the agency in the first place)
- Goals you have for the person and the person's own stated goals
- Any other pertinent circumstance or information about the person that you feel is relevant

Using the material you have assembled, you (and presumably a team) will develop in this meeting a plan for services, and for treatment if needed, that matches the client's problems and expectations.

Benefits of Conference Planning

Agencies use conference planning or team planning in order to have more than one person take a look at the case. Sometimes the people comprising the team come from different disciplines and thus they bring to each case a different perspective. A physician or nurse might be concerned with health issues or see clues in the history that call health issues into question. A psychiatrist would be interested in the diagnosis and medications, among other things. A therapist might offer ideas about programs and approaches that would be most beneficial. When children are involved, teachers, parents, a pediatrician, minister, scout leader, and other significant people in the child's life can raise their concerns and ideas about this child and become part of the child's support team.

When a team sits down together to look at a particular case, each team member contributes ideas and perspectives to the overall planning and to the overall understanding of the person's situation and needs. This is enormously beneficial in making sure that the individual receives the best care and all the important issues are addressed. A team will usually bring a richer approach to the case than a single person would be able to do.

Collaboration

Treatment or service planning, when it is done with a team, is a collaborative activity. While you come with background and some firsthand information about the person being presented and ideas for a plan, everyone contributes to the discussion in order to be sure the best plan is ultimately created.

Some case managers become defensive as others suggest changes or additions to the tentative plan. One case manager was "upset" when the tentative diagnosis she had in mind was questioned by the psychiatrist. Feeling that a change in the diagnosis would reflect on her and imply incompetence, she resisted any change and became angry.

Another case manager was sure that the man whose situation he was presenting was not experiencing medical problems as a result of his drinking. This case manager became sarcastic in the planning meeting when others asked that a medical examination be part of the plan. "I think I've seen enough alcoholics to know when someone needs medical care and when they don't," he sneered at the others on the team.

In a third situation, a woman presented a case of a client suffering from severe anxiety. The case manager thought the problem was a generalized anxiety disorder. Another case manager questioned the diagnosis, stating that it seemed more like posttraumatic stress disorder. The client's symptoms and complaints needed further evaluation by the team, and a discussion ensued. The case manager was asked questions as the team attempted to assemble more details in order to make the diagnosis. The case manager answered these questions, but she did so in a petulant manner and later reported that her "feelings were hurt" because "people I thought were my friends just ganged up on me in there."

It is important to ask questions of your colleagues in a collaborative and respectful manner. Do not grill the people presenting cases. Do not be dismissive of their interpretations of the situations or demean the conclusions the presenters have made. After all, the presenter is the person who actually saw the client, and her observations are extremely relevant to the decisions made on the person's behalf. Good collaboration is critical to quality exploration of the person's case and for positive outcomes. Further, when others question you about the case you have presented treat it as a team effort to clarify certain points about the case and not as a personal attack on you.

Preparing to Present Your Case

In the planning meeting, you will give the pertinent details that will help the group make decisions about where the person will be referred and the type of treatment the individual will need. Your presentation is a short, oral summary of your case to the group. It should be given in an orderly manner.

Before the meeting, review the details of the case. Bring the intake and assessment material with you, and refer to it if you do not know the answer to a question. In general, however, you should be able to answer most questions without reference to the material.

The elements of your presentation will include the following information:

1. *Why the person came to the agency? What were the presenting problems?* Discuss why the person came to the agency seeking assistance. What were the outstanding problems she talked about during the phone intake and at the assessment interview? Mention any other outstanding problems that you feel should be addressed.
2. *How the client presented in the assessment interview:* Talk about any unusual, inappropriate, or bizarre behavior. Describe any hallucinations or delusions the client might have had. Most clients will not have any of these. Simply describe how the person seemed during the interview.
3. *What the individual indicated are his goals and expectations:* What did the person say he wants to happen as a result of seeking services? Tell why the person sought help, and what he expects the outcome will be. Even though you will formulate the final specific goal plan for this person, be sure to indicate the client's input here. What did the individual want? What were his stated goals? What was his priority? Have a tentative goal plan ready that addresses these, but expect others to see and suggest additional goals or to suggest changes.
4. *Additional relevant information that would have bearing on the disposition of this case:* If there is any other information that the team should have in order to make a decision, be sure to discuss that. In addition, describe any unusual characteristics that might give a clearer picture of the person. Perhaps it is something the person said; provide her wording if possible. Perhaps she has worked for several years at the Humane Society and has four dogs she rescued. During the interview, she talks about her dogs and shows you pictures of them. This fact may not directly influence the treatment plan created to deal with her depression, but it will give the team a clearer picture of who this woman is and her interests and the goals she might have talked about, either short term or long term.
5. *Your impressions and recommendations:* What is your impression of this person, and what do you recommend that might address the problems the person felt were most pressing? This would be the information you put together for the impressions and recommendations after you took the social history.

Making the Presentation

By following the format discussed here, you will make a good presentation. Be prepared to talk about the case to the team for about 3 to 5 minutes. You should not have to read through notes or shuffle papers. Simply describe your client. After you are finished, the team will ask you questions about the person. You should be able to answer these without reference to your notes, but that may not always be so. If you do not have the information, turn to your notes when you need to.

Sample Presentation

Following is the presentation on Larry McCune by a case manager in the planning conference:

> Mr. McCune is a 34-year-old male, an engineer. He came seeking help for symptoms resulting from a severe traffic accident about 4 years ago. He said he was driving and was in the intersection when the light turned red. He was hit and his 4-year-old daughter died in the accident. His marriage broke up about a year ago, and the remaining child, a 9-year-old boy, is currently with the wife.
>
> Mr. McCune complained of severe headaches, and says he has nightmares that involve the accident. He also said he thought he had a phobia (his word) about riding in cars and, more recently, in other forms of transportation. He said he doesn't feel able to work consistently and talked about being irritable, especially at work.
>
> During the interview, Mr. McCune seemed to have a flat affect and sometimes he was tearful when he discussed the loss of his daughter and later when he was describing the divorce. He seemed to me to have some slowed motor responses. He sat quietly during the interview. He looked sad, and there was no animation in his speech. There were times he appeared not to be focused on the interview.
>
> He is asking for help that will allow him to return to work. He says the two biggest obstacles to that are his problems with transportation and also his irritability with coworkers. I talked to him about the need to find out the origin of the headaches and he agreed with that.
>
> Right now he lives alone. He has stopped attending church, and he said he has few friends. He accepted all my suggestions without much discussion. He seemed pretty passive.
>
> I see this man as depressed and anxious in the sense that he seems to have developed a strong fear about using a car or other transportation. The other impression I had was that he is dealing with a lot of guilt, which he agrees needs to be addressed too. I am recommending that he be evaluated for any residual neurological problems resulting from the accident, just to check on the headaches, and I would like to have him evaluated for depression and possible antidepressant medication. He asked specifically for counseling, and I recommend that. I think he could benefit from that. Mr. McCune might actually benefit from a grief-support group in time. His goal for coming, he told me, is to get back to feeling better and being able to work consistently. Does anyone have any questions?

In this presentation, the case manager has briefly addressed all five parts of the presentation. Others at the meeting might want additional information. For example, someone might ask this case manager about the client's relationship with the wife and son. Here is how the case manager might answer that question:

> We really only talked about that a little bit. He pays support but not consistently. He said his ex-wife is working. He did say he sees his son, but I understood that this is not like a regular visitation schedule or anything. He doesn't describe the relationship with his ex-wife as extremely hostile. My understanding was that she left because she couldn't take his irritability and not going to work regularly.

Other issues the team might want to know more about could be whether Mr. McCune expressed a desire to see more of his son; whether he has had any other health issues or health issues related specifically to the accident besides the headaches; what kind of engineering does he do; what has the response been from his employer to his inconsistent work habits; and what living arrangements he has now since his wife and son have moved out.

In Appendix C is a form titled "Service Planning Conference Notes." You may use this form to make sure you have notes on the five points you want to cover in a planning conference on your client. You would not give this form to anyone, but it helps you to make sure that all the points are covered and gives you a place for the notes.

Follow-Up to Meeting

After the treatment planning conference or disposition meeting, you will write up a formal service plan for your client using the "Treatment or Goal Plan" form we looked at in Chapter 20. When your goal plan is completed, you will:

1. *Meet with the individual and discuss the plan:* Your first contact note, for our purposes in the classroom, should be written on your first meeting with the person after developing the goals and referral options and having those confirmed in the treatment planning conference or disposition meeting. At this meeting, you will go over the plan with the client, or the client's parents in the case of a child, and note their response and any changes you make to the plan as a result. When you and the individual feel comfortable with what is planned, you can move on to referring the person to the place where he will receive the treatment or service. In Mr. McCune's case, he will probably be seen by a neurologist for evaluation of his headaches, by a psychiatrist for evaluation of his depression and possible medication, and by a counselor at a counseling service to work on issues of guilt and loss.
2. *Make referrals* for your client to the agencies that will actually carry out the services. This happens when your agency does not give services directly.

Summary

After you have met with a client and heard the person's concerns and issues, and after you and the person have developed a plan that appears to meet those needs and seems satisfactory to both of you, often you will be asked to sit down with others to go over the individual's situation and the plan for that person. This meeting is meant to offer support in the belief that several people looking at a person's case together can refine the plan somewhat. For that reason, keep in mind that the meeting is a collaborative one. You can expect to answer questions and hear ideas about your client, and you will be expected to ask questions and offer suggestions to others.

Once you have completed this meeting you will:

- Write a formal service plan and have your supervisor (in this text, your instructor) sign it
- Meet with the client to review the final plan
- Write a contact note about the meeting with the client for the person's file
- Refer your client to the people and agencies indicated in the plan

Video Examples

To view the videos that accompany this book, go to CengageBrain.com.

You can see part of an actual service planning conference online.

Exercises: Planning

These exercises can also be filled out online at CengageBrain.com.

Exercise I: Developing a Service Directory

Instructions: Before you go to a planning meeting, you need to know what services are available in your community. Gather information on various social service agencies. Some regional phone books contain special pages of social services. Some counties publish a directory you may be able to purchase. The library may have lists of agencies you can copy, or a local agency may have compiled a directory.

Be sure the directory you compile or obtain is relevant to the area in which you intend to work following completion of your courses. In this way, you will begin now to become familiar with the services that are available.

Exercise II: A Simulated Planning Meeting

Instructions: To simulate a planning meeting, form groups of no more than five students (otherwise it will take too long for everyone to present a case). Present one of the cases you have developed to the group, following the instructions for presentation provided in this chapter. Be sure to bring the information on the client and the material you will need to plan services and give a diagnosis.

You can have your instructor act as the senior specialist who can confirm with you the proper diagnosis, or you and the team can arrive at your own decision through discussions very similar to the discussions that take place in agencies among professionals.

Sign the revised planning form, and have your instructor also sign it, as your supervisor.

CHAPTER 22

Making the Referral and Assembling the Record

Introduction

After the service planning conference or disposition meeting, you rewrote your original, tentative service plan to develop a formal service plan for your client, and you had your instructor, who is acting as the supervisor in this case, sign it. You then discussed this final plan with your client, and now you will refer the person to the unit within your agency or outside agencies that will carry out the treatment or service.

Use the "Referral Notification Form" in Appendix C to make referrals to these other agencies. Referrals are generally faxed or sent electronically to the agency to save time, although they can be sent by mail.

1. All referrals are coming from the case management unit for which you work, in this case the Wildwood Case Management Unit.
2. Write the name of the agency to which you are referring the person after the word "To."
3. Note the date the **referral** was made.
4. Write the person's name, address, and phone number after "Re" in the box.
5. Write the goals the referral is to address in list form.
6. The **"Target Date"** is the date you expect the goals to be met.
7. After **"Review Date,"** place the date on which you intend to review this case plan to see if the plan is working for the person. The review date comes before the target date.
8. Write your name in the blank for case manager.

Determining Dates

Two dates must be determined: the target date and the review date.

The Target Date

When you set a target date, you are stating clearly to the providers how long you expect it will take for the service to obtain the goals you and your client have worked out for your client. You are, however, clarifying something else. You are informing the provider how long you will allow the service to be given without getting the desired result. If the target date is reached and the goal has not been accomplished, baring extenuating circumstances, it is time to stop this intervention and seek a more useful approach.

When setting a target date, decide how long you are willing to continue to try this approach without seeing any result. Beyond that date, you will not continue funding, and at that point you will evaluate with the client other options for the person.

Physicians routinely expect to see results from medications they prescribe for their clients in a specified amount of time. They know that if the medication has had little or no effect in a specified number of weeks, it is time to switch to another medication. Neither you nor the physician in this example can afford to administer a particular treatment plan for as long as it takes, no matter how long that might be.

The target date is influenced by two factors:

1. *Funding:* The amount of money available to spend on the service will influence how long the individual can stay in the service. If you are using public funds or insurance, there may be a cap on the amount of money you can pay for a particular service.
2. *Goal:* The goal is a factor in the length of time needed. Some goals are short term, such as a 6-day detoxification program, whereas others are a substitute for or to prevent inpatient hospitalization and may require weeks or months of service.

The Review Date

The review date comes before the target date and is the date you expect to review this specific service in the plan to see if the plan is actually working. In most cases, plans are reviewed at least every 90 days. Therefore, a person in a partial hospitalization program for 6 months would be reviewed in 3 months. As noted, however, some services are short term. In the case of a 6-day detoxification service, you might make a quick review on the third day.

The purpose of the review date is to make sure that the client is getting what the client needs and what the referral stipulated the client would receive and to be sure that if the plan is not working, no further money will be spent on trying to make it succeed. In most cases, the plan will be revised to better achieve the person's goals, and in some cases the goals may be reevaluated.

Sample Referral Notification Form

Figure 22.1 contains a simplified version of what you are likely to see in a referral form in actual practice. Many provider agencies have their own referral forms they want the case management unit to use when making a referral to that provider agency. These forms ask for more information about the person, usually in a detailed summary. We will use a simplified referral form for our purposes. Note that on our form we use both the *DSM 5* ICD-9 Code and the ICD 10 Code which will take effect on October 1, 2015. For a time you are likely to see both until everyone has switched over to the new coding system.

The referral in Figure 22.1 is for Paul Bittinger, a person with chronic schizophrenia. He has been in and out of hospitals for acute episodes of hallucinations during which he believes the voices he hears are giving him the power to walk on water. These voices, and his subsequent delusion that he is omnipotent and can walk on the surface of a nearby river, have caused him to jump in the river at times when it was high or there were huge, swiftly moving ice chunks. In an effort to help him manage his own illness and medications better, the case manager is referring him to a partial hospitalization program where there will be more extensive monitoring of his medications and triggers.

FIGURE 22.1 Referral notification form

Wildwood Case Management Unit

Referral Notification Form

Client *Paul J. Bittinger*   # *12365*

Address *4692 Caroline Drive, Apt B, Conradville UT 00012*

Home phone *555-5555*   Work phone *555-5555*   Date of referral *9/2/14*

Diagnosis *295.90 (F20.9) Schizophrenia*

Provider *Grandon River Hospitalization Program*

Type of service *Partial hospitalization*

For the purpose of

Learning to manage medications

Learning to manage the symptoms of his illness

Review date *10/5/15*   Target date *12/15/15*

Referring Case Manager *Brenda Walker-Poloski*

The Face Sheet

Your clients are now in your agency's system, and a record is assembled from the various forms and contacts you have had with these people and those concerned with them. All charts on people who have entered the system have a *face sheet*. (See the sample of a completed face sheet in Figure 22.2.) This sheet lies on top of all the other information and contains essential information for anyone who might need access to it quickly.

You will find a blank face sheet in Appendix C. To fill out the face sheet:

1. Place the name of the individual, the agency number you have assigned the person, and the person's address and phone numbers at the top of the form. (For our purposes here, give the client any agency number you wish. It should be at least five digits and begin with 0.)
2. If the person is a child, someone with a severe intellectual disability or mental illness, or an elderly person in need of a guardian, put the guardian's information in the next section. Most people will not need a guardian, in which case you can leave this area blank or write N/A (not applicable) in the space.
3. Nearly everyone has someone they wish to have notified in case of an emergency. It may not be a blood relative; a friend or a neighbor is acceptable. The person the client indicates he feels closest to should go in the section under next of kin. Indicate the relationship of that person to the client (e.g., mother, sister, or friend) in parentheses.
4. If the client has indicated there are numbers they do not want the agency to call place those numbers next. Generally people ask that certain numbers not be called in order to maintain confidentiality.
5. Below the top box, from "Date of First Contact" on down, fill in only the information that applies. Not all clients will be on medication or have a physician involved in their case. Where information does not apply, use N/A to indicate that.

    a. The top row of boxes is fairly self-explanatory. The date of the first contact is listed followed by who took the contact. The person's date of birth and gender complete this row.
    b. In the second row of boxes, give the person's marital status, the last grade completed or diploma received, the current employment if the person is working, and veteran status.
    c. The next row deals with issues related to pregnancy so that medications and treatments do not jeopardize the mother or the child.
    d. In the fourth row, the first box asks you to state the reason the person came to the agency. The next box requires information on current medical conditions and who is treating the person for those. The box next to that requires medications prescribed for that condition and the person doing the prescribing. In most cases it would be the same physician. The last box refers to any legal or incarceration issues the client is experiencing. All of these can affect treatment decisions and decisions about goals.
    e. In the fifth row of boxes, the first box asks about substance abuse issues and requires a diagnosis if substance abuse is present. The second box asks for the same information related to mental health issues the person may have. In the

FIGURE 22.2 Face sheet

Wildwood Case Management Unit

Face Sheet

Name *Lucinda Harris*   Agency # *02487*

Address *1235 Pleasant Street, Anytown, PA 01234*

Home or cell Phone *555-555-5555*   Work Phone *555-555-5555*

Guardian _____

Address _____

Home or cell Phone _____ Work Phone _____

Next of Kin *Jamal Harris (Husband)*

(If different from guardian)

Address _____

Home or cell Phone _____ Work Phone _____

May not call these numbers Home _____

Work _____ Cell _____

Date of first contact 8/19/15	Taken by Peter Van Voories	DOB 4/8/84	Gender Female
Marital status Married	Educational level Associate's degree	Employment status Secretary Emco, Corp	Veteran status N/A
Currently pregnant N/A	Rec Prenatal care N/A	Given birth last 28 days N/A	Pregnancy complications N/A
Reason for Visit Depression, loss of energy, loss of interest in life, poor job performance	Current medical conditions Fibromyalgia Treated by Carl M. Anchor, MD	Current medications Advil 400 mg TID Prescribed by: Carl M. Anchor, MD	Legal status/ incarcerations
Substance abuse problems Dx. Case Mgr	Mental health problems Dx. 296.22 (F32.1) Case Mgr Keyanna Lefevre	Referrals 1. Linden Counseling Center 2. 3.	Primary provider of service, counselor, therapist Michael Cedaneo Counselor
Psychotropic medications Wellbutrin 100 mg BID Prescribed by Carlos Cremara, MD	Psychiatric evaluation done Carlos Cremara, MD 8/24/15	Psychological evaluation No	Court ordered? No

Assigned case manager *Keyanna Lefevre*   First review date *2/19/15* third box enter the primary referrals you made. There may be other referrals but here you want to enter the referrals where most of the treatment or service will be provided. In the last box place the name of the person at the primary referral agency who will be seeing your client or managing the person's case.

    f. In the sixth row of boxes note first if the person has been prescribed psychotropic medications—that is, medications that treat mental and emotional symptoms—those medications are entered in the first box along with the name of the prescribing physician. Psychiatric and psychological evaluations are noted next, if these were done, along with the name of the person who conducted each of these and the date. There is also a place to note whether the person was ordered by the court to seek services.

6. On the last line, you sign your face sheet and place the date for the first review on the sheet.

Every agency has a specific order for the material within client charts. By maintaining a specific order with charts you might create in the classroom, you get used to the idea of following agency guidelines for organizing material in **clients' records**. When you place the face sheet in the front of the chart, you have an organized collection of documents. Using a manila file folder, place the other forms on your client in the following order, with the face sheet on the top and the referrals on the bottom:

1. Face sheet
2. Inquiry and referral form
3. Verification letter
4. Assessment form and/or social history
5. Release of information forms
6. Service plan
7. Referrals

Now you can add to the chart all further contacts, letters, and monitoring activities that take place for this person.

Summary

With the referral of your client to the agencies and people who will provide service and treatment, the person has become a formal client of your agency. When you make the referral, plan the target date and the review date carefully. The target date is the date beyond which you would not want to continue the service if there has been no improvement. The review date is the date you plan to check to see how well the service is going for the individual. When you review the plan, be prepared to modify the service if that seems useful. At the target date, the person still may not have reached the goal, but there may be improvement. Again, this would be a time to modify or extend the plan.

Once the individual has a plan and is a part of your case management caseload, the person becomes a formal client of the agency. A file or record must be kept on the contacts and changes that occur while the person is being served by your agency. Set up a file on the individual using a manila folder, and place a completed face sheet on the top of your documents.

Exercises

These exercises can also be filled out online at CengageBrain.com.

Exercises: Assembling the Record

Instructions: Complete the following exercises:

1. Look at the completed forms on the clients you have developed. Fill out referral forms for each agency to which you intend to send your clients for services. In each client's chart, clip these together and place them at the back of the chart. Use a separate referral form for each agency.
2. Next, develop a face sheet for the front of each chart.
3. Now assemble each chart as indicated in this chapter, using a manila file folder and placing the forms on your client in the following order, with the face sheet on the top and the referrals on the bottom:

    a. Face sheet
    b. Inquiry and referral form
    c. Verification letter
    d. Assessment form and/or social history
    e. Release of information forms
    f. Service plan
    g. Referrals

Once you have written them, your case notes will follow the referrals when they are put into the chart. Be sure the client's name and agency number are on the folder.

CHAPTER 23

Documentation and Recording

Introduction

If it isn't documented, it didn't happen. That is the most important thing to remember about **documentation**. You can say that you talked to the person about her diet but if you didn't document it the conversation never took place. You can insist that you and the person discussed suicide and she promised to call you if she felt the need to take her life. If that discussion isn't documented that conversation did not take place.

Once a person is in the case management unit's system and is receiving services from a provider, it is your responsibility to keep a record of all contacts relevant to this case. These contacts will be with the client or with those connected to the client in some way, such as providers of service, family, doctors, teachers, and counselors. You will document these contacts on the form titled "**Contact Notes**" (found in Appendix C and on CengageBrain).

While you are a person's case manager, you will see this person for many reasons. Sometimes people will come to your office because something upsetting has happened in their life or because they need a prescription. At other times, people will call on the phone because of a problem or need. You might see people at provider agencies where they are receiving services when you make site visits. In addition, you may talk to teachers, family members, and others related to the person, something called a **collateral contact**. Every contact related to the person is documented.

Keeping accurate records and documenting all contacts with or related to clients is needed primarily for legal and administrative purposes. Legally you need to be able to show that the service for which you are being paid is being given to the client. Administratively you need a record that documents the activities on behalf of the client and all contacts related to the client so that case managers are not relying on memory to reconstruct what has happened before.

These notes should focus on your client, and not on you. The treatment plan begins to go out of date soon after it is written due to the changes in programs and people's lives. The purpose of your notes is to keep the record current.

The Importance of Documentation

As noted above, if you do not document contacts and events related to the client these never took place when it comes to supporting what you did. You can say you made a home visit, that you discussed his suicidal feelings, that her mother called and you agreed to help with her school attendance, but if it is not in the record you cannot point to the record to back you up. The client, for whatever reasons, can say you never were at his home, never discussed with him his suicidal feelings, never offered to help with her school attendance. Therefore, the record functions as legal documentation of your work with the client.

In addition, documentation is money, particularly in fee-for-service agencies where the agency is reimbursed for the services the staff have provided. Every service and contact must be noted in the record in order for the agency to be paid for the service you gave.

Documentation is often an annoying task. However, many agencies track how much documented service case managers are doing and do not tolerate case managers who have poor documentation habits. These case managers are costing the agency money even if they are good case managers. For these reasons, it is a good idea to begin a habit of documenting all your activities and contacts with your clients. This means that you will set aside time to do this every day. In some agencies there is an online form that allows you to document your contacts on the computer as you move through your day, making it easier to get documentation completed. Whatever arrangements are used in the agency where you go to work, be sure that you begin by making a habit of documenting your work and maintaining your records.

Writing Contact Notes

Your **contact notes** in the chart should *always* include the following:

1. The focus of the interview or contact
2. Your assessment based on a concise summary of behavior, appearance, and affect
3. Any resolution that takes place
4. The reason for the next contact or follow-up that will occur

Figures 23.1 and 23.2 illustrate contact notes. Figure 23.1 identifies the four parts of a contact note. See if you can identify the four parts in the example shown in Figure 23.2.

FIGURE 23.1 Sample contact note broken into four parts

Focus of the Interview

Linda came into the office today to discuss her medication.

Your Assessment

She appeared somewhat disheveled and tearful and indicated her belief that the medicine is "not strong enough."

The Resolution

An appointment was set up for her to see Dr. Wentworth on July 2. She was advised to remain in her program where her depression can be closely monitored.

Reason for the Next Contact

Linda will return July 2 after her visit with the doctor to let CM know what was done about her medications.

Digital Download Download from CengageBrain.com

FIGURE 23.2 Contact notes: Can you identify the four parts?

1/13/16 (Phone) Mark called today asking for a voucher for public transportation in order to get to Polyclinic Medical Center for kidney dialysis. Suggested he go by county transportation for more direct service. Mark seemed bright and pleased with the results of dialysis. He will call next week to confirm that county transportation has begun to pick him up.

5/3/13 (Office visit) Jill came in to the office this afternoon because she is out of medication and did not renew her prescription at the time she was given a voucher to do so. She states she wanted to try to go without the medication at the time. Jill denied having any symptoms of her schizophrenia but seemed agitated and impatient. She paced the office and at one point appeared about to remove her blouse because "Its so hot!" An updated voucher was given to her in order for her to refill her prescription for Seroquel. She will refill her medication today and call CM tomorrow to report how she is feeling.

11/22/16 (Site Visit) CM visited Peter at his sheltered workshop on Tuesday where Peter has been moved to a different job after folding the towels became too tedious for him. Workers at Margrave Linens felt that Peter could handle something more challenging. He will be folding uniforms now. Staff at Margrave seemed pleased with Peter's progress. Peter was working enthusiastically at his new responsibilities and seemed proud to have moved on to something more challenging. CM will see Peter in his group home in 2 weeks to see how things are going.

A case note of this type is never more than six or seven sentences. Write concisely to facilitate others having access to the information. You may in some agencies refer to yourself as CM for case manager. No one can sit down and read through lengthy, descriptive narratives. Your notes must be clear, and they must be concise.

Labeling the Contact

In the left-hand margin for every case note, place (1) the date and (2) the type of contact in parentheses (Collateral Contact, Office Visit, Phone, Site Visit, Group, Home Visit). The types of contact are as follows:

- *Collateral Contact:* A collateral contact is with someone other than the client such as the client's mother, minister, or nurse. Be careful in all collateral contacts that you have the client's permission to make that contact and, if the client is a child, that you have the parent's permission to talk with this person.
- *Office Visit:* The client was seen in the office.
- *Phone:* The client called you on the phone or you called the client.
- *Site Visit:* You went to the site where services are being given to your client and evaluated those services or discussed problems that have arisen.
- *Group:* You saw the client in a group, and you are noting what took place during that contact. Sometimes clients are seen in groups to use time more efficiently.
- *Home Visit:* The client was seen at home.

Here are examples of several contacts that are labeled:

- 4/3/15 (Office Visit)
- 9/8/16 (Phone)
- 12/6/15 (Group)

Documenting Service Monitoring

In addition to your direct contact with the client, document your efforts to monitor the delivery of service to your client. Those notes should be set up and labeled in the same way. For example, if you went to the provider agency and talked to the client and the client's therapist there, it would be labeled like this: 2/14/16 (Site Visit). If you went to a provider agency to attend a treatment planning meeting that focused on your client's treatment or services from the program, your contact note would be labeled the same way: 2/19/16 (Site Visit). Be sure to name the agency in your notes.

Following is an example of notes from a site visit:

1/6/16 (Site Visit) Met with Rob and social worker, Ginger Diamond, at Riverview Center to monitor Rob's progress toward job readiness. He seemed eager to begin job search and somewhat annoyed at the length of time the job-readiness classes are taking. It was agreed that he will begin his job search with support next week, while completing the remaining segments of the training. Rob will notify CM in 2 weeks regarding outcome of job search.

All your contact notes must reflect the fact that you have made efforts to monitor the services delivered to your client. If you spoke to a contact person at another agency by phone regarding your client's progress or services, or you talked to a child's parent or teacher, you would also identify this as a collateral contact:

> 2/24/15 (Collateral Contact) James's teacher, Mrs. Pike, called today to say that James is doing much better in school. She reported that he seems better able to focus on classroom assignments. Further, she has had two meetings with James's parents, and they have begun to inquire about homework and enforce the need to complete it each night. Overall Mrs. Pike is feeling better about James finishing fifth grade this year with his other classmates. She stated she will call again if there are signs of problems.

Documentation: Best Practice

There are a number of ways to make your case notes sound professional. The following sections contain some tips for writing better notes.

Avoid Hostility

Do not use your notes to release hostility. When we are angry with someone, it is easy to sound sarcastic, facetious, or even annoyed with the person. Make sure your notes do not reflect any negative feelings you might have about another person.

> Example: Abdul and CM agreed he will come in to pick up his prescription on Monday 8/15/15, if he can manage to get out of bed.

Document Your Interactions with the Client

Your interaction with the client may be the most important thing that occurred. This could be a verbal exchange or some other form of interaction. Document what happened and what you observed. Always use quotation marks to indicate a word-for-word verbal exchange.

Document Significant Aspects of the Contact

Some aspects of the contact with a person are extremely significant. These are clues to the person's state of mind or situation. You are giving the points you found important when you did your mental status exam. When you think it is significant, document these aspects of the client's behavior:

- Appearance
- Dress
- Facial expressions
- Mannerisms
- Responses to others or to activities
- Participation concerning interaction with you or participation in the services to which the client was referred

- Attitudes concerning interaction with you or participation in the services to which the client was referred
- Cognitive problems

Here is a sample statement in a contact note that documents significant aspects of an interaction with a client:

> Mrs. Peters seemed unable to understand exactly when we would be meeting again or what services would be provided in the meantime. She appeared confused and was somewhat unresponsive to her daughter and son-in-law who tried to clarify these for her.

Be Clear and Precise

One factor that can make your notes more professional is your precision and clarity. Many people write vague notes or notes that give only general descriptions. Be clear about what you are documenting. Do not use vague terms or indefinite statements. For example:

> *POOR:* Alice was friendly today.
> *BETTER:* Alice initiated the conversation, joking about her Christmas shopping.
> *POOR:* Bill was upset today.
> *BETTER:* Bill was concerned about the possibility that he could lose his job.
> *POOR:* Marcella got along well in the program today.
> *BETTER:* Marcella's affect was improved today, and she participated in preparing lunch for the group and in both group sessions.

Use Quotations

What the individual has said to you may be extremely important. You may feel that this information should go into the record. The rule is to place *only* the person's exact words in quotation marks. If you paraphrase what the person has said, do not use quotation marks. Do not use them for any reason other than indicating an individual's exact words.

Avoid Contradictions

Your progress note must not contradict other previous notes without explanation. If the plan is changed, that must be documented. If the client regresses or improves considerably, this should be documented. There should not be gaps that lead the reader to conclude that something happened that has not been documented.

Use Language the People You Serve Can Understand

You have learned a rather extensive vocabulary that is used by professionals in the field. To the person receiving services, these words can sound like jargon. Sometimes new workers use a lot of jargon in an attempt to sound knowledgeable. Avoid jargon. Write your notes in language the client or the child's family can understand because they have the right to have access to the record if they request access.

Accurately Describe Disabilities

When writing about a person with a disability, make sure to use language that accurately reflects the person's life circumstances and does not label the person in pejorative ways. Here are some guidelines from the Three Rivers Center for Independent Living in Pittsburgh*:

- *Person first:* Identify the person first, rather than the disability. Use *person with disability* or *a person who is deaf* rather than *disabled person* or *deaf person*.
- *Disability:* The terms *afflicted with, suffering from, cripple,* and *victim* are all unacceptable. They emotionalize and sensationalize, often to induce pity. The term *handicapped* is based on the image of a person with a disability on the street with a cap in his hand, begging for money. Except when citing laws (the college has handicap parking restrictions), regulations, or environmental conditions (such as the stairs are a handicap to her), always use *disability* rather than *handicap*.
- *Wheelchair:* People are not confined to their wheelchairs; they use them for mobility. Say she *uses a wheelchair*; not that she is *wheelchair-bound* or *confined to a wheelchair*.
- *Blind: This term refers to total loss of vision. Partial vision, partial sight,* or *visual impairment* are more accurate terms in some cases.
- *Deaf:* This term refers to total loss of hearing. *Partial hearing, hard of hearing,* or *hearing impairment* are more accurate terms in some cases.
- *Nonverbal: A person who cannot speak* is preferred over terms like *mute, deaf-mute,* or *deaf-dumb*. These terms imply that people who are deaf are also unintelligent. The inability to speak does not indicate intelligence or lack of intelligence.
- *Congenital disability:* This is a disability that has existed since birth. Do not use the term *birth defect*. *Defect* is derogatory and is not a synonym for disability.
- *Learning disability:* This term refers to a disorder affecting the understanding or use of spoken and/or written language.
- *Mental disorder:* This term describes any of the recognized forms of mental illness or other emotional disorders. Terms such as *neurotic, psychotic,* or *schizophrenic* (as in she is a schizophrenic) are pejorative labels.
- Be careful about using phrases such as *he overcame his disability* or *in spite of her handicap*. These terms inaccurately reflect the barriers people with disabilities face. They do not succeed in spite of their disabilities as much as they *succeed in spite of an inaccessible environment* or *a discriminatory society*. They do not overcome their disabilities so much as they *overcome prejudice*.

Digital Download Download from CengageBrain.com

---

* Adapted with permission from the Three Rivers Center for Independent Living, 7110 Penn Avenue, Pittsburgh, PA 15208–2334.

Government Requirements

You must follow state and the federal government requirements for documentation to be reimbursed for your services to the client. These may vary from state to state and from one type of service to another. These funding sources treat the record the same way blank checks are treated: Correction fluid, erasures, and blank spaces are not acceptable. This applies if your documentation is kept in a paper record. Computerized documentation automatically addresses many of the legal documentation requirements.

Here are some general rules for documentation that are often required by state and federal governments:

1. Use black ink. Blue ink does not copy well.
2. Never use either a pencil or correction fluid.
3. All notes must be legible.
4. The person must be identified by name on each page. Sometimes you can use an agency number instead. For children, a date of birth on each page is often required. Do not use nicknames or initials.
5. When **recording**, place the actual date of the contact note in the margin.
6. Sign (do not initial) every note.
7. After your signature, add the date the note was written. It should be on or as close to the date of service as possible.
8. If the person is in ongoing service, every note must end with the next scheduled service date. In some places, case managers are required to note the actual date and time of the next appointment and also the plan of action. For example, "John will return on March 6, 2017, for an appointment at which time we will discuss what he has done about housing."
9. To correct a mistake in a note:
    a. Draw a line through the error—whether it is a letter, word, phrase, or entire paragraph.
    b. Write the word *error* above the line.
    c. Write the correction next to the word *error*.
    d. Sign or initial the line.
    e. Date your signature or initials.
10. If any blank lines are left on a page once you have completed your notes, draw diagonal lines through the blank spaces.

Do Not Be Judgmental

Do not write notes that sound as if you are sitting in judgment of the client. You may be inclined to judge some aspect of your client's life or behavior in negative terms, but that is not helpful. When negative judgments are obvious in your case notes, they leave a legacy that can follow the client. Avoid judgmental words in your notes. Figure 23.3 contains a list of words that tend to sound judgmental and a substitute word that is more objective for each. Become familiar with these.

FIGURE 23.3 How to avoid sounding judgmental

Poor Words for Documentation (Judgmental)	Better Words for Documentation (More Objective)
Dirty	Unclean habits, poor hygiene
Nasty	Unpleasant
Lazy	Inactive
Stubborn	Refused, Uncooperative
Nervous	Anxious
Wild	Restless
Bad-mouthing	Argumentative
Sarcastic	Critical
Mean	Unpleasant, insulting others
Troublesome	Uncooperative
Whining	Complained of
Glum	Sullen
Just sat there	Passive
Jittery	Restless
Foolish	Used poor judgment
Slow	Had trouble completing
Pushy	Persistent
Aggravating	Irritated others

Digital Download — Download from CengageBrain.com

Distinguish Between Facts and Impressions

A fact is something you observed, whereas an impression is a clue you picked up from the person. Use words such as the following to introduce your impressions:

- Bart seemed …
- Agnes appeared …
- Staff felt …

For example:

> *POOR:* Mary acted pleasant but was putting on a front.
> *BETTER:* Mary was pleasant, but CM felt she was angry over losing her job.
> *POOR:* Manuel wasn't telling the truth when he said he was comfortable with the new group.
> *BETTER:* Manuel said he was happy in the new group, but he appeared uncomfortable in the first session.

Give a Balanced Picture of the Person

Do not paint the client as entirely positive or entirely negative. People have strengths and weaknesses, and they have assets and problems. Give a balanced picture of the person, noting strengths and weaknesses, positive gains and negative problems. As noted previously, your notes should not be simply a collection of problems.

Provide Evidence of Agreement

There should be evidence in the written record that you and the program to which the individual was referred agree on the plan for the person and that you have had interaction with each other regarding the plan. This interaction and agreement can be documented by reporting on team or staff meetings you attended at the other facility or meetings called specifically to discuss the person's goal plan or treatment. This does not mean that the individual is excluded from participation in developing the specific plan for himself at the provider agency. You should also see evidence that the client was a participant in developing his plan, and if that is not clear, you should inquire about how the client took part.

Making Changes to the Plan

Sometimes even the best plans must be changed for any number of reasons: the client gets sick, the provider is closed for snow or overcrowding, the client has a death in the family, the plan is too difficult or is not addressing the real issue. Sometimes other, more pressing problems arise in the person's life. If there is a lack of progress toward the original goals:

1. Note the lack of progress in the notes.
2. State why there is a lack of progress.
3. Note the revisions to the plan.
4. Revise the actual service or treatment plan.

See Appendix C for samples of contact notes.

Summary

The ability to write good contact notes is important. These notes constitute the written record about the service a person receives from you. In addition, these notes serve to document changes in the person, the person's life, and the service plan. Keep your notes brief. Too much information is difficult to read quickly and is easily taken out of context by lawyers or others seeking to distort the individual's care. On the other hand, do not make the notes so brief that they ignore significant information.

Become familiar with the government guidelines as well. The government views the record as a blank check and wants to be able to see clearly when corrections and revisions have been made to the notes. Memorize the common government guidelines for making corrections so that your notes are not subject to government censure.

Exercises: Recording Your Meeting with the Client

These exercises can also be filled out online at CengageBrain.com.

Instructions: Your contact note will be written about your first meeting with the client after the treatment planning conference or disposition meeting and after developing the final plan and referral options. In this meeting, you go over the plan with the client, guardian, or the client's parents, in the case of a child, and note the client's response to your proposed plan. Also note any changes you make as a result of this interview.

Write a note for the record of the client whom you are following. This note will indicate that you met with the client and discussed the service plan. Use the four elements for writing notes. This first note is generally a little longer than the others—perhaps 6 to 12 sentences in all. Here is an example of such a note:

> 3/24/15 (Office Visit) John came into the agency today to discuss his service plan. He continues to appear anxious and asked repeatedly for directions to the Susquehanna Counseling offices where he will be receiving services. CM went over the plan for 10 sessions of counseling at Susquehanna Center with John who agreed to this plan. He will begin at Susquehanna Counseling on Monday March 28, 2016, where he will have his appointment with the intake staff and meet his therapist. John will call in one week to let CM know how that meeting went and what specific plans he and the therapist developed for him. *Carly Jameson*

Practice writing your own first contact note. When you complete the note satisfactorily, rewrite it into the record, using the form in Appendix C entitled "Contact Notes."

Exercise I: Recording Client Contacts

Part I: Recording a Client Contact

Instructions: Read the following material, and then write a paragraph of no more than six sentences that covers the following:

1. The focus of the interview
2. Your assessment based on a concise summary of behavior, appearance, and affect
3. Any resolution that takes place
4. The reason for the next contact or the follow-up that will occur

Mrs. Pell is seen in the emergency room, after which you are called by the ER physician. He tells you that her friend brought her in and that she arrived complaining of chest pains and shortness of breath. She was extremely anxious, and during her physical examination, she confided in him that she is suffering physical abuse at home and is afraid. He is uncomfortable with discharging her from the ER until you have seen her. She is about 26 years old, intelligent, and a bit unkempt. You notice old bruises on her arms.

YOU: How are you?

MRS. P.: I guess Dr. Ingram told you—a little scared.

YOU *(sitting down beside her)*: He said that you were facing some problems at home.

MRS. P. *(avoids looking at you)*: I am.

YOU: Can you tell me a little bit about that?

MRS. P. *(nods)*: —

YOU: I can see how difficult this is. Where would you like to start?

MRS. P. *(speaking just above a whisper)*: Well, I, I can't seem to get along with my husband. We've had some really bad fights lately. Really bad. I seem to be on the losing end of those fights.

YOU: So things are pretty rough at home right now.

MRS. P.: Well, we seem to fight all the time *(tears well up)*. I love him. I really do, but he doesn't believe me! He accuses me of seeing other men or being attracted to other men, and when I deny that he blows up.

YOU: Do you feel like you could talk a little more about some of those fights?

MRS. P. *(nods and reaches for a tissue)*: Yeah. I'm going to have to. I just don't know how to begin. It's gone on so long.

YOU: Maybe you'd like to start with what's been going on recently.

MRS. P.: Well, recently things have gotten so much worse. I feel as though it's something I'm doing. I think my husband has a short fuse. He had a head injury as a child, and he blames his temper and his moods on that.

YOU: So he can be pretty moody.

MRS. P.: Oh yes! And I get the brunt of it. I try to remember that and be careful—not upset him. I just feel like I'm walking on eggshells all the time lately. He just goes off at me at the least little thing. I try to please him. I tell myself that I know what it is he likes and what he dislikes, but no matter how hard I try, there is a fight because I didn't do something or I did it but I didn't do it right.

YOU *(gently)*: Can you give me some examples?

MRS. P.: Well, 2 days ago I forgot to get a roast at the store. I made a meat loaf for dinner instead. He wanted roast, and he just went off when he got home. He wouldn't eat, threw the dishes on the floor, and turned over the table and then held my head under the tap in the kitchen sink. I could hardly breathe. He told me he'd let me up when I agreed to fix exactly what he asked for from now on.

YOU: That must have been so frightening.

MRS. P. *(nodding)*: Yes. I just don't have a choice with him. I have been thinking for years—well, we've been married 6 years—that I could help him with this. I even thought I could become so important to him that he would never hurt me. But in all this time things have gotten worse.

YOU: So, in other words, no matter how hard you try, things don't improve and have actually gotten worse.

MRS. P.: Much worse. This is the first time I started to think maybe I should leave him, although that is scary too. Who knows what he will do then?

YOU: So leaving has some good points and some bad points.

MRS. P.: Well, if he would leave me alone, then I think it might be pretty good. I'm just afraid he would come after me.

YOU: It sounds like you might be interested in a safe place to go for a while.

MRS. P. (nods): —

YOU: I notice your friend brought you in. Are there people you can turn to?

MRS. P.: My friend only suspects what goes on. We never talk about it directly, but I'm pretty sure she knows. She told me one time I didn't have to stay in a bad marriage, and she looked at me kind of funny when she said it. I just think she has her suspicions.

YOU: So you really don't feel comfortable opening up to her about this.

MRS. P.: I don't feel comfortable opening up to anyone. My family told me not to marry him. He was a loner. He didn't like to be around them, and they thought he would keep me from seeing them. They were right, but for all these years I've pretended they were wrong.

YOU: You felt bad that their predictions turned out to be so true.

MRS. P.: That's right. He doesn't like them. About a year after we were married, he told me I couldn't see them again, and I've been sneaking around ever since. I make excuses to them—why we're not there at Christmas or why he didn't come along. They probably know.

YOU: It sounds like your world is getting tighter and tighter.

MRS. P.: Well, as I try to do everything he demands, it is. I have no friends except Suzanne (*nods toward the waiting room*), and I spend a lot of my time doing everything for him.

YOU: Do you work outside the home?

MRS. P.: Well, see, I used to, but he put a stop to that. He said the company I worked for was putting ideas in my head after the second promotion I got there. I think he was jealous, but he made me quit and said no wife of his was going to have a better job than he has!

YOU: It sounds like you are under a lot of stress. Can you tell me a little bit about what Dr. Ingram said about your stress?

MRS. P.: Yeah. He said he thinks there is nothing seriously wrong. But he could see these bruises, and he asked about them and then he called you. He was very nice, and he said I needed to look at the way things were going so I didn't keep having these episodes. [blurts out] I wish I didn't have to go home!

YOU: You don't have to. I can easily arrange for you to go into a shelter, and that would give you time to think some of this through.

MRS. P.: If I did that, I could never go home again. He's likely to kill me for seeing the people here. He doesn't like it when I see people he doesn't know. If I actually left and stayed somewhere for a while, he'd never forgive me.

YOU: So in many ways you feel it would be pretty dangerous to stay at a shelter.

MRS. P.: He's all I have. I, I, I've just started to think maybe I should leave. I'm just so afraid to. Afraid of what he'd do and afraid to be on my own (*looks pensive*). He won't find out from the hospital that I saw you?

YOU: He shouldn't. I'll speak to them out there before I leave. But the hospital and my agency have very strict confidentiality guidelines, and your time with me is absolutely confidential.

MRS. P.: I better get out of here. I'm afraid he'll find out I was here and—thanks. Really, this has made me feel better.

YOU: I think you should go, too, if you are concerned. Let me ask you one thing, just because I am concerned about your situation at this point. How do you think I can continue to help you?

MRS. P. (pulling her clothes on, looks over at you startled): You want to stay in touch?

YOU: I'd like that. My agency stays in touch with any number of women in your situation, and we have a support group for women and a hotline if you need us quickly. Is there some way you think we could help you?

MRS. P. (brightening): I didn't think about staying in touch with you. I could do that. I could even come in. Suzanne is bringing me here on Tuesday morning for an EKG. Could I see you then?

YOU: I would be happy to see you then. When is your appointment?

MRS. P.: I have to be here at 10:30. You could meet me in the waiting room out there at 10:20 and go up with me. We could talk more about what I should do—sort it out.

YOU: I'll be there. Could you think a little bit about our shelter in the meantime, and some of the other services I described? See if you think there is something you might want to do if things get too bad.

MRS. P.: They just might. I, well, it can't continue like this. I've told him I love him. He doesn't hear me. Sometimes I think he'll kill me someday, and I say to myself, "What are you waiting for? Run. Get out of here." But I have no place to go, and like I told you, I can't turn to my family. Oh, they'd help me in a minute, but he'd look there first, and they are no match for his anger. (*Mrs. P. has her coat on and her hand on the door handle.*)

YOU: I will be here Tuesday morning. If you need me in the meantime, or the agency, call us (*handing her a business card*). Tuesday we'll talk some more and see if there are some other things we might be doing to help you with this.

MRS. P. *(smiles, extends her hand):* Thanks! *(calling over her shoulder)* I promise to call if I need anything!

Write your contact note here:

Part II: Recording a Client Contact

Instructions: Read the following material, and then write a paragraph of no more than six sentences that covers the following:

1. The focus of the interview
2. Your assessment based on a concise summary of behavior, appearance, and affect
3. Any resolution that takes place
4. The reason for the next contact or the follow-up that will occur

Mr. Dudley comes to your office in a rumpled plaid shirt. He has oily hair. He appears to have neglected his appearance since he stopped drinking, which is not typical. He has been in an outpatient alcohol treatment program for 4 weeks. He sits in the chair beside your desk and appears sad and ready to weep.

YOU: How are you?

MR. D.: Not so hot.

YOU: Can you tell me a little bit about what's going on?

MR. D.: I quit the treatment program.

YOU: Tell me something about what happened.

MR. D.: Well, I got tired of going. I figured I could do this myself.

YOU: So you wanted to try not to drink on your own.

MR. D.: Yeah. I thought most of the people in that program were whiners and complainers—all the time whining about how hard life had been, and I got tired of listening to it.

YOU: So you went out on your own.

MR. D.: Yes.

YOU: How did that work?

MR. D.: Not so hot.

YOU: Could you tell me a bit more about that?

MR. D.: What does it look like? I started drinking again. There isn't a whole lot more to tell.

YOU: So, in other words, you left the program and immediately started to drink?

MR. D.: Well, no. I was sober for about 4 days. Went to work and everything. Stuff was happening at work. Some guys got laid off, and I thought I might be next.

YOU: It sounds like you got off to a good start on your own, but then things began to happen at your work and you decided to drink.

MR. D. (hesitantly): Well, yeah.

YOU: And the drinking helped you deal better with the possibility that you might be laid off.

MR. D.: Well, yes, it did. I was upset. I thought I might be next. I don't know. It all happened so fast. I just thought I'd stop with the fellows for one drink on the way home from work, and one thing led to another, and I've been drinking …

YOU: For several days?

MR. D.: Yeah, I can't stop.

YOU: Tell me a little more about what's going on with your job.

MR. D.: The first day I called in sick, but today I didn't do anything. I came here because I'm afraid I'm heading right back where I was before.

YOU: So you really want to stop this binge.

MR. D.: I do. I didn't know where else to go.

YOU: It looks to me like you could use a detox unit. What do you think of that as a place to start?

MR. D.: I could use getting away from bars and getting cleaned out. I would really like to start over—try again.

YOU: Well you had 4 days where you did fine. We can begin by getting you into detox, and from there we can talk about a plan for when you are sober. What do you think of going into detox?

MR. D.: Yeah. Well, I suppose the missus and the boss would appreciate that.

YOU: What about you?

MR. D.: That's why I came here. I can't stop drinking. I'm a loser.

YOU: You must be feeling pretty bad about yourself.

MR. D.: Well, I screwed up again. Wouldn't you?

YOU: It's hard to feel like a loser. We can't address that while you are drunk, but after you sober up I will be happy to talk to you about making some changes. There are some things you did that tell me that you can handle things. First you had those 4 days on your own where you didn't drink at all, and then when you started to drink you knew where to go for help in stopping the binge you were headed on.

MR. D.: Haven't we been through all this before?

YOU: I don't think we have. We only started this 5 weeks ago. It might be helpful for you to know that people often have relapses as they are moving toward healthier choices. It might be that you learned something valuable here. That's just my opinion. What do you think?

MR. D.: I hadn't thought of it as a learning experience or anything. Hmm. I don't know. I just wouldn't have thought of it that way.

YOU: I would like to see you go to the detox unit for 7 days, and then we could sit down and look at this together.

MR. D.: Yeah, I will.

YOU: Good. I'm not willing to talk to you about changes when you've been drinking. I'll call First Step Detox, and then I will see you there during your stay so that we can plan what happens next.

MR. D.: Thanks.

Write your contact note here:

Exercise II: Using Government Guidelines to Correct Errors

Instructions: Using the section in this chapter on the common requirements of government funding sources, correct (or change) the following case notes:

Case Note 1: Winnie was at the hospital for tests on 3/17/16. Those tests were done to determine whether she has a tumor on her thyroid gland. Homemaker assistance has been arranged.

(The date for the tests is wrong. It should read 2/13/16. Correct the note accordingly in the space provided.)

Case Note 2: Carmela is attending the partial hospitalization program 3 days a week. Today she appeared brighter and more talkative. Our interview focused on her need to find a more independent living arrangement. Partial hospitalization staff will assist her in looking at supported living programs in the community.

(Carmela is actually attending the partial hospitalization program 5 days a week. Correct the note accordingly in the space provided.)

Exercise III: Spotting Recording Errors

Instructions: Tell what is wrong or left out of each contact note and rewrite an improved contact note in the space provided. Be sure the contact note has all four parts.

Case 1: Jim came into the office today looking depressed. He said he wants another job. He sat slumped over in a chair and was unkempt. The worker wrote:

> 1/17/17 (Office Visit): Jim came into the office today to see about getting a different job. He is not working at present. Will call Goodwill to see if they can place him temporarily at the bakery where he was before.

Case 2: Alice has been asking to be relocated to another group home since December when another client, Cheryl, moved in. Alice and Cheryl have fought ever since. The staff is not sure which client should move, and they have communicated their concerns to the case manager. The worker wrote:

> 3/8/11 (Phone): Alice is carrying on again about her housemates. She is trying to get a better housing assignment. Will call the house where she is staying and see if something can be done.

Case 3: Kitsu is attending an intensive outpatient rehabilitation program for his drinking. He sees his case manager at the site about once every week. Recently it was decided that Kitsu is not making the progress he was expected to make. Part of this is due to his job, which he says prevents him from coming to outpatient meetings regularly. To accommodate his night schedule at work, his services will now be given in the early evening before he goes into work. The worker wrote:

> 8/6/16 (Site Visit): Met with Kitsu and his therapist at the rehab program. Therapist is concerned about Kitsu's lack of attendance. Changes will be made in his program to facilitate attendance.

CHAPTER 24

Monitoring the Services or Treatment

Introduction

Monitoring is one of the four case management tasks. Case managers monitor the cases on their caseloads and the referral agencies to which their clients have been sent. This responsibility is important for several reasons. First, you are generally required to document that you have monitored your cases in order for your agency to be reimbursed for your time. But why do funding sources insist on monitoring? These funding organizations recognize that with careful monitoring the individual is more likely to receive the service you are requesting. In addition, regular contact with your client gives that person a certain amount of reassurance and support. In other words, monitoring allows you to check to see that the services being given to your client at the provider agency are addressing the goals you and the client developed. Also, when done on a regular basis, monitoring allows you to spot and address problems early and prevent crises. This way, a case manager can take steps to prevent costly hospitalization or relapse.

Case managers bring new clients into the agency, assess their needs, listen to their desires and expectations, develop plans to address clients' most outstanding goals and needs, and make referrals of clients to the agencies that can best respond to the clients' plans. That is the first half of what a case manager does. The second half is to monitor clients regularly (in some cases, frequently) to make certain there is movement toward the desired goals you and the individual developed for him or for her. During this process, the case manager may make suggestions for midcourse corrections, encourage the individual to participate in revising the plan, and discuss with providers any changes in focus or goals.

What Is Monitoring?

Monitoring is two things:

- An ongoing review of people's participation in the services to which they were referred
- An ongoing review of the services being provided and the service provider's activities with regard to the clients referred to them

Reviews must be documented in the person's file and will be part of the case notes of each file. Reviews are carried out by doing the following:

1. Talking to clients regularly to see if they feel they are making progress and to learn whether they are satisfied with the services. If a person points out needed revisions in the plan, this must be in your notes, as well as what was done in response.
2. Contacting people in the agencies or programs that are primarily responsible for your clients meeting their goals in their service plans. In talking with such a person, find out what this person thinks of the client's progress toward completion of her goals. Should the service be continued or discontinued? Should the service be modified in any way?
3. Contacting other people, agencies, and services involved with the client. For instance, if the client is a child, contact the parents. If the child is in school, contact the teacher to learn whether there have been noticeable changes.

Remember that all monitoring is subject to the rules of confidentiality and the Health Insurance Portability and Accountability Act (HIPAA) standards. These will be clear to you at the agency where you work. For now, however: Do not question individuals or organizations about people unless you have the express permission of the person or the person's parent or guardian.

The Financial Purpose of Monitoring

Let's study the purpose of monitoring more closely. Many of the people with whom we deal are receiving services paid for by public funds. These funds are limited and must be spent wisely. As the case manager, you are responsible for deciding what services a person receives, based on the goals you have developed with the individual. Then you must estimate how long it will take to reach these goals.

You might have a client with intellectual disabilities who needs to develop a marketable skill. You refer that person to Goodwill Industries. You authorize services there for 1 year. This means that you have allowed for payment to be made to Goodwill Industries for 1 year of job training for this particular individual. The expectation is that at the end of the year, the person will have a marketable skill.

You would then go to Goodwill Industries at regular intervals to see whether the goal is being approached. Is the person moving toward a marketable skill? Are there obstacles and problems you did not foresee or that are new in the person's life? Rather than referring the individual to Goodwill Industries and leaving him there for the year without ever checking up on how well things are going, you visit with him at the site, talk to him and to his supervisor, and get a feel for how well the plan is working.

Angelica provides a good example. She needed a supportive environment in order to be able to live with minimal supervision in the community. She had been hospitalized four times for severe depression, and it was only during the last hospitalization that the medical staff was able to combine medications for her in such a way that she really felt free of her depression symptoms. The severity of her depression and the number of acute episodes had prevented her from working or from living on her own successfully. Now, as the depression cleared, the staff was looking for a program in which Angelica could begin to work toward an independent living arrangement and other goals Angelica saw for herself.

The case manager in a mental health case management unit placed her in a small program called the Yellow House, so named for the color of the house. There, three women lived for a period of 90 days, preparing to move out on their own. The time in the house was spent in normal activities such as cleaning, cooking meals, decorating for the holidays, and going to the movies. During the day, the residents engaged in activities to help them secure a job and a place to live.

Case management chose the Yellow House program specifically for Angelica, believing this was the best place for her to receive the support she needed to move toward independence. Her improvements and her accomplishments were reported to the case manager, and as Angelica began to make plans to move to her own apartment, the case manager collaborated in the planning, coming to the Yellow House for meetings with Angelica and the staff.

Let us look at another example. Suppose you are working with a person who came into your transitional housing program after suffering years of marital difficulties and homelessness. You would develop a service plan for this person that addresses his most urgent needs. Perhaps you and he decide that one of his needs is for a community college degree in order to be self-sufficient. In addition, he may need professional help in parenting his children, who appear to be out of his control. His children may need academic and recreational programs as well, and there may be a medical problem requiring attention. After you develop the plan with him, you would refer him to the appropriate services. In this case, your program would not pay the actual costs of each of these services. A Pell grant and other financial aid will pay for college; the children's programs are supported by community funds and are free to the participants; and the parenting help is obtained through the public child protection agency's parenting classes.

Nevertheless, your agency is receiving public money to develop good service plans for people like this person to help them move from dependence to independence. The funding source expects you to develop plans that are usually successful. Too many poor plans and resulting failures might cause you to lose your funding. The plan you devise, therefore, must be in the best interests of the individual. Thus, for these reasons, you would carefully monitor your client's participation in the services and his progress toward the goals. Very often, in a situation like this, you would talk more directly with him about his progress or obstacles rather than talking to the providers of the service. For instance, it would be an invasion of his privacy to go to the community college to check on his grades, but you would want to keep in touch with him so you would know about problems that might arise and might cause him to fail if they are not addressed. From time to time, you would ask him about his parenting course and how his studies are progressing.

Follow-Up

"Follow-up" is a term used to mean the case manager went back to check on things after a contact with the client. Follow-up is an integral part of the case management process and specifically a part of the monitoring process. Doing a follow-up is particularly important after a **crisis**. It may be a call to see if the new medication is working, a home visit after a suicide attempt, or a visit to a partial hospitalization program after the person begins a program there. Slightly different from the usual monitoring, follow-up refers to going back to make sure things are going well for people after what was for them considerable change or an emergency. Follow-up occurs when the situation was unstable in the first place and the case manager is seeking to stabilize things for the person. Follow-up is done to make sure changes are working and not destabilizing.

Generally case managers follow up for two reasons. First, they want to be sure that the changes or the introduction of new services they made with the client are actually working for the person. Second, they want to be sure that the client is doing well since the last contact when changes were made. In these situations, the sooner follow-up is made, the more likely it is that problems can be avoided and relapses prevented.

Collaboration with Other Agencies

Collaborating with other agencies is not always as simple as it seems. There are times when people at other agencies begin to think they know the person better than the case manager does, and they may develop goals and objectives without collaboration with the case manager. Sometimes they discharge people or switch them to other services within their agency without telling the case manager.

You may also run into situations in which an agency is willing to take anybody referred to them, but gives very poor care or minimal service. Generally agencies are motivated to do this to collect the funds that come with the person for the person's care. Although the agency appears to be an ideal place to refer an individual for whom there is no other service, this may not be in the best interests of the person.

In these situations, it is tempting to become angry with the agency or with specific individuals working there. Anger and unpleasant disputes can permanently spoil relations between entire groups of professional people, curtailing the system's effectiveness and creating a hardship for clients. There are other, better ways to handle disagreements.

Try going to the other professionals and expressing your concern. If that does not work or the same difficulties continue to occur with other clients, ask your supervisor to talk to their supervisors, or ask the head of your agency to work with the head of the other agencies, to create a positive agreement on how these issues will be handled in the future.

For your part, make every effort to be responsive to the concerns of the provider agency. Be available for planning meetings and reviews. Support decisions the agency wants to make that really are in the best interests of the client. In this way, all the professionals involved with the individual present a united approach to the person's problems. If your client is creating a problem at the provider agency, help the agency develop a plan to deal with it, talk to your client if that will help, and invite the staff there to join you in looking for the causes of the problem.

Do not blame the staff in the provider agency—not to them personally or behind their backs. Refrain from gossiping about how poor the program is or the limited abilities of staff members. Instead, use the communication skills you have perfected to raise your concerns and listen to the responses. By following these guidelines, you will serve your clients better and maintain the smooth operation of the delivery system.

Advocating

As noted in the previous section, there are times when the agency to which you referred your client is unable or unwilling to give the services for which you are paying. Goals may be changed or weakened; services may be sporadic; changes within the agency itself may impede service to clients. When you are monitoring your clients' progress and their treatment or service at other agencies you are likely to discover these issues.

This is the time to advocate on behalf of the client. Questions to ask first are:

- How can we together help this client get more from the services?
- What can I do to help you better serve this person?

If you can work with people at the provider agency to get the services needed by your client, that is useful. If you cannot get services for the person that will actually help that person meet his goals, you may need to look for another provider.

The issue for case managers is always whether or not the client's best interests are being served. You need to stand your ground on behalf of the client, but not in a manner that creates a lot of animosity. When things do not seem to be as they should be, here are some ways to advocate:

- Ask open questions about how things are going for the agency and your client. ("Tell me about Jim's time so far with the agency." "Describe some of the things you and Jim have worked on so far.")
- Express an "I message" if you feel things are not working well or if you have concerns. ("I think I am most concerned that the goals we sent over and your goals don't look the same. Can you walk me through how the goals were changed so I can understand where we are with Jim right now?")
- Finally, if changes in treatment need to be made and the agency is unwilling to do this, discuss the need for changing providers. ("I think what we have decided to do at this point is transfer Jim to Oak Creek Rehabilitation Center. I think for us this would be a better fit for Jim until some of these issues are resolved." This is said matter-of-factly and without sarcasm.)

There are other ways to advocate for your client. For example, Ralph's client, Homer, was living in a very large apartment complex in which a number of low-income individuals resided. The complex had numerous code violations and the city was moving to close it down. There had been a legal settlement in which the court ruled that the company, which owned the complex, was responsible for helping people relocate. Some people had been relocated to other properties owned by that company. Others had moved and found their own housing. However, the company had made no effort to help Homer and Homer did not have good interpersonal skills that would have allowed him to be assertive about his right to be relocated.

In this situation after countless attempts to help Homer work with the company and after observing that nearly all the clients living in the complex were having similar difficulties, Ralph took Homer to visit his state representative. The representative was helpful, listening to Homer and to Ralph and then making calls to make sure the court decision was enforced for clients of the agency.

Rita's client, Clare, inherited an apartment building when her parents died. Clare simply did not know what to do with it and after her parents' death Clare ignored the building. One of the things she ignored was the tax notices that came on the building. When the letters from the tax office became more threatening and the person in the tax office informed Clare that, "the building will be sold for taxes," Clare became extremely upset.

When Rita talked to Clare it was obvious that Clare was not able to be a responsive landlord or handle the issues related to owning the building. In this case Rita went with Clare to meet with the lawyer who promulgated Clare's parents' will. Together they decided that it would be in Clare's best interests to sell the building and invest the money to support Clare whose ability to live independently was very tentative at best. Rita also went to the tax office with Clare. The employee there who had been so stern before softened with Rita present and Rita was able to advocate for a plan that would allow Clare to sell the building and pay the back taxes. **Advocacy** often involves going with people to handle things they cannot handle themselves, speaking for people when they cannot speak for themselves, and leveraging authority on behalf of your client when that is needed.

Leave the Office

Too many case managers want to monitor their clients' progress from the comfort of their offices. They find it to be too much trouble to drive around town, find parking, go out in all kinds of weather, or visit in homes located in areas that make them uncomfortable. Talking to providers and clients on the phone and insisting that people and their families always come to you appears both arrogant and somewhat lazy.

When your client is admitted to the hospital, visit her there. When your client begins a job-readiness training program, drop in to check with him there. When your client is in a program to gain independent living, stop by to see how she is doing. If the neighborhood makes you uncomfortable, go in pairs to see clients.

People live and work and strive toward their goals in a community. Only by visiting them in the settings in which they are living and functioning can you get a true picture of who they are and their real obstacles or accomplishments. The telephone is certainly useful as a support to your monitoring efforts, but relying on that exclusively cuts off important information and opportunities to strengthen rapport.

Responding to a Crisis

From time to time, people may experience an unsettling life event or escalation of their condition. For all of us, there are events that throw us off balance for a while, and for our clients this is true as well. However, if a person is already handling problems or an illness, these events can precipitate a serious breakdown in functioning. A person in such a situation may not use good judgment, may become overwhelmed and immobilized, or may disregard personal safety in an attempt to feel better.

Wherever possible, foresee and plan for destabilizing events. For example, if a person's mother is dying of cancer and you know that she will die soon, begin to work with the individual about handling this event and be prepared to give additional support when it happens.

We can't, however, foresee everything. A person loses his job, another person is evicted, the state closes a group home for corruption by the workers there and your client is suddenly without housing. People who have lived together for years now find themselves being separated and going to strange new environments. A person's liver fails, and there is emergency hospitalization. Another person's wife and children leave; a third person calls seeking shelter after a domestic dispute.

Life events and problems can throw people off balance, and so can illness and emotional issues. The desire or intention to commit suicide is one such emergency. Another might occur when an individual becomes so ill that she must be placed in a safer environment because she is suffering from delusions that endanger her or others around her. Another might be people who have become so ill from their addictions that their condition is medically life threatening.

Here are four steps for handling crisis events effectively:

1. *Respond immediately.* The faster you are able to make contact with your client, the more likely it is that you can stop the crisis from escalating. Your presence alone can have a calming effect, particularly if the person already knows you.
2. *Construct the best course of action under the present circumstances.* Consult your client if possible, but be ready to act. A person may have to be hospitalized for her own protection or to address a serious medical condition. Another may have to go to another home where he knows no one. Your being there for these changes and supporting your client can make the interventions more effective. When Jason's mother died unexpectedly, his case manager went through the entire funeral with him, including the planning at the funeral home. When Meg was hospitalized following liver failure, her case manager went immediately to the emergency room and spent that evening with Meg. After that the case manager visited daily to give support even when she could not stay long.
3. *Listen to the person.* By now you are probably a very good listener. What did the relapse or the move or the abuse mean to them? Let people talk to you about this event or change in their life and help them to put it all in perspective through reflective listening and open questions. This is part of accepting what is and being able to look at the event in a more practical and realistic light.
4. *Help the person to begin to look ahead.* Where do they want to go from here? Where would they want to be a year from now? What do they want to see in their future

following this? What is their thinking about where they should go from here? In this phase encourage people to brainstorm, look at all the options, talk about possibilities, discuss real obstacles and new considerations. In this way you are supporting a return to the person taking back control of his life.

Crises are difficult for everyone in the sense that they require an immediate response and concentrated effort. On the other hand, we don't want to lose sight of the fact that people can grow as a result of these crises, and our skillful participation can facilitate that growth.

Summary

Monitoring is more than checking in with your client from time to time and more than checking with providers every so often to see if services are being given correctly. Monitoring involves numerous methods for staying in touch including home visits, **site visits**, office contacts and phone calls. When we talk about case management monitoring we are talking about:

- Talking to people connected to your client and his or her care such as parents, teachers, counselors, doctors, and ministers
- Working closely with providers to make sure the services that were outlined and are being paid for are the ones being given
- Helping providers revise the plan for your client if the need arises
- Attending meetings related to the care of your client
- Being available in a crisis
- Following up after a change in plans or a crisis or some other change of direction for the client
- Giving support when your client is facing a difficult time or decision
- Arranging for services that are best suited for your client

The way we monitor our cases prevents crises and hospitalizations and gives the kind of support people need to reach recovery or to make meaningful changes. It is our monitoring that helps people move forward and it is our monitoring that keeps the plans and changes on track.

Video Examples

To view the videos that accompany this book, go to CengageBrain.com.

Online, you can see a site visit taking place. Keyanna is visiting a site where Michelle is receiving services. The discussion that takes place among Michelle, Kathryn, her worker in the program, and Keyanna, the case manager, is typical of most site visits.

CHAPTER 25

Developing Goals and Objectives at the Provider Agency

Introduction

It is important to note at the outset that **goals** and **objectives** like the ones we will be discussing in this chapter are done in many different settings. Case managers in many agencies develop these more specific goals and objectives with clients. This would be particularly true if the agency did their own intakes and provided the services as well. For example, it might be that a program for domestic violence would do the intake, learn the goals the woman wanted for herself, and develop a specific goals and objectives plan with her. Perhaps in an agency that served individuals with substance abuse issues, the case manager would learn that abstinence was the goal the person had set for himself and the case manager might go on to develop with the person the specific goals and objectives to get him there.

For our purposes in this textbook, we have case managers outlining the general goals with people that these clients have for themselves. These goals are then sent to the provider agency where the actual service will be given and a very specific plan with goals and objectives is developed there with the person. In other words, these more specific goals and objectives become the steps or the plan to achieving the general goal.

Therefore, at the provider agency they have received your broad general goals for the client. That person has arrived there, and the people at the provider agency have read over the general goals you wrote on your referral sheet. Now they will sit down and develop with the person very specific goals and objectives to address the larger goals you put on the referral form. In other words, they develop the goals you sent over in much greater detail.

In this chapter, you will step out of your role as case manager at the case management unit and step into the role of the person primarily responsible for implementing the client's service plan at the provider agency. In the agency where the service is actually given (the provider agency), goals and objectives are written very specifically and in greater detail. Here the broad general goal supplied by case management is broken down into more specific goals and objectives. The objectives tell us how the goal will actually be met. The objectives are a plan or blueprint for reaching each goal. This enables the staff at the case management agency to know exactly what the plans are for the client.

When the referred person arrives at the treatment or service agency, that agency's staff takes their turn looking at the stated goals on the referral form that were worked out with the individual. They then decide with the person just how to meet those goals in the time allotted by the case manager. Completion of the more specific goals is expected to take place during the time for which the case manager has authorized payment of services for the client. Sometimes the client cannot meet the goals in that time or needs more time because of other issues that have surfaced or new problems that have occurred. For example, a person who has periodic difficulty with asthma was hospitalized on a pulmonary unit for a week and missed several weeks of services, necessitating an extension to the agreement. In another case, a person did not do well in the program where she went 4 days a week to learn more about independent living. Although she appeared to make progress, her progress was slower than anticipated, so the case manager extended the authorization for 6 more weeks. In these cases, the case manager authorized additional time for the person in that agency.

Much of the material in this chapter is based on the work of Arnold R. Goldman (1990), from his newsletter *Practical Communications*.

Client Participation/Collaboration

We would not make goals for people without collaborating with them. If you are working with a child, you want to note that the parent or parents were involved in the decisions. For an adult, you need to include the fact that the adult has participated in determining his own goals. If the person is unable to participate at the time due to her mental or physical condition, try to learn who the person would want to participate in the planning on her behalf. For example, Ardith assisted her mother in developing a plan with the worker because her mother was in the beginning stages of Alzheimer's disease.

Assessment and evaluation forms usually have places to record people's answers when you ask them what they see as the main issues to be resolved and what they expect of services. Each of these forms addresses this issue in a different way, but look at this material when developing goals with people. The information you collect from the person should indicate that you and the person developed the goals and objectives together. Someone reading your plans should see it clearly indicated that the client participated and agreed with the direction the goals tend to lead.

Make Objectives Manageable

Goals and objectives can overwhelm people. Sometimes when working with people to develop goals and objectives, case managers develop objectives that are too difficult. It is not always possible to foresee that what you and the client have planned will overwhelm the client when she attempts to meet the objectives. It is best to choose small objectives, small attainable steps that you know the person will be able to accomplish. Meeting these objectives shows progress toward the goal much faster and gives people a sense of having accomplished something important and a sense of moving forward. You can stress to them that they have accomplished something important. When you develop your objectives, work with the tasks you are sure people will be able to attain.

In addition, be careful not to overwhelm people with too many goals and objectives to accomplish. Try only two or perhaps three to start so that people do not feel buried in to-do lists right from the start.

Here is an example of a goal and the objectives that were developed for a woman who wanted to leave welfare. In this example, the case manager overwhelmed the person.

Goal: Larita will become self-sufficient financially by 2019 as evidenced by:

> Objective 1: Larita will go to college in nursing by fall 2015
> Objective 2 Larita will attain a nursing degree by June 2017
> Objective 3: Larita will attain a position in the local hospital by July 2017

In this example, Larita has a plan but the goals are large and could be overwhelming. In addition, she would not accomplish anything before the fall of 2015. Here is a better way to plan with Larita.

Goal: Larita will go to college in nursing by fall 2015

> Objective 1: Larita will get the college catalog
> Objective 2: Larita will choose a course of study
> Objective 3: Larita will enroll at the college
> Objective 4: Larita will choose her courses for the semester
> Objective 5: Larita will complete her first semester

In this example, your service interventions might be to assist her with these tasks and support her as she tackles her first semester. Larita, however, has a clear step-by-step plan to begin her degree. As she checks off each task she will sense she is moving forward. This example has a greater chance of being successful and a greater opportunity to instill confidence.

Expect Positive Outcomes

Goals are actually the outcomes you expect to occur as a result of the treatment, service, or intervention you have chosen (Goldman, 1990). Goals are written, therefore, in the positive—what *will* happen, rather than what will *not* happen or what *might* happen.

The Elements in a Goal

- The goal is the result of the treatment or intervention.
- The client is the subject of the goal. (The therapist, case manager, or treatment team is never the subject.)
- There is only one condition for the goal.
- The goal is written in one sentence.
- Goals are written in the positive.

Writing Goals: Best Practice

Positive Outcomes

POOR: *Kimberly will try to find an apartment by June 1.* (Here the goal does not have a clear, positive outcome. To write a positive goal, state what will happen as a result of the intervention or service being provided, not what might happen. Use the word *will* in constructing your goals.)

BETTER: *Kimberly will find an apartment by June 1.*

POOR: *Alice will no longer frequent bars.* (Here the goal tells what Alice will not do.)

BETTER: *Alice will attend AA meetings 4 times a week.* (Here the goal tells what Alice will do.)

POOR: *Paul will go to public housing and hopefully find housing there.* (Here the goal sounds doubtful and contains an editorial note that has no place in the goal.)

BETTER: *Paul will obtain housing for his family from public housing.*

The Client Is the Focus

POOR: *Johnny's grandfather will work with Johnny to get reacquainted.* (Johnny's grandfather is not the identified client. Instead, make Johnny the focus of the goal.)

BETTER: *Johnny will spend time with his grandfather once a week.*

POOR: *Therapy to help Anne relate better to others.* (This makes the therapy team the object of the goal. Instead, the therapy team needs to talk about what intervention will be used to support this goal in another place and write the goal making Anne the focus of the goal.)

BETTER: *Anne will work with the therapy team to learn to control her anger.*

The Goal Has Only One Outcome

POOR: *Alice will communicate verbally and attend AA meetings.* (Here there are two goals that are unrelated in the same sentence.)

BETTER: *Alice will communicate verbally. Alice will attend AA meetings. (Two separate goals.)*

Written in One Sentence

POOR: *Horace will attend work consistently. He will accomplish this by August 6, 2016.* (Two sentences are cumbersome.)

BETTER: *By August 6, 2016, Horace will attend work consistently.*

Writing the Goals

Most clients will identify more than one goal. Be careful that the person does not have so many goals that they are impossible to complete or overwhelming to the person. First write or study the broad, general goals for the person. In these broad general terms, state what it is you intend to bring about. For example:

- *Phyllis will be able to play cooperatively with the other children.*
- *John will be able to abstain from drinking.*
- *Gladys will be able to prepare her own meals.*
- *Harriet will obtain reliable housing.*
- *Paul will complete job-readiness training.*
- *Michael will complete the outpatient drug treatment program.*
- *Agnes will have regular medical checkups for her asthma.*

Objectives

Goals are often quite similar for similar populations. Many people seek help for similar reasons and, therefore, may have similar goals. *What individualizes a person's plan are the goal objectives*, often called "treatment objectives" or "service objectives". Every goal has objectives. The objectives either are concrete, observable, and measurable manifestations of a treatment goal (Figure 25.1) or the individual steps the person will take to achieve the treatment goal (Figure 25.2). The figures provide some examples of these two types of objectives for three different clients.

Objectives need to be observable and measurable. For example, you cannot see or measure whether or not Skip likes his grandmother. You can see and you can measure whether or not they work together on the issues related to his homework. Skip working with his grandmother on his homework is something you could see. If you say Skip will work with his grandmother 4 nights a week you can also measure the objective. Does Skip meet the 4 nights a week or does he only work with his grandmother once or twice a week?

Skip's mother died and his grandmother is raising him. You cannot see if he has dealt with the death of his mother, but you can see and you can measure how many times he goes to a Caring Place, an agency that helps children deal with the loss of their parent. Look at Figures 25.1 and 25.2. Here you can see the person meeting the objectives. These objectives are measurable. They contain the number of times something will happen.

FIGURE 25.1 Examples of treatment goals and objectives

(Objectives are concrete, observable, and measurable)

Written with Objectives You Expect to Achieve on the Way to Achieving the Goal

Client: Billy
Goal: Billy will play cooperatively with other children as manifested by:

Objectives (concrete, observable, measurable manifestations):

- *Billy will talk to at least three other children every day without hitting.*
- *Billy will* eat *lunch with the other children at least 4 days a week.*
- *Billy will help other children carry out a classroom chore without yelling.*

Client: Karen
Goal: *Karen will take steps to curb her alcohol intake as evidenced by:*

Objectives (concrete, observable, measurable manifestations):

- *Karen will attend four AA meetings a week.*
- *Karen will work collaboratively with her AA sponsor at least once a week.*
- *Karen will attend her* church *on Sunday mornings.*
- *Karen will not miss work for the next month.*

Client: Katherine
Goal: *Katherine will use her local senior center to meet new friends as demonstrated by:*

Objectives (concrete, observable, measurable manifestations):

- *Katherine will go to the senior center twice a week.*
- *Katherine will attend at least one special event at her senior center during the month.*
- *Katherine will participate in one activity each time she is at the center.*

Digital Download Download from CengageBrain.com

© Cengage Learning®

Combining Goals and Treatment Objectives

Think of goals and objectives as being part of a continuum ranging from abstract to concrete. Figure 26.3 shows where goals and objectives would fall along such a continuum.

To check on whether you have written a goal or an objective, apply the "See Billy" test. Read what you have written, and ask yourself if you actually will be able to see or hear the client doing that. If you cannot, you have a goal; but if you can, you have an objective (Goldman, 1990). You can see or hear the client achieve the objectives. Here are some things you could not hear or see, and so you would not use these as objectives:

1. *The client will gain insight into his problems with his mother.*
2. *The client will understand the importance of AA.*
3. *The client will work well with other children at school.*

FIGURE 25.2 Examples of treatment goals and objectives

(Objectives are individual steps to attain the goal.)
Written with Objectives as Individual Steps to the Goal

Client: Billy
Goal: Billy will play cooperatively with other children as manifested by:

Individual Steps:

- *Billy will observe the activities without hitting other children.*
- *Billy will participate in part of at least one activity daily without hitting other children.*
- *Billy will complete one activity daily without hitting.*
- *Billy will participate in more than one activity without hitting.*
- *Billy will participate in an entire day of camp without hitting other children.*

Client: Karen
Goal: Karen will take steps to curb her alcohol intake as evidenced by:

Individual Steps:

- *Karen will determine where AA meetings are held.*
- *Karen will choose the site most convenient for her.*
- *Karen will attend one AA meeting as an observer.*
- *Karen will participate in an AA meeting.*
- *Karen will attend two AA meetings a week.*
- *Karen will secure a sponsor*

Client: Katherine
Goal: Katherine will use her local senior center to meet new friends as demonstrated by:

Individual Steps:

- *Katherine will call her local senior center for hours they are open.*
- *Katherine will choose the dates and times she prefers to go to the center.*
- *Katherine will arrange for transportation with the case manager's help.*
- *Katherine will attend the senior center twice a week.*

Digital Download Download from CengageBrain.com

FIGURE 25.3 shows where goals and objectives would fall along such a continuum

Goals are more abstract, and objectives more concrete

Abstract   *Concrete*

Most goals fall here   Objectives fall here

These would work better as goals. Objectives in these areas might be written as follows:

1. You might hear the person has gained insight by what he talks about. So you might write a treatment objective that reads: *Leonard will be able to talk about the problems he and his mother have.*
2. You could see the person understands the importance of AA by the way she attends meetings. So you might write a treatment objective that reads: *Keyanna will attend AA meetings twice a week.*
3. You could see that the child worked well at school. So you might write a treatment goal that reads: *Janet will work cooperatively with other children on the playground garden.*

In this last set of objectives we can see that they are written specifically to fit the client's circumstances and can be observed and measured.

Finishing Touches

Following is information on other things to consider when writing goals.

Proper Endings

End every goal statement with one of these phrases:

- "as manifested by..."
- "as evidenced by..."
- "as demonstrated by..."
- "as indicated by..."

Then write the objectives after that phrase. For example:

>Goal A: Anita will understand the importance of AA *as evidenced by* ...
>Objective 1: Attendance at AA meetings twice a week for 6 weeks.
>Objective 2: Securing a sponsor by week 2

Numbering System

Develop a numbering system for easy identification so you can tell which objectives go with which goals. For example, you could use this system: Goals: A, B, C, D, and so on; Objectives: 1, 2, 3, 4, and so on. Goals are given capital letters while the objectives for the goals are given numbers. Thus, A1 would represent goal A, objective 1. In this way, you do not have to write out everything in your notes; you have an easy reference system with which to refer to goals and objectives. For instance, in the example of Anita, you might want to note progress in Anita's consistent attendance at AA meetings. You would say that Anita is meeting A1 by having attended AA meetings twice weekly for the past 2 weeks.

Every goal must have at least two objectives. It can have as many more objectives as you see fit.

Target Dates

Give each objective a target date. Remember to set the date for the amount of time you are willing to try this particular intervention without getting the desired result. In this way, as the worker, you will be able to monitor whether the person is progressing. With no target date, it is easy to leave people in programs and treatment for unnecessarily long periods of time. Even if the individual has not reached the goal by the target date, you can assess any progress that has been made and determine whether the program is working and needs a new target date or whether the program is not working and other interventions need to be tried or other arrangements need to be made for the person.

If the client meets the goal or the objective before the target date, simply record this in the progress notes. Also note the new goals and objectives and the new target dates for these as the client continues along an improving continuum.

When a person has a chronic mental illness, some form of intellectual disability, or problems related to aging, a provider of service may be asked to provide an intervention to simply prevent the person's condition or circumstances from growing worse. In this case, you set the target date according to the date at which you would expect to see deterioration if your intervention was not working. When you reach the target date, if the patient's behavior or situation has remained stable, you can extend your target date. Review the intervention and the person's stability at each target date before extending it, and note your review and extension in the chart.

Review Dates

Choose a date between your referral date and your target date to check in and review the person's progress toward meeting the goals. Review dates are generally set for halfway between referral and the expected date of completion. Some large case management units will make all review dates 6 months after the initial intake. Case managers receive a list of people due for a review of their services and each case on that list is carefully reviewed to be sure the goals are being met and to make revisions if those are needed.

Treatment Interventions

A treatment intervention is something you or another professional does to enable the client to reach the goal. If you give the individual an assignment or some task to do that will help her complete one of the objectives, this is a treatment intervention. It is not a goal or objective. It is a treatment intervention because it is what you (the worker) are doing to help the client complete an objective and work toward her ultimate goal. As an example, you might ask Clara, whose goal is to become a better parent, to read some material about a particular parenting technique. Or if Clara attended a parent-training workshop and you were the trainer, your training of Clara would be a treatment intervention.

Notice that a *treatment objective* gives some anticipated change in the client's behavior (e.g., client will attend a parenting workshop twice a week), whereas an *intervention* tells what you, the worker, will use to bring about this change (e.g., worker will give client testimonials to read). Another intervention by the worker, in Clara's case, might be for the worker to contact and get Clara into the parenting workshop. When case managers support their clients by offering information, finding good programs, making effective arrangements, or setting up services, the case managers are practicing treatment interventions.

Long Term and Short Term

Sometimes you are asked to state what your targets are for the long term and for the short term. This may happen if you are in a meeting regarding the client or receive a letter from the client's insurance company. Simply state your goals as your long-term targets and your objectives as your short-term targets. Figure 25.4 provides examples of goals and objectives for two people.

When case managers make a referral to another program, the program generally sets the objectives with the individual. Goals are developed in broad general terms at the case management unit with the person, and the service provider develops the more specific goals and objectives—or the staff of a service provider can take the broad general goals from case management and develop objectives for them with the client. For example, a person referred to a partial hospitalization program would be referred there because of the need to attain a particular goal, such as learning to control anger. The objectives would be designed by the staff there with the client. A person referred to AA would work out the objectives with his case manager because there is no paid staff at AA. Exactly who designs goals and objectives will depend on the program chosen for the individual and the extent to which that program has in-house case management services. However the goals are developed and with whom, this is always done with the client's input and collaboration.

Vocabulary

Let's review the vocabulary.

- A **broad, general goal** is the long-term place a person hopes to be in time. *Harriet wants to become financially independent.*
- A goal is sometimes the same but can be more specific.

    a. Harriet will begin college training in the pulmonary technician program by August 2016.
    b. Harriet will graduate from college with a pulmonary technician's degree by June 2017.

- **Objectives** are the observable and measurable steps to reach the goal. Objectives function as the plan that will be used to meet the goal. Here are the objectives:

    A. Harriet will begin college training in the pulmonary technician program by August 2016, as evidenced by:

    1. Harriet will contact the college for information on the pulmonary technician program by June 1, 2015.
    2. Harriet will discuss the program with CM by June 15, 2015.
    3. Harriet will apply to enter the college by July 1, 2015.
    4. Harriet will begin studies toward her pulmonary technician's degree by August 19, 2015.

FIGURE 25.4 Examples of goals and objectives for two clients

Client 1: Shanika is a young girl who is participating in a therapeutic camp program. Here are her goals and objectives as a camper in this program:

Goal A: Shanika will play cooperatively with the other children as evidenced by:

1. Shanika will be able to engage in one group activity daily without yelling and pushing the other campers by the end of camp.
2. Shanika will be able to follow her team leader in games each day by the end of camp.

Goal B: Shanika will demonstrate her understanding of leadership as evidenced by:

1. Shanika will lead one activity with her team every week from the third week of camp to the end.
2. Shanika will talk at least once in group about what it takes to be a good leader.
3. Shanika will assist younger children in a team activity at least once during the camp session.

Goal C: Shanika will express anger verbally as evidenced by:

1. Shanika will talk in group about her anger at least twice during the camp session.
2. Shanika will be able to tell people what she wants without anger by the end of camp.

Client 2: Chapter 22 contains a referral notification form for a client named Paul J. Bittinger. Paul has been diagnosed with schizophrenia and has been referred to the Grandon River Hospitalization Program where there are mental health technicians as well as psychologists, social workers, and psychiatrists. Here are his goals and objectives at this provider agency.

Goal A: Paul will learn how to manage his medication by February 15, 2016, as evidenced by:

1. Paul will attend Medication Management group once a week.
2. By January 15, 2016, Paul will ask for his medication without prompting.
3. By February 15, 2016, Paul will take his medications during the day without assistance.
4. By February 15, 2016, Paul will be able to discuss in Medication Management group the importance of his medications to managing his illness.

Goal B: Paul will be better able to manage the symptoms of his illness by March 15, 2016, as evidenced by:

1. By January 15, 2016, Paul will be able to discuss in group his behaviors that contribute to an exacerbation of his symptoms.
2. By February 15, 2016, Paul will smoke only one cigarette during the day.
3. By February 15, 2016, Paul will have only one drink with caffeine a day.
4. By February 15, 2016, Paul will sleep 8 hours each night.

- **Target dates** are the dates you anticipate the goal will be met. These are also the dates beyond which you will not continue treatment or services if they are not supporting the person in meeting their goals. In the example above, Harriet's target dates appear at the end of the goal and each objective.
- **Review dates** are the dates on which you will review the entire plan. These are usually halfway between the intake and the target date but this should be flexible given the specific circumstances and activities planned for each person.

Summary

Goals and objectives give more specificity to the plan you and your client have agreed upon. This gives everyone an outline for reaching the goals important to the person. As you work together on these steps, you get to understand better what your clients can handle, how they view their futures, what they want for themselves, and what they think they can handle. Your task is to assist in building the kind of goals and objectives that support success and make people feel a sense of competence.

Remember to write each goal with only one result, to write the goals in the positive, and to provide objectives that are observable and measurable. Include people and their families, when that is useful, in planning these goals. This also helps you to create goals with individuals that they believe are obtainable. It is not helpful to people, nor does it garner their cooperation, if goals and objectives are made up by workers and imposed on them.

Be flexible in reworking goals that the person has not been able to accomplish. In this way, the client has more successes than failures.

Goal development is sometimes difficult as you look for just the right exercise or activity for people to use to get to their goals. With practice, however, you will begin to write goals easily, and thereby provide your clients with the means to move forward.

Exercises: Developing Goals and Objectives

These exercises can also be filled out online at CengageBrain.com.

Exercises I

Instructions: Use the following vignette to develop a detailed treatment plan for the client.

> Helen is 89 years old. She has lived independently in her own home until now. Last week she fell off a chair in her home while reaching up to clean out the top shelves in one of her closets. She sat with her arm aching all night until morning when she called her daughter to take her to the hospital. There an X-ray was done, and the arm was set. The hospital staff sent her home, and her daughter, who works, called the Office of Aging to help her plan so that her mother could remain at home. The Office of Aging set the main goal, "Helen will continue living in her own home with assistance" and referred the case to your agency to provide the service. Helen is unable to make meals or bathe herself, and the pain medication is making her dizzy. The Office of Aging has called your agency, which provides homemaker and in-home care, to provide the actual service to Helen. As the case manager for the agency, you go out and evaluate Helen's situation.

1. The broad goal is *to keep Helen in her own home.* List two very specific goals your agency will work on with Helen. Next, place the objectives for each goal under that goal. Remember the phrase "as evidenced by...."

Goal A:

   Objective 1:

   Objective 2:

Goal B:

   Objective 1:

   Objective 2:

2. Identify Helen's strengths, and show how these strengths will help her meet the goals.

   Helen's strengths:

   Explain how these will help her meet the goals:

3. Helen broke her arm on November 9. What are your target dates for your goals?

4. Describe a service intervention you will use with Helen to help her meet her goals.

Exercise II

Instructions: Use the following vignette to develop a detailed treatment plan for the client.

> Art has been chronically ill with schizophrenia for many years. It started when he was in college, and he has been unable to hold a job. His family has just moved to your area. Art's sister, with whom he lives, has gone into the mental health center seeking services for her brother. Art's parents are deceased. The sister states the move has upset Art, and he appears ready to have another acute episode. She is also concerned that he is not taking his medication as he should, further jeopardizing his health. The sister hoped the case manager could find a place to send Art for treatment following the move. In college Art majored in engineering, and he is quite good at math. Art has been referred to the partial hospitalization program. There you are responsible for devising goals and objectives for Art.

1. The broad goal is *to help Art adjust to the move.* List two very specific goals your agency will work on with Art. Next, place the objectives for each goal under that goal. Remember the phrase "as evidenced by. . . ."

    Goal A:

    Objective 1:

    Objective 2:

    Goal B:

    Objective 1:

    Objective 2:

2. Identify Art's strengths, and show how these strengths will help him meet the goals.

    Art's strengths:

    Explain how these will help him meet the goals:

3. Art and his sister went to mental health case management on January 12. What are your target dates for your goals?

4. Describe a service intervention you will use with Art to help him meet his goals.

Exercise III

Instructions: Use the following vignette to develop a detailed treatment plan for the client.

> Lester has been in prison and has recently been released on parole. He has family in the area, but he distanced himself from them after high school. They did visit him while he was in prison. Lester has been referred to your agency, which works with ex-convicts to help them turn their lives around. Lester has indicated that his juvenile crime spree was "crazy" and that he would like to become a useful citizen. As the case manager, you interviewed

Lester and learned that he would like to become a chef. He tells you that he had some cooking classes in the prison and that he worked mainly in the kitchen there and enjoyed the work. He feels, however, that he could have learned more and taken on more responsibility. At present Lester is staying in a halfway house and is charged with getting a job and a place to stay.

1. The broad goal is *to help Lester get acclimated to the community.* List two very specific goals your agency will work on with Lester. Next, place the objectives for each goal under that goal. Remember the phrase "as evidenced by. . . ."

Goal A:

   Objective 1:

   Objective 2:

Goal B:

   Objective 1:

   Objective 2:

2. Identify Lester's strengths, and show how these strengths will help him meet the goals.

   Lester's strengths:

   Explain how these will help him meet the goals:

3. Lester came into your office on May 6. What are your target dates for your goals?

4. Describe a service intervention you will use with Lester to help him meet his goals.

Exercise IV

Instructions: Use the following vignette to develop a detailed treatment plan for the client.

> You are a case manager in an alternative school for teenagers who have behavioral and emotional problems, and Rick is referred to you. Rick's mother is addicted to cocaine, and his father is not in the picture. Only recently Rick was moved to his grandmother's home. His grandmother seems to be very intent on helping Rick. She works as a domestic in a large office building downtown and is not home before 9:00 P.M. You are concerned that Rick is home alone after school, and you are also concerned about his getting his homework done. In addition, Rick has been having problems with other teens: hitting and punching, name-calling, and cutting up in class, all behaviors that need to be curbed if Rick is going to get anything out of the alternative school.

1. The broad goal is *to help Rick succeed at the alternative school*. List two very specific goals your agency will work on with Rick. Next, place the objectives for each goal under that goal. Remember the phrase "as evidenced by. . . ."

Goal A:

   Objective 1:

   Objective 2:

Goal B:

   Objective 1:

   Objective 2:

2. Identify Rick's strengths, and show how these strengths will help him meet the goals.

Rick's strengths:

Explain how these will help him meet the goals:

3. Rick came into your office on December 10. What are your target dates for your goals?

4. Describe a service intervention you will use with Rick to help him meet his goals.

Exercise V

Instructions: Use the following vignette to develop a detailed treatment plan for the client.

Twelve-year-old Christina has come to the attention of your agency because her parents have abused her physically. The case was reported to Child Welfare, where you work, by a teacher at Christina's school and Christina was removed from her home. The teacher noticed that Christina always seemed to have bruises, and when she asked Christina about these, Christina was reluctant to discuss them. You have met the parents and talked at length with Christina. All three say they want the family reunited in time. For the time being Christina is being placed with an aunt who offered to provide foster care for Christina. Christina is a good student, particularly in math and science. She is extremely shy and quiet and has few friends. You are not sure whether this is because she was forbidden to bring friends home or whether she has always been this way.

1. The broad goal is *to help Christina and Christina's family reunite successfully*. List three very specific goals your agency will work on with Christina and her family. Next, place the objectives for each goal under that goal. Remember the phrase "as evidenced by. . . ."

Goal A:

   Objective 1:

   Objective 2:

Goal B:

   Objective 1:

   Objective 2:

Goal C:

   Objective 1:

   Objective 2:

2. Identify Christina's strengths and any strengths you see within her family, and show how these strengths will help Christina and the family meet their goals.

   Christina's and Christina's family's strengths:

   Explain how these will help them meet their goals:

3. Christina came into your office on April 4. What are your target dates for your goals?

4. Describe a service intervention you will use with Christina and her family to help them meet their goals.

CHAPTER 26

Terminating the Case

Introduction

We have seen how people enter the human service system and how their services and treatment are determined and monitored. There is usually a point at which people leave the system, moving on in their lives. Here are some of the reasons a case may be terminated.

1. *The individual and the case manager agree the individual is ready to move on.* This is the ideal. The service or treatment has been successful and is no longer needed. Many people do leave for this reason, feeling that their original issues and problems are less significant than they once were or that these problems have been resolved.
2. *The individual dies or moves away.* When people die or move to another jurisdiction, their cases are closed. If they formally request that their records be sent to the new jurisdiction, this should be done immediately to facilitate a smooth transition to the new program.
3. *The funding source will no longer finance services.* Managed care has introduced limitations on care that you and the client may find unrealistic. It is important that people know the limitations during the first interview so that they can prepare for **termination**. When termination is imposed by a funding source before you or the client feel it should, work with the person to find alternatives to your service. Support groups or specialized programs funded by other sources may not give the level of service you have provided, but they may help the person make the adjustment. It is not ethical to drop someone with no plan or referral when the insurance or funding runs out.
4. *The individual no longer wants the services.* People may be dissatisfied with the services being offered and request that their cases be terminated. In situations like this, sit down with the dissatisfied individuals and learn why they are not

pleased with the service. This may provide you with valuable information about how you or a provider agency is perceived by people who are served there, and it may facilitate people leaving with the feeling that they can come back if they need to do so. Others decide to leave because they no longer find working with you a priority and simply wish to move on.

5. *You cannot find the individual.* Sometimes people indicate they are not interested in our services by disappearing. It may be a person you feel is really in need of support, medication, or treatment of some sort; but often people feel case management is either intrusive or a nuisance, and just disappear. You may make attempts to find such people, and you may even track some of them down, but they have every right to refuse services. When you encounter this situation, make sure that both your contact notes and the **termination summary** reflect your attempts to contact the person who has disappeared.

Not all clients will leave case management. A child with autism may need services all his life. A woman with severe developmental disabilities may require case management during her entire lifetime in order for her to live in her community successfully.

A Successful Termination

Cases should not be closed without the person, or the person's family, if they are involved, knowing that this is about to occur and why. A letter by itself cannot convey warmth and concern for the individual and often comes across as bureaucratic and unfeeling. A phone call is not much better. You might convey warmth, but there is no meaningful exchange or documentation.

All clients whose cases are being closed, except those who have moved away or died, should receive two things from the case management unit:

1. An opportunity to meet with you to discuss the termination
2. Follow-up by letter outlining the main points in your **final interview** and inviting the person to return if the need arises

Feelings about Termination

Leaving anything can be difficult, and leaving services can be particularly difficult for people who have grown fond of the workers at the agency or who feel uncertain about how they will handle life on their own. Sometimes terminations are milestones. The people have reached a new level of independence, emotional health, or sobriety. But the accomplishment still can be tinged with misgivings.

In some cases, people may regress in order not to have to leave your services. Will struck a pose as a very independent person. He came with obvious reluctance to the case management unit after his supervisor demanded he get help for his drinking problem or he would lose his job. Throughout his time with his case manager, Will boasted that he could manage all this on his own and did not need to receive help. He wanted to be sure his case manager knew he had come only to please his boss and to keep his job.

Will did well and at one point confided to his case manager that he was surprised at how he had managed to go to AA and "stay off the bottle." When it was time for Will's case management services to end, he suddenly began drinking on a Saturday afternoon. When he came in for his final interview he announced that he had been "on a bender" although his wife said he "only had a couple of beers." The situation was perplexing and Will insisted he had strong urges to drink again and was concerned that he could lose his job if he reverted to his old behavior. His case was not terminated and Will continued to come in and report abstinence, but he wanted the case manager to know "it is a struggle."

Again when termination was about to take place, Will reported drinking again. Contrite and embarrassed, he came into the office on a Monday morning to tell his case manager that he "tied one on at the wedding Saturday." He reported being extremely intoxicated and said he needed more time in case management to be able to "get a better handle on this." It was at this point that the case manager and his supervisor sat down with Will and addressed what seemed to be his reluctance to leave case management, even though he had shown remarkable progress most of the time. Will agreed to terminate his case and was told he could certainly return if he had further problems. He seemed cooperative and willing to leave the services. He would continue in AA and he would work with his sponsor. His job was now secure and he appeared to be well prepared to move on. However, 3 weeks following termination while on vacation Will showed up drunk at the case management unit and his case was reopened. This episode, timed to take place on vacation when it could not interfere with his work seemed coincidental. Will reported that he had been having a "whale of a battle, I tell you," trying to remain sober. Clearly, Will is able to control his drinking and has made progress, but leaving a case manager he has come to lean on for support has been extremely difficult for him.

Try to recognize the underlying feelings your client is experiencing in the final interview and respond empathically. Some people resent being terminated. They may have grown accustomed to the support, enjoyed having a person in their lives who cared for them, or they may just be feeling that they are being "shoved out the door," as one person put it.

As the person's case manager, you may have feelings as well about the termination. Perhaps you and your client have had a particularly good working relationship. Together a lot was accomplished. It is sometimes hard to say goodbye to clients when we are fond of them or have enjoyed our work with them.

Terminations work better if clients have a chance to get used to the idea that the relationship will end and can in some ways prepare for that. This is particularly true if you and the person have had a long relationship and been through the ups and downs of the person's life. Encourage people to talk about their feelings about leaving and respond empathically to what they have to tell you about what the change will mean to them. Acknowledge your own sense that you have worked hard together, if that applies.

While this chapter is about the client leaving the services, much of what we have said here applies to times when you are leaving and the client will receive a new worker. Feelings of uncertainty, loss, even grief are not uncommon. Sometimes life intrudes inconveniently and we need to leave our positions with very little notice, but wherever possible plan your leave-taking so that clients have an opportunity to know and prepare for the change.

The Final Interview

Use the interview to summarize the work that has been accomplished and the reasons for the termination. If there are gains, review how far the person has come since first seeking help some time before. For those who have requested termination of their cases themselves, ask them to tell you more about their reasons for making the request and be open to their suggestions for change. Invite questions. People often want to know where they should turn should former problems resurface. Give them information they can use, particularly information that will help them prevent a relapse or regression.

There should be a sense of reassurance during the interview that clients are not being cut off or dumped and that they are welcome to return should they need services again. These feelings are especially likely when people have run out of funding or they were funded for only a specific amount of time. They may not tell you all of that explicitly. Nevertheless, people may feel dumped or dropped and be unable to express that to you. Speak to these concerns in the final interview.

It is useful to go over in the final interview the gains and accomplishments that the client has attained. Talking about these specifically is a good way to summarize for the person where you and he started, what you have accomplished together, the gains he has made, and where he is now. This is a good time to point out specifically the strengths you have come to see in this person and the specific gains made in reaching goals you set together some time before.

In addition to strengths and achievements, talk about areas the person still sees as weaknesses or areas she feels still need to be addressed in her life. This too is part of summarizing for people where they have been, where they are now, and perhaps where they feel they need to go from here. This summary gives people a feeling of tying up the loose ends, relating the tasks and work, and looking positively toward an end to the relationship.

Finally, talk about where people will go after the relationship ends. Will they be referred to another service? Will they join community self-help groups? Are there things they can now do on their own to continue their progress and prevent future issues?

Many agencies seek an evaluation of their services when clients leave the agency. If your agency does that explain the process to people and prepare them to receive evaluation materials or a call.

Digital Download Download from CengageBrain.com

The Letter

The letter is a follow-up to the final interview. It should summarize the main points of the interview, recapping briefly what was discussed. There should be a brief statement about why the termination took place so the person has documentation of it. In addition, questions that seemed particularly important to the person during the interview should be answered again in the letter, especially if the individual needs addresses, names of resources, and other supportive information. Figure 26.1 provides a sample termination letter. In this letter there is a summary spelling out why Mrs. Warren is leaving the agency, the positive gains she made, and her plans for the future.

Digital Download Download from CengageBrain.com

FIGURE 26.1 Sample termination letter

> Dear Mrs. Warren,
>
> It was good to talk to you on Monday and to hear about your plans to move. As we discussed on Monday you have been able to consistently manage your medication and work reliably at your job as an artist. These have been real gains that you have made in the last 6 months. As we planned in our meeting, we will be terminating your case here as you requested because of your move to another state. We will forward those records you designated to the new case management unit when you are settled.
>
> We have given you a prescription for medications for one month that can be filled at a pharmacy where you will be living. If you have been unable to see a doctor before the prescription expires, call us and we will issue a prescription for 1 more month of your current medication.
>
> The case management unit where you will be living is at 3132 Green Road. The unit's phone number is (555) 555-5555.
>
> If there is any way we can be helpful in the future, please don't hesitate to get in touch with us again. It was a pleasure working with you, and we wish you all the best in your new home.
>
> Sincerely,
>
> *Cindy Parker,*
>
> Case manager

Documentation

Like all contacts with clients, this last contact and letter should be documented in the person's chart. In the note, the focus of the interview would be the termination of the case. You would note the highlights of the discussion, any follow-up arrangements that were made for the person, and his or her response to the interview.

The Discharge Summary

In most cases, your agency will ask for a termination summary or discharge summary. This is not the same thing as the final case note discussed earlier. Although you include the information from that contact note in the discharge summary, you are actually summarizing the most important information about what took place while the individual was working with you. It is wise to think of your summary as a document that may go to other professionals in other agencies who will see the person in the future. Asking for summaries is common practice when beginning work with a new person who was seen before in another agency. For that reason, your summary functions in part as an indication to other professionals of how thoroughly your agency addresses the needs of clients. This is a major way that other programs and professionals can learn about the quality of the care given in your case management unit. Sloppy summaries containing little useful information or those that indicate little organized effort on the client's behalf can make your agency look unprofessional.

What you include in your summary should be helpful to a new case manager or therapist developing a strategy to help the person. For this reason, your summary should discuss what was tried, what worked, and what was less successful and why. Figure 26.2 provides a sample discharge or termination summary.

Your agency may have a standard format for discharge summaries that you can use as a guide to writing good discharge summaries. If not, you can use the "Discharge Summary" form provided in Appendix C.

Here are important items to include in the discharge summary:

1. Name, date of birth, date of admission, and date of discharge
2. Diagnosis (if there is one)
3. Any medication that was prescribed by physicians who were working with the case management unit and whether it has been discontinued. (Make sure to note the name of the medication, the dosage, the frequency, and any adverse reactions.)
4. The reason for discharge
5. The major presenting problem that brought the person to you
6. Your goals and objectives for the individual
7. The extent to which the client participated in formulating these goals and objectives
8. Progress that was made or goals that were accomplished
9. Problems that were identified but were not addressed
10. How the individual appeared to be at intake and how the individual appeared to be at termination
11. Attempts to locate the person if she has disappeared

Examples

We can look at several cases and how they were terminated. First is Alex, whose insurance stopped payment for services, which stopped Alex from having further contact with the agency. Alex did not want to leave and felt he had gained support he needed to continue to work through his difficulties on his job. The reason for discharge was "insurance no longer available." The goal for Alex to begin a course that could have led to a promotion was not met. This was noted and the need for further support to obtain this advanced training was noted under problems that had been identified but not addressed. Because Alex seemed as agitated about his work situation when his insurance ran out as he did the day he came in for assistance that too was noted and in fact, the case manager noted that the loss of insurance coverage was contributing to his difficulty. Before discharge the case manager worked with Alex and his minister to give Alex some support for attending the advanced classes. This gave Alex someone to work with before the insurance was reinstated the following year. This was not an ideal solution, but the case manager made sure there was some support system in place before the final termination.

FIGURE 26.2 Sample discharge or termination summary

Wildwood Case Management Unit

Discharge Summary

Name *Juan Gonzales*   Date of birth *1/17/1958*

Date of admission *10/24/16*   Date of discharge *3/5/17*

Reason for Admission *House fire and death of his wife and 2-year-old daughter*

Diagnosis on Admission

309.81(F43.10) Acute Post Traumatic Stress Disorder

Asthma following smoke inhalation during fire

Medication:

Welbutrin 100 mg TID

Reason for discharge:

Client is moving to Cambridge to be near relatives

Presenting problem:

Juan Gonzales is a 48-year-old male, a draftsman with a large engineering company, who came to the case management unit suffering from anxiety and depression following a house fire 4 months before in which he lost his wife and 2-year-old daughter. He was referred by his family physician after his physician had seen Juan on several occasions. At the time Juan came to the agency, he was complaining of depression, feeling tired and "numb," and feeling intense anxiety about the status of other loved ones, particularly his surviving child. He reported recurrent, intrusive memories about the fire and an unsuccessful struggle to diminish these memories. His affect was flat. He stated that others in his life were concerned about his irritability. At work he was unable to concentrate and felt intense irritability with coworkers.

Goals:

The case manager and Mr. Gonzales developed two goals. The objectives were developed with Mr. Gonzales by his therapist, Kelly Marcuson.

GOAL I: To feel less depressed by 3/24/17 as demonstrated by:
  A. Attending 10 session of psychotherapy with Dr. Kelly Marcuson
  B. Meeting with Dr. Pederos to evaluate his need for medication
  C. Attending at least one session of the Market Square Presbyterian Church's grief support group

GOAL II: To be able to work consistently by 3/25/17 as demonstrated by:
  A. Obtaining information from Dr. Pederos as to how much he is able to work at present
  B. Meet with his employer and discuss his recent absences
  C. Discussing with his employer a schedule for his gradual return to work full time
  D. Keeping the devised schedule for at least 1 month

These were the two most important concerns when Mr. Gonzales came to the case management unit, and he participated in developing the goals and objectives.

(continued)

FIGURE 26.2 *(continued)*

Progress:

At intake Mr. Gonzales was unable to attend work regularly, was experiencing intrusive memories of the fire, and had a flat affect. At discharge he appeared more animated and expressed hope that being near his family would offer support to both him and his son. He was referred to Family and Children's Services where he saw Kelly Marcuson.

Mr. Gonzales attended 10 sessions with Kelly Marcuson, psychologist, from November 2016 through March 25, 2017. He reported the sessions were helpful in relieving some of his more acute symptoms and in helping him to "sort out what happened to all of us." In November 2016 he was evaluated by Dr. Marcella King, who prescribed Wellbutrin, 100 mg TID, which he tolerates well. He remains on this medication at discharge. He reports the memories of the fire are less intrusive, and he is more animated in contacts with CM.

The client followed up with a meeting with his employer and began to attend work 3 days a week. At discharge he was attending work full time but still concerned about some lack of concentration and some irritability with coworkers.

In addition, client attended the grief support group at a local church where he met two individuals who became close friends. He credits them with helping him to think more about the effect the fire had on his son. His one regret in leaving is that he will be leaving this support group.

Client is leaving the area to move closer to relatives in the Cambridge area. He is seeking employment with the Massachusetts Department of Highways as a draftsman and has taken a civil service examination, on which he obtained a high score.

Additional issues:

Not addressed in our time with Mr. Gonzales were his concerns for his surviving 7-year-old son, now in second grade. He indicated a willingness to discuss his son's problems in school that started after the fire. Client intends to seek help for his son in Cambridge.

Impressions and recommendations:

Juan Gonzales is a 48-year-old man, widowed as the result of a house fire, and currently caring for his 7-year-old son. He sought help for debilitating symptoms of posttraumatic stress disorder, primarily depression and anxiety, and an inability to work consistently. At discharge his symptoms are less acute and he remains on 100 mg of Wellbutrin, TID, which he tolerates well. He met both goals. His depression is diminished and he has returned to working consistently.

The reason for termination was a move to be closer to relatives in the Cambridge area. Recommend he seek further assistance with remaining symptoms and also seek assistance for his son.

Will need a medication re-evaluation and continuation of his Wellbutrin until that evaluation.

Prepared by
Angela Carter, Case Manager

Serina had been homeless and very independent. Case managers had visited her on the street where she frequently stayed and made certain she had her medications. However, Serina was found dead one morning of an apparent homicide. No one was ever charged in her death. The discharge summary noted the reason for termination was that she was deceased. There were, however, issues for case managers compiling a discharge summary for Serina. Had there been goals for her? The one goal was to get Serina into a single-room occupancy shelter, particularly given her age, but there were few notes in the record that indicated that case managers had helped Serina pursue this goal or even that Serina agreed with such a move. In addition, the record begged the question: Had case managers done enough to protect Serina? The discharge summary, while well written, left many unanswered questions. In part, Serina's unwilling cooperation made the case management work in this case difficult. Nevertheless, the discharge summary could not ethically show a vigorous plan and work in support of Serina.

Finally, Augustine had met all his goals. Diagnosed with a bipolar disorder, he had learned to take his medications consistently, obtained a job in the local symphony orchestra, and had worked successfully to stabilize his marriage. His reason for termination was that his goals had been met and the case manager could list his goals and the progress that was made. There were no apparent issues left unaddressed and Augustine remarked during his last interview with his case manager that while he hated to leave he felt like "I'm more normal doing this from now on, on my own, rather than needing someone to help me stay on track." Unlike Serina, Augustine had worked hard on his goals, cooperated with the work put forth by his case manager to help him meet the goals, and even came up with some ideas of his own for meeting his goals.

Summary

Termination should be approached as skillfully as all other aspects of case management. When cases are closed well, both clients and agencies benefit. People feel reassured and supported as they take leave of your services. In the community, the perception of your agency as a caring and professional place is strengthened.

Termination involves good documentation as well. The discharge or termination summary you prepare outlines concisely the history of the person's relationship with your agency. This is enormously useful to others who may work with your client later.

Video Examples

To view the videos that accompany this book, go to CengageBrain.com.

Terminating a Case

Online, you can watch as Danica conducts a termination interview with Alison.

Exercises

These exercises can also be filled out online at CengageBrain.com.

Exercises I: Termination of a Middle-Aged Adult

Instructions:

1. Record in your contact notes a termination interview, indicating the reason for termination and any follow-up arrangements you made for the client after he leaves your services. Refer to the goals for such an interview in this chapter.
2. Write your client a letter as a follow-up to this interview, summarizing for the client the discussion and resolutions that took place in your last contact.
3. Prepare a discharge summary on your client. It will be the last item in the client's chart. Use the guidelines for a discharge summary found in this chapter.

Exercises II: Termination of a Child

Instructions:

1. Record in your contact notes a termination interview, indicating the reason for termination and any follow-up arrangements you made for the client after she leaves your services. Be sure your termination interview includes the parents, and indicate if the child was present. Refer to the goals for such an interview in this chapter.
2. Write your child's parents a letter as a follow-up to this interview, summarizing for them the discussion and resolutions that took place in your last contact.
3. Prepare a discharge summary on this child. It will be the last item in the client's chart. Use the guidelines for a discharge summary found in this chapter.

Exercises III: Termination of a Frail, Older Person

Instructions:

1. Record in your contact notes a termination interview, indicating the reason for termination and any follow-up arrangements you made for the client after he leaves your services. Indicate if others were present, such as relatives or a worker from a new service where your client will be going. Document where you conducted the interview. Refer to the goals for such an interview in this chapter.
2. Write your client a letter as a follow-up to this interview, summarizing for the client the discussion and resolutions that took place in your last contact. If you have the client's permission to do so, indicate that you are sending a copy of your letter to the person who was present for the termination interview.
3. Prepare a discharge summary on your client. It will be the last item in the client's chart. Use the guidelines for a discharge summary found in this chapter.

Exercises IV: Organizing the Record

Instructions: Organize the record you have compiled on any client you have been following. The record should be organized according to Appendix C.

APPENDIX A

Ten Fundamental Components of Recovery

Self-Directed Consumers lead, control, exercise choice over, and determine their own path of recovery by optimizing autonomy, independence, and control of resources to achieve a self-determined life. By definition, the recovery process must be self-directed by the individual, who defines his or her own life goals and designs a unique path towards those goals.

Empowerment Consumers have the authority to choose from a range of options and to participate in all decisions—including the allocation of resources—that will affect their lives and are educated and supported in so doing. They have the ability to join with other consumers to collectively and effectively speak for themselves about their needs, wants, desires, and aspirations. Through empowerment, an individual gains control of his or her own destiny and influences the organizational and societal structures in his or her life.

Non-Linear Recovery is not a step-by-step process but one based on continual growth, occasional setbacks, and learning from experience. Recovery begins with an initial stage of awareness on which a person recognizes that positive change is possible. This awareness enables the consumer to move on to fully engage in the work of recovery.

Individualized and Person Centered There are multiple pathways to recovery based on an individual's unique strengths and resiliencies as well as his or her needs, preferences, experiences (including past trauma), and cultural background in all of its diverse representations. Individuals also identify recovery as being an ongoing journey and an end result as well as an overall paradigm for achieving wellness and optimal health.

Holistic Recovery encompasses an individual's whole life, including mind, body, spirit, and community. Recovery embraces all aspects of life, including housing, employment, education, mental health and health care treatment and services, complementary and naturalistic services (such as recreational services, libraries, museums, etc.), addictions treatment, spirituality, creativity, social networks, community participation, and family supports as determined by the person. Families, providers, organizations, systems, communities, and society play a crucial role in creating and maintaining meaningful opportunities for consumer access to these supports.

Strengths-based Recovery focuses on valuing and building on the multiple capacities, resiliencies, talents, coping abilities, and inherent worth of individuals. By building on these strengths, consumers leave stymied life roles behind and engage in new life roles (e.g., partner, caregiver, friend, student, employee). The process of recovery moves forward through interaction with others in supportive, trust-based relationships.

Peer Support/Mutual Support Including sharing of experiential knowledge and skills and social learning—plays an invaluable role in recovery. Consumers encourage and engage other consumers in recovery and provide each other with a sense of belonging, supportive relationships, valued roles, and community.

Responsibility Consumers have a personal responsibility for their own self-care and journeys to recovery. Taking steps toward their goals may require great courage. Consumers must strive to understand and give meaning to their experiences and identify coping strategies and healing processes to promote their own wellness.

Respect Community, systems, and societal acceptance and appreciation of consumers—including protecting their rights and eliminating discrimination and stigma—are crucial in achieving recovery. Self-acceptance and regaining belief in one's self are particularly vital. Respect ensures the inclusion and full participation of consumers in all aspects of their lives.

Hope Recovery provides the essential and motivating message of a better future—that people can and do overcome the barriers and obstacles that confront them. Hope is internalized; but can be fostered by peers, families, friends, providers, and others. Hope is the catalyst of the recovery process.

Source: U.S. Department of Health and Human Services, Substance Abuse and Mental Health Services Administration, Center for Mental Health Services, *Consensus Statement Defines Mental Health-Recovery*, March/April 2006, Vol. 14, Number 2.

APPENDIX B

Vocabulary of Emotions

	Happiness	Caring	Depression	Inadequateness	Fear	Confusion	Hurt	Anger	Loneliness	Remorse
Strong	Delighted Ebullient Ecstatic Elated Energetic Enthusiastic Euphoric Excited Exhilarated Overjoyed Thrilled Tickled pink Turned on Vibrant Zippy	Adoring Ardent Cherishing Compassionate Crazy about Devoted Doting Fervent Idolizing Infatuated Passionate Wild about Worshipful Zealous	Alienated Barren Beaten Bleak Bleeding Dejected Depressed Desolate Despondent Dismal Empty Gloomy Grieved Grim Hopeless In despair Woeful Worried	Blemished Blotched Broken Crippled Damaged False Feeble Finished Flawed Helpless Impotent Inferior Invalid Powerless Useless Washed up Whipped Worthless Zero	Alarmed Appalled Desperate Distressed Frightened Horrified Intimidated Panicky Paralyzed Petrified Shocked Terrified Terror-stricken Wrecked	Baffled Befuddled Chaotic Confounded Confused Dizzy Flustered Rattled Reeling Shocked Shook up Speechless Startled Stumped Stunned Taken-aback Thrown Thunderstruck Trapped	Abused Aching Anguished Crushed Degraded Destroyed Devastated Discarded Disgraced Forsaken Humiliated Mocked Punished Rejected Ridiculed Ruined Scorned Stabbed Tortured	Affronted Belligerent Bitter Burned up Enraged Fuming Furious Heated Incensed Infuriated Intense Outraged Provoked Seething Storming Truculent Vengeful Vindictive Wild	Abandoned Black Cut off Deserted Destroyed Empty Forsaken Isolated Marooned Neglected Ostracized Outcast Rejected Shunned	Abashed Debased Degraded Delinquent Depraved Disgraced Evil Exposed Humiliated Judged Mortified Shamed Sinful Wicked Wrong
Medium	Aglow Buoyant Cheerful Elevated Gleeful Happy In high spirits Jovial Light-hearted Lively Merry Riding high Sparkling Up	Admiring Affectionate Attached Fond Fond of Huggy Kind Kind-hearted Loving Partial Soft on Sympathetic Tender Trusting Warm-hearted	Awful Blue Crestfallen Demoralized Devalued Discouraged Dispirited Distressed Downcast Downhearted Fed up Lost Melancholy Miserable Regretful Rotten Sorrowful Tearful Upset Weepy	Ailing Defeated Deficient Dopey Feeble Helpless Impaired Imperfect Incapable Incompetent Incomplete Ineffective Inept Insignificant Lacking Lame Overwhelmed Small Substandard Unimportant	Afraid Apprehensive Awkward Defensive Fearful Fidgety Fretful Jumpy Nervous Scared Shaky Skittish Spineless Taut Threatened Troubled Wired	Adrift Ambivalent Bewildered Puzzled Blurred Disconcerted Disordered Disorganized Disquieted Disturbed Foggy Frustrated Misled Mistaken Misunderstood Mixed up Perplexed Troubled	Annoyed Belittled Cheapened Criticized Damaged Depreciated Devalued Discredited Distressed Impaired Injured Maligned Marred Miffed Mistreated Resentful Troubled Used Wounded	Aggravated Annoyed Antagonistic Crabby Cranky Exasperated Fuming Grouchy Hostile Ill-tempered Indignant Irate Irritated Offended Ratty Resentful Sore Spiteful Testy Ticked off	Alienated Alone Apart Cheerless Companionless Dejected Despondent Estranged Excluded Left out Leftover Lonely Oppressed Uncherished	Apologetic Ashamed Contrite Culpable Demeaned Downhearted Flustered Guilty Penitent Regretful Remorseful Repentant Shamefaced Sorrowful Sorry

452 Appendix B Vocabulary of Emotions

	Happiness	Caring	Depression	Inadequateness	Fear	Confusion	Hurt	Anger	Loneliness	Remorse
Light	Contented	Appreciative	Blah	Dry	Anxious	Distracted	Let down	Bugged	Blue	Bashful
	Cool	Attentive	Disappointed	Incomplete	Careful	Uncertain	Minimized	Chagrined	Detached	Blushing
	Fine	Considerate	Down	Meager	Cautious	Uncomfortable	Neglected	Dismayed	Discouraged	Chagrined
	Genial	Friendly	Funk	Puny	Disquieted	Undecided	Put away	Galled	Distant	Chastened
	Glad	Interested in	Glum	Tenuous	Goose-bumpy	Unsettled	Put down	Grim	Insulated	Crestfallen
	Gratified	Kind	Low	Tiny	Shy	Unsure	Rueful	Impatient	Melancholy	Embarrassed
	Keen	Like	Moody	Uncertain	Tense		Tender	Irked	Remote	Hesitant
	Pleasant	Respecting	Morose	Unconvincing	Timid		Touched	Petulant	Separate	Humble
	Pleased	Thoughtful	Somber	Unsure	Uneasy		Unhappy	Resentful	Withdrawn	Meek
	Satisfied	Tolerant	Subdued	Weak	Unsure			Sullen		Regretful
	Serene	Warm toward	Uncomfortable	Wishful	Watchful			Uptight		Reluctant
	Sunny	Yielding	Unhappy		Worried					Sheepish

Source: Tom Drummond, North Seattle CC, tdrummon@me.com

APPENDIX C

Wildwood Case Management Unit Forms

Arrangement of the Client's Chart

1. **A Clean File Folder**   Use a clean file folder and place client's name and number on the tab of the folder, last name first.
2. **Face Sheet** on top
3. **New Referral/Inquiry Sheet**
4. **Request/Release of Information Form(s)**
5. **Intake Assessment Form**   Your instructor may ask you to provide either a social history or an assessment form, or both, for practice purposes. Be sure you show collaboration with client on goals and expectations.
6. **Social History**   Your instructor may ask you to provide either a social history or a brief social history, or both, for practice purposes. Be sure you show collaboration with client on goals and expectations.
7. **Notes for Service Planning Conference**   These may be handwritten notes you made for yourself. In an agency these would not be part of the record, but your instructor may want to see these.
8. **Treatment or Goal Service Plan**
9. **Referral Notification Form(s)**
10. **Contact Notes**   Be sure these reflect monitoring.
11. **Provider Agency Goals and Objectives**   This should be on a letterhead from the provider agency, informing you of the goals for the client in response to your referral. Students usually make up a letterhead for the provider agency on their computers. Have someone else sign as the worker at the provider agency.
12. **Termination Letter**
13. **Discharge Summary**

Wildwood Case Management Unit
Face Sheet

Name _____ Agency # _____

Address _____

Home or cell Phone _____ Work Phone _____

Guardian _____

Address _____

Home or cell Phone _____ Work Phone _____

Next of Kin _____
If different from Guardian

Address _____

Home or cell Phone _____ Work Phone _____

May not call these numbers Home _____

Work _____ Cell _____

Date of first contact	Taken by	DOB	Gender
Marital Status	Education Level	Employment Status	Veteran Status
Currently Pregnant	Rec. Prenatal care	Given birth last 28 days?	Pregnancy Complications
Current Medical Conditions Treated by	Current Medications Prescribed by	Legal status/ Incarcerations	Reason for visit
Substance Abuse Problems Dx. Case Mgr	Mental Health Problems Dx. Case Mgr	Referrals 1. 2. 3.	Primary provider of service, counselor, therapist
Psychotropic medications Prescribed by	Psychiatric evaluation done	Psychological evaluation	Court Ordered?

Assigned case manager _____ first review date _____

Wildwood Case Management Unit
New Referral or Inquiry

CLIENT _____ SEX _____ DOB _____

ADDRESS _____

_____ ZIP _____

HOME TELEPHONE _____ WK TELEPHONE _____

PARENT OR SPOUSE _____

EMPLOYER _____

SCHOOL _____

REFERRED BY _____

CHIEF COMPLAINT &/OR DESCRIPTION OF PROBLEM _____

_____

_____

_____

_____

_____

_____

_____

_____

_____

_____

PREVIOUS EVALUATION, SERVICES, OR TREATMENT

_____

_____

_____

_____

TAKEN BY _____ DATE _____

DISPOSITION FOR INTAKE _____

VERIFICATION SENT _____

Wildwood Case Management Unit
4600 Wildwood Drive
Harrisburg, PA 17110
255-5555

Verification of Appointment

Date _____

Dear

This letter is to inform or remind you that you have an appointment scheduled:

Date: _____

Time: _____

Staff: _____

Location: _____

Please contact me if you have any questions or if you need to reschedule.

Sincerely,

Case Manager

Wildwood Case Management Unit
Request/Release of Information

RE: _____   DOB: _____

To Whom It May Concern:

[ ] I hereby authorize the Wildwood Case Management Unit to release information about services rendered to the above-named, for the purpose of:

_____

[ ] I hereby authorize the Wildwood Case Management Unit to receive information about services rendered to the above-named from:

_____

for the purpose of _____

Such information may be transmitted under the conditions stated below, and/or as required by Federal or State statute or order of the court. This release will be effective for a period of ninety (90) days from the date signed below and will expire on _____

Information to be released/received may include

- ( ) Medical records
- ( ) Discharge summary
- ( ) Psychological evaluation
- ( ) Vocational evaluation/summary
- ( ) Treatment summary
- ( ) Behavior management services
- ( ) Psychiatric evaluation
- ( ) Educational record
- ( ) Substance abuse treatment history
- ( ) Social/developmental history
- ( ) Personal information including Social Security No(s) address(es) and telephone No(s)
- ( ) Other _____

To the agency or professional person receiving this release:

THIS INFORMATION HAS BEEN DISCLOSED TO YOU FROM RECORDS WHOSE CONFIDENTIALITY IS PROTECTED BY STATE LAW. STATE REGULATIONS PROHIBIT YOU FROM MAKING ANY FURTHER DISCLOSURE OF THIS INFORMATION WITHOUT PRIOR WRITTEN CONSENT OF THE PERSON TO WHOM IT PERTAINS.

THIS CONSENT TO RELEASE OF INFORMATION CAN BE REVOKED AT THE WRITTEN REQUEST OF THE PERSON WHO GAVE THE CONSENT.

I have read this form carefully and I understand what it means.

_____   _____
Authorized Signature   Date   Staff Person Signature   Date

I have read this carefully and I understand what it means and as I am not physically able to give my written consent, I am giving my verbal consent to release these records.

_____   _____
Witness Signature   Date   Staff Person Signature   Date

_____
Witness Signature   Date

Wildwood Case Management Unit
Release of HIV/AIDS-Related Information

RE: _____   DOB: _____

To Whom It May Concern:

[ ] I hereby authorize the Wildwood Case Management Unit to release information about services rendered to the above-named, for the purpose of:

[ ] Such information to be released includes information regarding my HIV/AIDS status and/or treatment.

To the agency or professional person receiving this release:

THIS INFORMATION HAS BEEN DISCLOSED TO YOU FROM RECORDS WHOSE CONFIDENTIALITY IS PROTECTED BY STATE LAW. STATE REGULATIONS PROHIBIT YOU FROM MAKING ANY FURTHER DISCLOSURE OF THIS INFORMATION WITHOUT PRIOR WRITTEN CONSENT OF THE PERSON TO WHOM IT PERTAINS.

THIS CONSENT TO RELEASE OF INFORMATION CAN BE REVOKED AT THE WRITTEN REQUEST OF THE PERSON WHO GAVE THE CONSENT.

I have read this form carefully and I understand what it means.

Authorized Signature	Date	Staff Person Signature	Date

I have read this carefully and I understand what it means and as I am not physically able to give my written consent, I am giving my verbal consent to release these records.

Witness Signature	Date	Staff Person Signature	Date

Witness Signature	Date

Wildwood Case Management Unit
Intake Assessment Form

Client Name _____ Agency# _____

D.O.B. _____ Date of Assessment _____

1. PRESENTING PROBLEM (Functional impairment, symptoms, background)

_____
_____
_____
_____
_____
_____
_____
_____
_____
_____
_____
_____

2. CURRENT CLIENT INVOLVEMENT WITH OTHER AGENCIES

Agency/Person	Phone	Service	Date

3. ASSESSMENT OF LIFE CIRCUMSTANCES OR CHANGES IN THE FOLLOWING AREAS

Family
_____
_____
_____

Social
_____
_____
_____

Support
_____
_____
_____

Legal
_____
_____

Education
_____
_____

Occupation
_____
_____

Finances
_____
_____

Psychosocial & environmental problems
_____
_____

4. CURRENT MEDICAL CONDITIONS

Condition	Physician	Treatment
_____	_____	_____
_____	_____	_____
_____	_____	_____
_____	_____	_____
_____	_____	_____

5. PREGNANT ( ) YES   ( ) NO

Receiving prenatal care? ( ) YES   ( ) NO

Prenatal Care from

Due Date

Complications
_____

6. PRIMARY CARE PHYSICIAN

_____
_____

_____
Date of last physical examination

7. CURRENT MEDICATIONS

Name/Dosage   Prescribed by   Condition

_____
_____
_____

_____
Side effects

_____
_____

_____
Medication allergies

_____
_____

8. RELATIONSHIP RISK FACTORS;

Is client safe at home? ( ) YES   ( ) NO

Does client feel threatened in any way? ( ) YES   ( ) NO

If *YES* describe

_____
_____
_____
_____

_____

Has client been abused in any way? ( ) YES   ( ) NO

If *YES* check all that apply

( ) Physical   ( ) Emotional   ( ) Sexual

Relationship of perpetrator to client

_____

Any legal action taken?
_____
_____
_____
_____
_____

Does client have a safety plan? ( ) YES   ( ) NO
Needs shelter   ( ) YES   ( ) NO
Needs protection from abuse order   ( ) YES   ( ) NO

9. SUICIDE/HOMICIDE EVALUATION

Client's self-rating of suicide risk _____

Client's self-rating of becoming violent _____

Client's self-rating of homicide risk _____

(1 – none   2 – slight   3 – moderate   4 – extreme/immediate)

Previous attempts or episodes
_____
_____
_____
_____

Evaluation of suicide risk

( ) none ( ) slight ( ) moderate ( ) significant ( ) extreme ( ) no plan ( ) plan (describe)
_____
_____
_____

Evaluation of violence risk

( ) none ( ) slight ( ) moderate ( ) significant ( ) extreme ( ) no plan ( ) plan (describe)
_____
_____
_____

10. MENTAL STATUS EXAM

Appearance
( ) age appropriate   ( ) well groomed   ( ) disheveled/ unkempt   ( ) bizarre   ( ) other

Orientation
( ) person   ( ) place   ( ) time   ( ) situation

Behavior/Eye Contact
( ) good   ( ) limited   ( ) avoidant   ( ) none   ( ) relaxed/calm   ( ) restless   ( ) rigid
( ) agitated   ( ) slumped posture   ( ) tense   ( ) tics   ( ) tremors   ( ) other

Motor Activity
( ) mannerisms   ( ) motor retardation   ( ) catatonic behavior

Manner
( ) appropriate   ( ) trusting   ( ) cooperative   ( ) inappropriate   ( ) withdrawn   ( ) seductive
( ) playful   ( ) evasive   ( ) guarded   ( ) sullen   ( ) passive   ( ) defensive   ( ) hostile
( ) manic   ( ) demanding   ( ) inappropriate boundaries

Speech
( ) normal   ( ) incoherent   ( ) pressured   ( ) too detailed   ( ) slurred   ( ) slowed
( ) impoverished   ( ) halting   ( ) neologisms   ( ) neurological language disturbances

Mood
( ) appropriate   ( ) depressed   ( ) irritable   ( ) anxious   ( ) euphoric   ( ) fatigued
( ) angry   ( ) expansive

Affect
( ) broad   ( ) tearful   ( ) blunted   ( ) constricted   ( ) flat   ( ) labile   ( ) excited
( ) anhedonic

Sleep
( ) good   ( ) fair   ( ) poor   ( ) increased   ( ) decreased   ( ) initial insomnia
( ) middle insomnia   ( ) terminal insomnia

Appetite
( ) good   ( ) fair   ( ) poor   ( ) increased   ( ) decreased   ( ) weight gain   ( ) weight loss

Thought process
( ) logical and well organized   ( ) illogical   ( ) flight of ideas   ( ) circumstantial
( ) loose associations   ( ) rambling   ( ) obsessive   ( ) blocking   ( ) tangential
( ) spontaneous   ( ) perseverative   ( ) distractible   ( ) other

Thought content
( ) delusions   ( ) paranoid delusions   ( ) distortions   ( ) thought withdrawal
( ) thought insertion   ( ) thought broadcast   ( ) magical thinking   ( ) somatic delusions
( ) ideas of reference   ( ) delusional guilt   ( ) grandiose delusions   ( ) nihilistic delusions
( ) ideas of inference   ( ) other

Perception/hallucinations
( ) illusions   ( ) hallucinations   ( ) depersonalization   ( ) derealization

Judgment
( ) intact   ( ) age appropriate   ( ) impulsive   ( ) immature   ( ) impaired   ( ) mild
( ) moderate   ( ) severe

Insight
( ) intact   ( ) limited   ( ) very limited   ( ) fair   ( ) none   ( ) aware of current disorder
( ) understands personal role in problems

Sensorium
( ) alert   ( ) drowsy   ( ) stupor   ( ) obtundation   ( ) coma

Memory
( ) intact   ( ) impaired   ( ) immediate recall   ( ) remote   ( ) amnesia

Type of amnesia

Intelligence
( ) average   ( ) above average   ( ) below average   ( ) unable to establish

Interviewer summary of findings (add details where appropriate)

11. **SUBSTANCE USE/ABUSE**

Type	Amount used	How taken	Duration	Frequency	Date of last use
Tobacco					
Alcohol					
Illicit Drugs					
Prescription Drugs					
OTC Drugs					
Other					

Experiencing:
Withdrawal	( ) YES	( ) NO
Blackouts	( ) YES	( ) NO
Hallucinations	( ) YES	( ) NO
Vomiting	( ) YES	( ) NO
Severe Depression	( ) YES	( ) NO
DTs and Shaking	( ) YES	( ) NO
Seizures	( ) YES	( ) NO
Other	( ) YES	( ) NO
Describe		

Patterns of use
- Uses more under stress ( ) YES ( ) NO
- Continues use when others have stopped ( ) YES ( ) NO
- Has lied about consumption ( ) YES ( ) NO
- Has tried to avoid others while using ( ) YES ( ) NO
- Has been drunk/high for several days at a time ( ) YES ( ) NO
- Neglects obligations when using ( ) YES ( ) NO
- Usually uses more than intended ( ) YES ( ) NO
- Needs to increase use to become intoxicated ( ) YES ( ) NO
- Has tried to hide consumption ( ) YES ( ) NO
- Sometimes uses before noon ( ) YES ( ) NO
- Cannot limit use once begun ( ) YES ( ) NO
- Failed to keep promises to reduce use ( ) YES ( ) NO

Describe attempts to stop

_____
_____
_____

Describe circumstances that usually lead to relapse

_____
_____
_____

Is client involved in AA/NA?   ( ) YES   ( ) NO

12. CLIENT REQUESTS, GOALS, EXPECTATIONS

_____
_____
_____
_____

13. CLINICAL SUMMARY (Using information you have collected summarize—identifying possible relationships—conditions and causes that may have led to current situation)

_____
_____
_____
_____
_____
_____
_____
_____

14. IMPRESSIONS

_____
_____
_____
_____
_____
_____
_____
_____
_____
_____
_____

15. RECOMMENDATIONS

_____
_____
_____
_____
_____
_____
_____
_____
_____
_____
_____

16. DSM 5 DIAGNOSTIC IMPRESSION

_____
_____
_____
_____
_____
_____
_____
_____
_____
_____
_____
_____

Case Manager Signature   Date

Wildwood Case Management Unit
PERSONAL SAFETY PLAN

FOR _____

INSTRUCTIONS: Check all the things you will do and fill in the blanks of those items where required _____

SAFETY AT HOME

____ If I need shelter the domestic violence hot line is _____

____ I will tell _____ about the violence and ask them to call the police if they hear signs of violence coming from my house

____ I will teach my children how to contact the police and fire department

____ I will be sure my children know their address and can give it to the police and fire department

____ I will get a programmable phone, or if I have one, I can program emergency numbers into it

____ I will teach my children to auto dial

____ I will avoid making long distance calls from the home phone if my partner can learn whom I called by looking at the phone bill

____ I will make long distance calls, if I need to, from _____

____ When I expect my partner and I are going to have an argument I will try to move to a place where I am less at risk, such as _____ (avoid the garage, bathroom, kitchen, rooms where there are weapons, rooms with no outside exit)

____ I will use my judgment and intuition to determine if the situation is serious. I will, where possible give my partner what he wants until the situation is under control and I/we are out of danger.

____ I will teach these protections to some/all of my children, where appropriate

PLANNING AHEAD FOR LEAVING

____ If its not safe to talk openly I will use as the code word/signal to my children that we are going to go or to family and friends that we are coming or leaving home

____ My safety route out of my home will be _____
_____
_____
_____

____ I will rehearse my escape plan and, if appropriate, practice it with my children

____ My purse and car keys are kept _____
_____

____ I will leave money and an extra set of keys with _____
_____
_____

____ I will leave extra clothes with _____

____ I will keep important documents, copies of important documents at

_____

_____

____ If I have to leave I will go to _____

_____

____ If I can't go there I will go to _____

INDEPENDENT MEASURES BEFORE I LEAVE

____ I will obtain a post office box

____ I will ask for help in planning from the domestic violence program

____ I will keep change and important numbers with me at all times

____ I will open my own bank account in my name

____ I will seek credit in my own name

____ I will take classes or get job skills

____ I will increase my independence in other ways _____

_____

_____

_____

AFTER LEAVING

____ I will install deadbolts where I live

____ I will change the locks where I live

____ I will check all my windows for access from the outside and take steps to bar entry into my home through the windows

____ I will replace all wooden doors with steel/metal doors

____ I will install a security system in my home

____ I will install rope or chain ladders to be used for escape from upper stories

____ I will purchase fire extinguishers

____ I will install smoke detectors on every floor

____ I will teach my children how to reach me if they are concerned for their own safety

____ I will inform the school or others who take care of my children as to who has permission to pick up my children and I will make sure the school personnel can recognize them

____ I will give those who take care of my children copies of custody and protective orders and emergency numbers

____ I will tell _____ and _____ that I am no longer with my partner and ask them to call the police it they believe I or my children are in danger

AT WORK AND IN PUBLIC PLACES

____ I will inform these people at my job

      ____ My boss

      ____ The security supervisor

      ____ The Employee Assistance Program

The Employee Assistance number is _____

____ I will ask _____ to screen my calls at work

____ I will protect myself when arriving at work by _____
_____
_____

____ I will protect myself when leaving work by _____
_____
_____

____ When traveling to and from work if there is trouble I will _____
_____
_____

____ I will carry a cell phone and carry it with me at all times

____ I will vary my routines

____ I will vary my route to and from usual destinations

____ I will avoid places my partner is likely to find me (laudromat, banks, stores, doctor's offices I used before leaving)

MY PROTECTION FROM ABUSE ORDER (PFA)

____ I will keep my PFA on or near me

____ If my partner destroys my PFA or if I lose my PFA I will get another copy from the court that issued it

____ I will give a copy of my PFA to the police where I live

____ I will give copies of my PFA to the police in communities I frequently visit

____ I will give a copy of my PFA to my employer or my employer's security department

____ I will give a copy of my PFA to my religious advisor

____ I will give a copy of my PFA to my closest friend

____ I will give a copy of my PFA to my children's school

____ I will give a copy of my PFA to my children's day care center

____ I will report violations of my PFA to the police

____ I can call my advocate, my attorney, or my case manager if my partner violates the PFA

My attorneys number is _____

My Advocate's number is _____

My Case manager's number is _____

____ I will call a domestic violence program if I have questions about how to enforce an order

____ I will call a domestic violence program if I have trouble getting my PFA enforced

RECOVERING MY EMOTIONAL HEALTH

____ I will call _____ if I feel lonely, blue or confused about what to do next

____ I will get a checkup at a doctor's

____ If I have left my partner and I am considering returning I will call _____ or spend time with _____ before I make the decision

____ I will remind myself daily of my strengths

They are _____

_____

_____

____ I will remind myself daily of my good qualities

They are _____

_____

_____

____ I will attend a support group to help build a support system

____ I will go to workshops and take classes on domestic violence and recovering from domestic violence to get the information I need to grow beyond this experience

____ I will take precautions not to drink or take drugs in a way that jeopardizes my safety

____ If I drink or take drugs at all I will do it with people who are committed to my safety

____ I will read one or more books written for battered women recommended by my domestic violence program

____ I will take other steps to feel stronger and more independent

These steps are

_____

_____

_____

WHAT TO TAKE WITH ME WHEN I LEAVE

Check these off as you get them together in a safe place

____ Personal identification _____

____ Drivers License

____ My birth certificate _____

____ Car registration

____ Social Security card

___ Public assistance ID/ Medicaid cards _____

___ Passports, green cards, work permits

___ Keys to House

___ Keys to car _____

___ Medication

___ Keys to office

___ Money _____

___ Children's toys

___ Children's security blankets

___ Checkbook _____

___ ATM cards _____

___ Children's social security cards

___ Credit cards

___ Insurance papers _____

___ Sentimental items

___ Divorce and separation papers _____

___ Personal safety plan ___ Lease, rental agreement, house deed

___ Car payment book _____

___ Other _____

___ Mortgage payment book _____

Wildwood Case Management Unit
Peer Support Services Referral Form

Peer support is a therapeutic system based on self-help founded on the principles of respect, shared responsibilities, and mutual agreements between peers.

GENERAL DEMOGRAPHICS

Name _____

Agency Number _____

Date of Birth _____

Address _____

_____

Case Manager _____

Date _____

SSN _____

Gender: ☐ Male   ☐ Female

Phone Number _____

Phone Number _____

CURRENT LIVING STATUS   ☐ Lives independently   ☐ Lives with family   ☐ Lives with others
   ☐ Other

HEALTH   Serious health condition _____ Physician's name _____ Phone _____

Individual must meet all of the following criteria in order to be eligible for Peer Support Services:

( ) 18 years of age or older with one or more serious challenges to recovery.

( ) Has a moderate to severe functional impairment that interferes with or limits role performance in at least one of the following areas:

   ( ) Educational   ( ) Social   ( ) Vocational   ( ) Self-maintenance

( ) Individual agrees to Peer Support Services.

REASON FOR REFERRAL

_____
_____
_____

REFERRAL SOURCE

_____   Phone _____

Peer Signature _____   Date _____

☐ Accepted   ☐ Not Accepted   ☐ Reason _____

_____

Appendix C   Wildwood Case Management Unit Forms

Wildwood Case Management Unit
Peer Support Mutual Agreements, Outcome Report, and Renewal

Peer Name _____	Agency Number _____
Certified Peer Specialist _____	_____
Date of Original Agreement _____	Review Due by _____
Review Completed _____	Next Review Date _____

Domain: ☐ Educational ☐ Social ☐ Vocational ☐ Self-maintenance

Overall goal for a _____ month period _____

_____

Objectives (Action steps toward goal)

Action	Target Date	Person Responsible

Certified Peer Specialists Role _____

_____

_____

_____

_____

Progress Made/Obstacles Encountered

_____

_____

_____

_____

Domain: ☐ Educational ☐ Social ☐ Vocational ☐ Self-maintenance

Overall goal for a _____ month period _____

_____

Objectives (Action steps toward goal)

Action	Target Date	Person Responsible

Certified Peer Specialists Role _____

_____

_____

_____

_____

_____

Progress Made/Obstacles Encountered

_____

_____

_____

_____

Domain: ☐ Educational ☐ Social ☐ Vocational ☐ Self-maintenance

Overall goal for a _____ month period _____

_____

Objectives (Action steps toward goal)

Action	Target Date	Person Responsible

Certified Peer Specialists Role _____

_____
_____
_____
_____

Progress Made/Obstacles Encountered

_____
_____
_____
_____

Domain: ☐ Educational ☐ Social ☐ Vocational ☐ Self-maintenance

Overall goal for a _____ month period _____

_____

Objectives (Action steps toward goal)

Action	Target Date	Person Responsible

Certified Peer Specialists Role _____

_____
_____
_____
_____

Progress Made/Obstacles Encountered

_____
_____
_____
_____

Domain: ☐ Educational ☐ Social ☐ Vocational ☐ Self-maintenance

Overall goal for a _____ month period _____

_____

Objectives (Action steps toward goal)

Action	Target Date	Person Responsible

Certified Peer Specialists Role _____

_____

_____

_____

_____

Progress Made/Obstacles Encountered

_____

_____

_____

_____

Signatures and Agreement with Plan and Review/New Goals Developed

Plan _____ Date _____ Review _____ Date _____
Consumer _____
Certified Peer Specialist _____
Practitioner/Case Manager _____

Plan _____ Date _____ Review _____ Date _____
Consumer _____
Certified Peer Specialist _____
Practitioner/Case Manager _____

Plan _____ Date _____ Review _____ Date _____
Consumer _____
Certified Peer Specialist _____
Practitioner/Case Manager _____

Plan _____ Date _____ Review _____ Date _____
Consumer _____
Certified Peer Specialist _____
Practitioner/Case Manager _____

Plan _____ Date _____ Review _____ Date _____
Consumer _____
Certified Peer Specialist _____
Practitioner/Case Manager _____

Plan _____ Date _____ Review _____ Date _____
Consumer _____
Certified Peer Specialist _____
Practitioner/Case Manager _____

Plan _____ Date _____ Review _____ Date _____
Consumer _____
Certified Peer Specialist _____
Practitioner/Case Manager _____

Wildwood Case Management Unit
Service Planning Conference Notes

Client _____   Date of conference _____

PRESENTING PROBLEM

_____
_____
_____
_____
_____

HOW CLIENT PRESENTED IN INTERVIEW

_____
_____
_____
_____
_____

CLIENT'S EXPRESSED GOALS, EXPECTATIONS, REQUESTED SERVICES

_____
_____
_____
_____
_____

ADDITIONAL RELEVANT INFORMATION

_____
_____
_____
_____
_____

IMPRESSIONS AND RECOMMENDATIONS

_____
_____
_____
_____
_____

Treatment or Goal Service Plan

CLIENT _____ # _____ Next of Kin _____ Review Date _____

Initial plan [ ] Updated plan [ ] Date _____

Developed with _____

Level of case management _____ Case Manager _____

Provisional DX: _____

TYPE	STRENGTH/NEED	GOAL(S)	COMMENTS	REFERRAL
INCOME/FINANCIAL SITUATION	STRENGTH/ NEED			
HOUSING/LIVING ARRANGEMENT	STRENGTH/ NEED			
VOCATIONAL	STRENGTH/ NEED			
EDUCATIONAL	STRENGTH/ NEED			
TRANSPORTATION	STRENGTH/ NEED			
MEDICAL	STRENGTH/ NEED			

TYPE	STRENGTH/NEED	GOAL(S)	COMMENTS	REFERRAL
ACTIVITIES OF DAILY LIVING	STRENGTH/ NEED			
LEGAL	STRENGTH/ NEED			
RECREATION & LEISURE TIME	STRENGTH/ NEED			
MENTAL HEALTH	STRENGTH/ NEED			
SUBSTANCE ABUSE	STRENGTH/ NEED			
FAMILY RELATIONSHIPS	STRENGTH/ NEED			
SOCIAL SUPPORTS	STRENGTH/ NEED			
OTHER	STRENGTH/ NEED			

Case Manager Signature   Date   Supervisor's Signature   Date

Wildwood Case Management Unit
Referral Notification Form

Client _____ # _____

Address _____

Home Phone _____ Work Phone _____ Date _____

Diagnosis _____
_____
_____
_____
_____

Provider _____

Type of service _____

For the purpose of _____
_____
_____
_____
_____
_____
_____
_____
_____
_____
_____
_____

Review Date _____ Target date _____

Referring Case Manager _____

Wildwood Case Management Unit
Contact Notes

Client _____   Agency # _____

Wildwood Case Management Unit
Contact Notes: Children's Case Management Services

Client _____ DOB _____ Agency # _____

Wildwood Case Management Unit
Discharge Summary

Name _____ Date of Birth _____

Date of Admission _____ Date of Discharge _____

DIAGNOSIS ON ADMISSION:

_____

_____

_____

_____

_____

MEDICATION:

REASON FOR DISCHARGE:

PRESENTING PROBLEM:

GOALS:

PROGRESS:

ADDITIONAL ISSUES NOT ADDRESSED:

IMPRESSIONS AND RECOMMENDATIONS:

PREPARED BY:
_____
CASE MANAGER   DATE

APPENDIX D

Prochaska and DiClemente's Stages of Change Model

Stage of Change	Characteristics	Techniques
Pre-contemplation	Not currently considering change: "Ignorance is bliss"	Validate lack of readiness Clarify: decision is theirs Encourage re-evaluation of current behavior Encourage self-exploration, not action Explain and personalize the risk
Contemplation	Ambivalent about change: "Sitting on the fence" Not considering change within the next month	Validate lack of readiness Clarify: decision is theirs Encourage evaluation of pros and cons of behavior change Identify and promote new, positive outcome expectations
Preparation	Some experience with change and are trying to change: "Testing the waters" Planning to act within 1 month	Identify and assist in problem solving re: obstacles Help patient identify social support Verify that patient has underlying skills for behavior change Encourage small initial steps
Action	Practicing new behavior for 3–6 months	Focus on restructuring cues and social support Bolster self-efficacy for dealing with obstacles Combat feelings of loss and reiterate long-term benefits
Maintenance	Continued commitment to sustaining new behavior Post-6 months to 5 years	Plan for follow-up support Reinforce internal rewards Discuss coping with relapse
Relapse	Resumption of old behaviors: "Fall from grace"	Evaluate trigger for relapse Reassess motivation and barriers Plan stronger coping strategies

Pre-contemplation Stage

"Ignorance is bliss"
"Weight is not a concern for me"

Goals:
1. Help patient develop a reason for changing
2. Validate the patient's experience
3. Encourage further self-exploration
4. Leave the door open for future conversations

1. Validate the patient's experience:
"I can understand why you feel that way"

2. Acknowledge the patient's control of the decision:
"I don't want to preach to you; I know that you're an adult and you will be the one to decide if and when you are ready to lose weight."

3. Repeat a simple, direct statement about your stand on the medical benefits of weight loss for this patient:
"I believe, based upon my training and experience, that this extra weight is putting you at serious risk for heart disease, and that losing 10 pounds is the most important thing you could do for your health."

4. Explore potential concerns:
"Has your weight ever caused you a problem?" "Can you imagine how your weight might cause problems in the future?"

5. Acknowledge possible feelings of being pressured:
"I know that it might feel as though I've been pressuring you, and I want to thank you for talking with me anyway."

6. Validate that they are not ready:
"I hear you saying that you are nowhere near ready to lose weight right now."

7. Restate your position that it is up to them:
"It's totally up to you to decide if this is right for you right now."

8. Encourage reframing of current state of change—the potential beginning of a change rather than a decision never to change:
"Everyone who's ever lost weight starts right where you are now; they start by seeing the reasons where they might want to lose weight. And that's what I've been talking to you about."

Contemplation Stage
"Sitting on the fence"
"Yes my weight is a concern for me, but I'm not willing or able to begin losing weight within the next month."

Goals:
1. Validate the patient's experience
2. Clarify the patient's perceptions of the pros and cons of attempted weight loss
3. Encourage further self-exploration
4. Leave the door open for moving to preparation

1. Validate the patient's experience:
"I'm hearing that you are thinking about losing weight but you're definitely not ready to take action right now."

2. Acknowledge patient's control of the decision:
"I don't want to preach to you; I know that you're an adult and you will be the one to decide if and when you are ready to lose weight."

3. Clarify patient's perceptions of the pros and cons of attempted weight loss:
"Using this worksheet, what is one benefit of losing weight? What is one drawback of losing weight?"

4. Encourage further self-exploration:
"These questions are very important to beginning a successful weight loss program. Would you be willing to finish this at home and talk to me about it at our next visit?"

5. Restate your position that it is up to them:
"It's totally up to you to decide if this is right for you right now. Whatever you choose, I'm here to support you."

6. Leave the door open for moving to preparation:
"After talking about this, and doing the exercise, if you feel you would like to make some changes, the next step won't be jumping into action—we can begin with some preparation work."

Preparation Stage
"Testing the Waters"
"My weight is a concern for me; I'm clear that the benefits of attempting weight loss outweigh the drawbacks, and I'm planning to start within the next month."

Goals:
1. Praise the decision to change behavior
2. Prioritize behavior change opportunities
3. Identify and assist in problem solving re: obstacles
4. Encourage small initial steps
5. Encourage identification of social supports

1. Praise the decision to change behavior:
"It's great that you feel good about your weight loss decision; you are doing something important to decrease your risk for heart disease."

2. Prioritize behavior change opportunities:
"Looking at your eating habits, I think the biggest benefits would come from switching from whole milk dairy products to fat-free dairy products. What do you think?"

3. Identify and assist in problem solving re: obstacles:
"Have you ever attempted weight loss before? What was helpful? What kinds of problems would you expect in making those changes now? How do you think you could deal with them?"

4. Encourage small, initial steps:
"So, the initial goal is to try nonfat milk instead of whole milk every time you have cereal this week."

5. Assist patient in identifying social support:
"Which family members or friends could support you as you make this change? How could they support you? Is there anything else I can do to help?"

Source: www.stepupprogram.org

APPENDIX E

Work Samples

Examples of Progress Notes or Contact Notes

1/5/16 (Office Visit) Clementine came in today to discuss her medications. After her last hospitalization, she has been concerned about running out of medications. She was given a county prescription signed by Dr. Horace Merkle. Clementine seemed bright and eager to return to work. She will call next week after her first day at work.

4/9/16 (Phone Contact) Larry called today stating he was intoxicated and requesting CM's help. He stated he had been drinking for 2 days since his boss gave him a bad evaluation that he feels he did not deserve. Larry states that since rehab he has worked hard to reverse his work record at the plant where he works. He sounded depressed and was tearful at times. CM will visit the home this afternoon.

4/9/16 (Home Visit) CM went to Larry's residence today after he called sounding intoxicated and requesting assistance. He and his wife were home and both expressed anger over the evaluation Larry received and concern over the fact that it led to renewed drinking on Larry's part. CM learned that Larry has not gone to AA meetings or outpatient treatment since he was discharged from inpatient rehabilitation. CM offered 3 days of inpatient detox followed by a meeting to decide what would be the best course of action regarding Larry's job. Larry agreed to this plan and arrangements were made at Madison Detoxification Center. Client seemed relieved to have a plan and resigned to the need for further treatment. Larry will come into the office when discharged.

4/13/16 (Office Visit) Larry and his wife came to the office today to decide how Larry should work on both his employment situation and his treatment. After listening to Larry's concerns with his present position and his concerns about relapse, he and CM worked out a plan for Larry to meet with his boss to better understand how he can improve in his present position. In addition, Larry was referred at his request to

Madison's outpatient rehabilitation program. Larry appeared more confident today and seemed interested in developing a plan. Larry will call after he makes an appointment to meet with his boss.

8/3/16 (Site Visit) CM visited Marvel at sheltered workshop where she works. Marvel sat in on a meeting with CM and her worker at the workshop. She seems to enjoy her work at the workshop and to relate well to her worker, Karen Pillsbury. Karen stated that Marvel works hard and is careful about her work. During the visit Marvel took CM to see her workstation and introduced CM to her friends. CM requested that both Marvel and her worker call if they need anything. It was agreed that this placement is working well for Marvel and that she seems to have developed more of a social life as a result. CM will make another follow-up visit in 3 months.

Dating Your Forms

Students often ask about the proper way to date forms. The following table provides examples for dating various types of forms.

Type of Form	Example	Description
Initial Referral or Inquiry Form	The form was filled out March 6, 2016. The disposition is for March 12, the verification letter went out on March 6 after the client called.	Enter the date on which the form is filled out or on which the inquiry is received. For the disposition date, enter the date on which the client is seen for the first time. Date the disposition within 2 weeks of the date of inquiry. The verification letter should be dated for the day on which it is sent to the client and that date should be within the first few days of the initial inquiry.
Verification Letter	March 6	The verification letter should be dated on the day it is sent and that date should be the same date you stated it was sent on your initial inquiry form.
Release of Information Form	The client comes in on March 12, so a few days later (e.g., March 14) you send for information.	The release of information form is sent after meeting with the client for the first time and after the client has signed the forms on the date of the first interview—the date that you presumably signed the release as well.
Treatment or Goal Plan	The planning conference was March 26, so date the plan for March 26 or later. Most review dates are 6 months later, but you might want yours sooner.	• Enter the date on which the plan was developed after the first interview and the Treatment Planning Conference. • Enter a review date (typically 6 months or later from the plan development date), which is the general date the agency will remind you to check on the plan again.
Referral Form	The treatment/goal plan is in order and you refer the client out on a date soon after the March 26. Decide how long you want the treatment to continue and give that as the target date. Make your review date halfway between March 26 and the target date.	• The *top date* is the date you enter upon filling out the referral form and after the service plan is decided. • The *target date* can be set as far into the future as you feel is reasonable for the goals to be achieved. • The *review date* (generally set about halfway to the target date) is the date on which you check up on this particular service to see how well it is going.

Contact Notes	Give the notes dates that follow March 26.	These dates can begin with the second contact with the client to go over the plan developed. Your initial inquiry is on the inquiry form; your first interview is documented in the social history and the assessment form.
Termination	Give the date well after the March 26 date so that the client has had time to go through the treatment and services you authorized.	• The date of the final interview should be well after you opened the case. • The termination letter is dated on the day it is sent, which is usually one or two days after the final interview.

Sample Cases with Service Plans

Beverly

Beverly is a second-grade teacher whose children had become increasingly unruly. The principal reported that on a recent visit to the classroom, children were jumping on the desks and chasing each other around the room. Beverly had been a good teacher for several years, and the situation in her classroom had not existed in former years. To the principal, Beverly looked exhausted and unable to organize her thoughts. She appeared unconcerned about the children's misbehavior and equally unable to handle it. The principal suggested a leave of absence, and a substitute took over Beverly's classroom. Beverly came into your office seeking help with a problem with prescription medications. She reported that she started taking prescription drugs after her mother had surgery and got a prescription for pain medications. Beverly was 14 and has been taking pain medications ever since. Recently she has been taking more than previously and noticed that she became less and less motivated to do anything. On the other hand, currently she doesn't feel that she can get through the day without using some form of prescription pain killer.

Here is Beverly's service plan:

1. Weekly individual and group counseling to gain an understanding of the addiction process, identify supports for prescription pain killer use, and develop appropriate coping skills to deal with relapse situations.
2. Weekly drug tests.
3. Attend at least three 12-step meetings weekly.

Hal

For years Hal has been working for a trucking company as a loader on the dock. He is divorced and has minimal contact with his three children. At a party about 4 years ago, someone gave him some meth to try. From then on he occasionally found more meth and was able to take it, gradually increasing his use of the drug. Hal is brought in by a coworker who tells you that Hal "hasn't been right for months." The coworker went to Hal's home after Hal missed a week of work and failed to call in to the company. The company tried unsuccessfully to reach Hal, and when they couldn't, the coworker went to Hal's home and found Hal had chills and was vomiting. When the coworker suggested that Hal go to the emergency room, Hal said he wanted to go to a substance abuse clinic instead. The coworker reports that Hal has been losing weight, "but there's so much to do at work I just noticed it—didn't say nothing though." He goes on to tell you that Hal stopped "hanging around with us after work and, you know, his work got behind. We covered for him some, and then he disappeared last week." Hal seems to be confused and is showing signs of moderate confusion.

Here is Hal's service plan:

1. Enter detox for 3 to 5 days stat.
2. Complete inpatient drug and alcohol treatment with group counseling, lectures on addiction process, recovery tools, and relapse process.
3. Attend 12-step meetings.
4. Possible halfway house referral.

Marrietta

Marrietta was in an abusive relationship over an 11-year period. Recently she was hospitalized with severe injuries following another domestic dispute in her home with her husband. Husband has been jailed and cannot make bail. Marrietta will not be discharged from the hospital for another 3 or 4 days. The worker from a domestic violence program comes to the hospital to meet with Marrietta to plan what will happen following her discharge from the hospital. Marrietta is adamant that she does not want to return to her home and the relationship with her husband. However, she fears that if she leaves she will be in even greater danger. She asks the worker to help her leave the relationship and help her remain safe. She is against getting a Protection from Abuse Oder as she feels this will make her husband even angrier when he comes out of jail.

Here is Marrietta's service plan:

1. Discharge to women's safe home temporarily. Seek one out of the area if feasible.
2. Complete 12 sessions in support group for survivors of domestic violence.
3. Evaluation of housing situation to determine whether shelter or new housing is required.
4. Legal support from attorney and worker in the program.

Angelina

Angelina is an 82-year-old grandmother of eight and a mother of four children. A widow, she has been living alone successfully for a number of years, but recently she has seemed confused and agitated. She is irritated with her children who try to help her, and her daughter reports that she has gone into the home and found her mother "almost normal" one time and regressed several hours later. The daughter has accompanied her mother to your office and gives the following history: Her mother has been active in her community and lived alone since her husband died 10 years ago. She is currently on medications for high blood pressure, but other than that she has no other health problems. In the office, she appears to be sweating profusely and seems agitated and irritable.

Here is Angelina's service plan:

Angelina's Strengths	Angelina's Needs	Angelina's Service Plan
• Financial stability	• Medical reevaluation	• Joseph Eberly Medical Center for complete gerontology workup
• Owns home	• Psychiatric evaluation	• Psychiatric evaluation at Eberly Medical Center
• Drives and has friends who drive for her	• Neurological evaluation	• Hargrave home health services for in-home care
• Moderately good health	• Temporary assistance in her home	• C M to work out visitation among family members for next 2 weeks
• Can manage activities of daily living		• Plan to be reviewed in 2 weeks
• Belongs to a number of organizations and her church		
• No substance abuse problems		
• Strong family relationships		
• Good friends		

APPENDIX F

Grading the Final Files

If you are turning in a file to your instructor on a hypothetical client, use this rubric to check yourself to ensure your work is in order.

The File	Is the file in order? Is the name on the tab? Is there a number on the tab? Is the folder clean?
Face Sheet	Does the personal information match? Are the appropriate boxes filled in? Is it signed?
New Referral and Inquiry	Is parent or spouse circled? Is N/A in appropriate places? Does the chief complaint adequately capture the reason the client called today? Is the narrative organized? Does it state how the client seemed? Do referrals go to the places mentioned in the file?
Verification Letter	Is the letter dated? Is there a name after Dear ___? Is the letter signed? Is the signature between Sincerely and Case Manager?
Brief Social History	Is presenting problem fully and concisely described? Is pertinent background to the presenting problem provided? Are case manager's impressions and recommendations included? Are subheadings used?

Assessment Form	Are interviewer comment spaces used appropriately? Is the form in order and stapled? Does the assessment fit the presenting problem?
Release of Information	Were appropriate documents sent for and recorded? Is the form filled out correctly? Is the authorized signature the correct one? Are there witness signatures?
Service or Treatment Plan	Are goals in the goals column, comments in the comments column, and so forth? Is the diagnosis correct? Were services consolidated in one or two places?
Referrals	Are they grammatically correct? Is the type of service the right one for this case? Are all five axes filled out? Are review and the target dates entered correctly?
Goals and Objectives	Are the goals and objectives doable for this client? Are they written properly? That is, are goals lettered and objectives numbered? Do goals and objectives meet the "See Billy Test"? Are the goals and objectives written up in a letter on a letterhead from the other agency? Did the other agency person sign the letter?
Case Notes	Is every case note dated? Is every case note labeled? Is every case note signed? Is there evidence that the case manager collaborated with the client? Is there evidence of client agreement? Does each note have all four parts? Are there at least 12 notes? Is information about how the client seemed, not just how the client says he feels, included in each note?
Termination Letter	Does the letter match the case note for the final interview? Were client's questions answered? Does the letter summarize major points discussed in the termination meeting? Is the client invited to contact the agency again in the future should the need arise?
Termination or Discharge Summary	Are all the headings addressed? Are the impressions and recommendations well written? Do impressions and recommendations match where the client is now?

General	Is written work free of spelling errors?
	Is written work free of grammatical errors?
	Are headings in bold?
	Are similar items stapled together (pages in the social history; all the release forms)?
	Do all the dates fall as they should?
	Is there a believable chronological progression?
	Is colored paper used for any of the forms?
	Is professional language used throughout (not "some guy," but "a man"; not "all shook up," but "anxious or agitated")?
	Are descriptions specific rather than generalities (not "gives mother a hard time," but "does not conform to rules mother has set for her"; not "she's been drinking," but "she began to drink heavily on Saturday and was still drinking this morning")?
	Are all diagnoses numerical and also written out?
	Is everything signed and dated where required?
	Is handwriting legible?

APPENDIX G

Information for Understanding DSM IV TR Diagnoses

The material in Appendix G will help you understand diagnoses you find that were made using the DSM-IV-TR.

> **Axis I**: All clinical syndromes listed in the *DSM-IV* are coded on this axis *except* personality disorders and intellectual disabilities. Axis I includes developmental disorders and other conditions that might be a focus of clinical attention.
>
> - V71.09   No diagnosis on Axis I
> - 799.9   Diagnosis deferred on Axis I (meaning too little time or information to establish a diagnosis)
>
> **Axis II**: Coded on this axis are personality disorders, intellectual disabilities, significant maladaptive personality traits, and habitual defense mechanisms.
>
> - V71.09   No diagnosis on Axis II
> - 799.9   Diagnosis deferred on Axis II
>
> **Axis III**: This axis is used for all general medical conditions that are relevant to planning and understanding the patient's diagnosis. *International Classification of Diseases (ICD-10)* codes can be used here.
>
> - None (meaning no medical conditions)
> - Deferred
>
> **Axis IV**: Psychosocial and environmental problems that affect the prognosis, management, or treatment of the case are coded here.
>
> **Axis V**: This axis is for the rating on the Global Assessment of Functioning (GAF) scale, which is usually a single number between 1 and 100 indicating the current level of functioning the patient possesses.

Axis I

All clinical syndromes listed in the *DSM* are coded on this axis *except* personality disorders and intellectual disabilities. The primary diagnosis, or reason a person is seeking treatment, generally is the diagnosis appearing on Axis I. However, if a person is seeking help for intellectual disabilities or a personality disorder, there may not be an Axis I diagnosis.

If there is more than one Axis I disorder, the primary diagnosis is listed first and often is qualified with the phrase "Reason for visit" or "Principal diagnosis" (unless the principal diagnosis is on Axis II; LaBruzza, 1994). If the clinician does not use one of these phrases, the Axis I diagnosis is always considered to be the primary diagnosis.

The following categories of disorders are coded on Axis I:

1. Disorders usually first diagnosed in infancy, childhood, or adolescence (except intellectual disabilities, which is coded on Axis II)
2. Delirium, dementia, amnesia, and other cognitive disorders
3. Mental disorders due to a general medical condition not elsewhere classified
4. Substance-related disorders
5. Schizophrenia and other psychotic disorders
6. Mood disorders
7. Anxiety disorders
8. Somatoform disorders
9. Factious disorders
10. Dissociative disorders
11. Sexual and gender identity disorders
12. Eating disorders
13. Sleep disorders
14. Impulse control disorders not elsewhere classified
15. Adjustment disorders
16. Other conditions that may be a focus of clinical attention

When there is no diagnosis on Axis I, the clinician writes V71.09 on the axis. If the practitioner suspects there may be an Axis I disorder but is not sure, the deferred code, 799.9, is used.

Axis II

Axis II is used to code personality disorders and intellectual disabilities. This was done to be sure relevant personality factors would be part of the entire diagnosis. It also makes intellectual disabilities a separate factor on a separate axis in the person's general diagnosis. *Of all the disorders listed in the* DSM, *only intellectual disabilities and the learning disorders require diagnostic testing before the diagnosis can be given.* If a person meets the criteria for more than one personality disorder, all of them should be coded on Axis II. In addition, the clinician can write in any significant maladaptive personality traits or habitual defense mechanisms on this axis. These items have no code number.

A habitual defense mechanism would be behavior a person uses that is maladaptive, but seems to the person to protect or benefit them in some way. Lying, cheating, and stealing when part of the usual way a person behaves or responds, would all be considered habitual defense mechanisms.

The absence of an Axis II disorder is coded V71.09. If the practitioner suspects there may be an Axis II disorder but is not sure, the deferred code, 799.9, is used. You will find that insurance companies are not pleased with Axis II diagnoses due to the amount of costly treatment required. This treatment can be intense and time consuming.

There are 11 categories of personality disorder.

1. Paranoid
2. Schizoid
3. Schizotypal
4. Antisocial
5. Borderline
6. Histrionic
7. Narcissistic
8. Avoidant
9. Dependent
10. Obsessive-compulsive
11. Personality disorder not otherwise specified

In addition, borderline intellectual functioning is also coded on Axis II (V62.89). When there is no diagnosis on Axis II, the clinician writes V71.09 on the axis.

Axis III

Often, individuals with severe mental disorders also have general medical conditions that affect the prognosis, treatment, and even the understanding of their situations. Any general medical condition that is relevant should be coded on Axis III, along with any medical history that may be relevant to the current problem.

Axis III is not meant to indicate the mind and body are entirely separate entities. Axis III is separate in order to be sure the full picture of mental and medical disorders is recorded. People with medical conditions often feel less well psychologically. Axis III also alerts the physician to the possibility that the person is on medications that might interfere or interact negatively with psychotropic medications that might be prescribed for the mental disorder.

In some cases, it is the medical condition that has caused the mental disorder. If this is the case, the medical condition is recorded on Axis III, and the mental condition is recorded on Axis I with the phrase "due to . . . " (e.g., Axis I might read "major depressive disorder, single episode, due to hypothyroidism," while Axis III would read "hypothyroidism").

If no medical condition exists, write "Axis III: None." If it is suspected that there may be a medical condition but there is not enough information, it is coded "Axis III: Deferred." The clinician can note in writing any significant symptoms or physical signs that were observed that need further evaluation.

Axis IV

Axis IV should list any psychosocial stressors or environmental problems that appear to have an impact on the conditions noted on Axes I, II, and III. The clinician looks at what has happened to the individual, particularly in the last year. Some stressors happened a number of years ago. For instance, a diagnosis of posttraumatic stress disorder (309.81) on Axis I may have Vietnam War on Axis IV as a stressor that is affecting the current condition of a Vietnam war veteran.

The *DSM* lists "other conditions" that may be the focus of clinical attention. These codes begin with the letter V and are listed on Axis I. They refer to problems people encounter in life that might cause them to seek professional help. It might be bereavement (V62.82), school problems (V62.3), or any number of situational and relationship issues that have come to clinical attention. After further treatment, a discrete mental disorder sometimes emerges and the diagnosis is changed at that point. At other times the person just needs help adjusting to the circumstances.

The V codes were developed to outline broad general categories of psychosocial stress a person might experience.

1. Relational problems
   - Childhood (V61.9)
   - Adult (V61.9)
   - Parent–child (V61.20)
2. Problems related to the social environment (V62.4)
3. Educational problems (V62.3)
4. Occupational problems (V62.2)
5. Housing problems
6. Economic problems
7. Problems with access to health care services
8. Problems related to interaction with the legal system/crime
9. Other psychosocial and environmental problems

If stress from a specific stressor is the main reason a person is seeking help, the V code for the stressor should go on Axis I, and a description of the exact stressor should go on axis IV. For example, if a person is having trouble on his job, you would put V62.2 occupational problem on Axis I, and on Axis IV you would write, "Difficulty adjusting to additional responsibilities" or "Problems with supervisor." (*Note:* These V codes should not be confused with V71.09, which is the code used on both Axis I and Axis II to indicate that no clinical condition is present on that axis.)

Much psychosocial stress has to do with problems in the person's support system. This can include births, deaths, separation, divorce, remarriage, abuse, neglect, and significant illness. When looking at the person's social environment, you might find social isolation, lost friendships, retirement, relocations, and cultural differences as contributors to the current difficulty.

The "other" category can be used to record natural disasters and catastrophes such as floods or earthquakes or problems related to professional caregivers, for example, a nursing assistant who is rough and seemingly threatening. The term *other* is used for any stressor that does not fit into the other categories. After the category is listed, the clinician should write out the specific stressors within each category (unemployed, mugged, best friend was killed, etc.).

There are positive stressors such as weddings or the birth of a child. These positive stressful events are not coded on Axis IV unless they create a clinical problem for the patient.

Axis V

This axis is reserved for the person's number on the Global Assessment of Functioning (GAF) scale. How well a person functions or does not function will affect any treatment plan developed and may affect insurance payment, research options, and the overall assessment of the person. A single rating of how the person is functioning at the time of the evaluation is given on Axis V. You will find the GAF in your *DSM*. The number assigned to the patient should reflect the person's "current level of psychological, social, and occupational functioning at the time of the evaluation," according to Anthony LaBruzza (1994, p. 79). He further pointed out that "impairment due to physical, and environmental limitations is excluded from Axis V" (p. 79). For example, if a person were impaired because of a stroke and therefore unable to get around, this reduced functioning would not be part of the assessment. Nor would circumstances such as poverty or a poor education be included in the assessment. The assessment looks at how well people are functioning beyond their environmental and physical limitations.

The number given on Axis V is a number from 1 to 100. Zero indicates that there has not been enough time to adequately assess the functioning. The scale is divided into 10-point segments, with 1 to 10 being the lowest functioning and 90 to 100 being the highest. Most individuals requiring inpatient hospitalization have a GAF score of 50 or less. The scale has been criticized for making it possible to a greater degree to abuse reimbursement, forensic, and disability situations.

Use of Codes V71.09 and 799.9

These two designations are used only on Axis I and Axis II. You need to know what these numbers mean when you see them. On both Axis I and Axis II, the codes indicate the following:

- V71.09 means there is no diagnosis on that particular axis.
- 799.9 means the diagnosis was deferred, presumably because there was too little time or information to make a diagnosis.

Making the Code

All the disorders in the *DSM* have a numerical code. The code has three whole numbers followed by a decimal point and one or two additional numbers. The form of the codes looks like this: XXX.XX. The last two numbers in the code give the diagnosis more specificity.

Let us take an example. A person comes in with an obvious depression. In *DSM* terms, this is called a "major depressive disorder, single episode" (296._ _). The number 296.21 indicates the person's condition is mild, 296.22 means moderate, 296.23 means severe but without psychotic features, and 296.24 is severe with psychotic features. A person in partial remission would get 296.25, and a person in full remission would get 296.26. We know that the Axis I diagnosis will be 296._ _. By choosing the proper digits to follow the decimal point, we create a more accurate clinical picture of this person.

The fifth digit is generally used to identify four things:

1. Subtypes
2. Specifiers (or Modifiers)
3. Course of the disorder and severity
4. Additional information such as reason for visit; provisional diagnosis; and the diagnoses of not otherwise specified, unspecified, or deferred.

Subtypes

When the clinician codes a diagnosis, the manual may require that the clinician "specify type." This refers to subtypes within a specific diagnosis. For instance, the diagnosis may be a delusional disorder (297.1), but there are seven distinct types of delusional disorders, including jealous, persecutory, grandiose, and so forth.

Supposing the person you are seeing believes his wife is seeing another man when all information points to the fact that this is not true. This man has believed this about his wife for over a month, affecting adversely his work and home life. His diagnosis would be written as: "297.1, Jealous Type" or "297.1, Delusional Disorder, Jealous Type."

Specifiers (Modifiers)

The manual may require that the practitioner "specify if" when coding a diagnosis. These are specifiers—also called "modifiers"—that allow the clinician to indicate if certain factors are present in this particular diagnosis. For instance, if the diagnosis is pedophilia (302.2), the practitioner will be asked to specify if the person is sexually attracted to males, females, or both (LaBruzza, 1994).

Thus if you are working with a person who is a pedophile you would write the diagnosis like this: "302.2, Sexually Attracted to Males" or "302.2, Pedophilia, Sexually Attracted to Males."

In another example, the person may have a diagnosis of 293.83, Mood Disorder Due to . . . . Here we are asked to indicate the general medical condition causing the mood disorder and then specify the features present in the disorder. There are four types of features from which to choose. Thus you might write: "293.83, Mood Disorder Due to Breast Cancer, with Manic Features."

Modifiers for Past and Present

It is always understood that the diagnosis is a present condition. Sometimes, however, it is useful to note a past history of a particular diagnosis. A diagnosis from the past is modified with the phrase "prior history." In this case, you might write: "309.81, Posttraumatic Stress Disorder, Prior History."

Modifiers for Course of the Disorder and Severity

The fifth digit is used to specify the status of the remission or the degree of severity in some diagnoses. Severity is usually either mild, moderate, severe, or psychotic. Each of these degrees of severity, specified by the fifth digit, indicate the intensity of the signs and symptoms and the degree to which the person's functioning is impaired. "Mild" means the person's symptoms just minimally meet the criteria for the disorder. Severe means the person meets the full criteria with intensity and is severely impaired in functioning.

Look once more at the example of Major Depressive Disorder under "Making the Code" to see how the fifth digit is used to indicate severity.

Remission is usually either "partial remission" or "full remission." Partial remission means the person meets some of the criteria for the disorder but is no longer showing the full criteria that were present when the original diagnosis was made. You might also use this specifier to indicate that a person who had been in full remission is now showing some of the criteria again. Full remission means all the signs and symptoms of the disorder have disappeared but the diagnosis is clinically significant at present.

Additional Information

Reason-for-Visit Modifiers

Beginning with the *DSM-IV*, there was a shift toward multiple diagnoses. If the person fits the criteria for more than one disorder, then all of the disorders are listed. This means that the practitioner needs to identify the disorder that brought the individual in for treatment or the disorder that will most likely be the focus of treatment. This is done by placing the phrase "principal diagnosis" or "reason for visit" beside the primary diagnosis.

List the primary diagnosis first with the modifier ("reason for visit" or "principal diagnosis"), and then list the other diagnoses under it in order of importance. Generally, the primary diagnosis is on Axis I. On occasion, it will be on Axis II, in which case you need to make sure to use the modifier on the Axis II diagnosis.

Here is an example:

Axis I: 296.23, Major Depressive Disorder, Recurrent, Severe Without Psychotic Features, Reason for visit

305.60, Cocaine Abuse

Provisional Diagnoses

When it is not clear what the diagnosis should be, the clinician needs to indicate that uncertainty. Some clients are uncooperative or impaired to the point that they cannot give much information. If the clinician has a strong idea of what the diagnosis probably is, a diagnosis is written with the term *provisional* after it.

There are some disorders that *must* initially be modified with "provisional." These would be disorders where a time lapse is needed between episodes in order to confirm the diagnosis. For example, panic disorder (300.01) requires at least one panic attack, followed by at least a month of persistent worry about having more attacks. A clinician who sees someone who appears to have panic disorder but had the panic episode only 2 days previously cannot give the client that diagnosis with a degree of certainty. Thus the word *provisional* is used until enough time has passed to confirm or revise the diagnosis.

Not Otherwise Specified

Sometimes a person comes in with most of the symptoms of a particular disorder and the practitioner thinks she knows what the diagnosis should be, but the signs and symptoms do not quite fit the criteria as they are outlined in the manual. All the major classes of mental disorders have a "not otherwise specified," or NOS, category. The practitioner would write NOS after the diagnosis in this sort of case.

For example, a person might have symptoms that do not meet the full criteria for any one of the mood disorders, but meets some criteria for several different mood disorders, giving a mixed version of the disorder. The diagnosis would be 296.90, Mood Disorder, NOS or not otherwise specified.

There are other reasons to use the NOS designation. You might encounter a person who has an obvious mental disorder, but it is not found in the *DSM*. Perhaps it is a characteristic of the person's culture, or it is something being researched that has not been given a code yet. In another situation, the clinician may not know the cause of the disorder. The mental disorder he is seeing might be due to the medication the person is on, the general medical condition the person has, or the stress of the individual's life. This may need to be sorted out before giving a certain diagnosis. Finally, clinicians who lack enough information to be sure of the diagnosis can use the NOS category. After getting good information, the diagnosis can be changed.

Unspecified and Deferred Diagnoses

There is an "unspecified mental disorder" category that cannot be used when psychotic symptoms are present. The number is 300.9. This is used when there is some certainty that a mental disorder exists, but there is inadequate information to make a clear diagnosis. Later it can be changed to a specific disorder, unless it is found that the symptoms do not meet the criteria for any specific disorder.

If there is inadequate information to make any diagnosis, the number on both Axis I and Axis II is 799.9, "diagnosis deferred." Today most private and public insurance agencies insist that a diagnosis, even if provisional, be given at the very first person-to-person contact. This means that in some locations it is case managers who are giving those provisional diagnoses.

You can see part of a service planning conference where the participants discuss what might be the appropriate diagnosis for Alison. Note the various considerations the group makes before a diagnosis is assigned.

APPENDIX H

Case Manager's Toolbox

Assessing Lethality

Numerous groups concerned about the safety of women have developed assessment tools to help workers and the women they serve assess the risks associated with abusive partners. Here are those signals that seem most suggestive of dangerous abusive behavior and the possibility that the abuser could kill his partner.

History

- ☐ Does the abuser have a history of violence toward others?
- ☐ Has law enforcement been involved in violent episodes in the past?
- ☐ Is there a history of arson or threats of arson?

Threats and Attempts

- ☐ Have there been threats to kill others or kill himself?
- ☐ Has the abuser a plan or a fantasy that includes killing him or herself or a victim? Has the abuser given explicit examples of what he or she intends to do and how the plans would be carried out? According to the National Center for Victims of Crime, "The risk is greater if the abuser is very specific about plans and intended methods."
- ☐ Has the abuser ever attempted to strangle or suffocate his or her partner?
- ☐ Have there been other forms of violence in the past?
- ☐ Have there been threats to kill the children or pets?
- ☐ Does the abuser intend to take a hostage? There should be considerable concern when someone threatens to take a hostage.
- ☐ Does the intended victim believe the threats?
- ☐ Has violence escalated recently?

Attitude Toward the Victim

- ☐ Does the abuser insist that the intended victim belongs to him and no one else can ever have this individual?
- ☐ Does the abuser indicate the intended victim is not entitled to any life apart from him?
- ☐ Does the abuser diminish and denigrate the victim's ability to be independent?
- ☐ Does separation or the hint of separation cause great rage?
- ☐ Does the abuser stalk the intended victim?
- ☐ Does the abuser demand accountability for every minute the victim is not in his presence?

Isolation

- ☐ Is the abuser virtually isolated from others?
- ☐ Is the victim effectively the abuser's only contact whom he depends upon to organize his own life?
- ☐ Does the abuser isolate the victim from the larger community?
- ☐ Has the abuser cut his partner off from family and former friends?

The Abuser

- ☐ Is the abuser depressed with little hope of improving the situation?
- ☐ Is the abuser unemployed?

Access

- ☐ If the intended victim has left, does the person have easy access to the intended victim?
- ☐ Does the person have access to weapons?

If any of these elements are present in the history you take you may conclude that the person is likely to kill his victim or himself. You need to take extraordinary precautions in working out the safety plan if the intended victim is your client. If the threat is posed by your client, you need to notify the potential victim and notify the police. You may need a mental health commitment, if this seems reasonable. If the intended victim is your client, always talk frankly about what these indicators could mean.

When working with an intended victim the National Center for Victims of Crime makes a number of useful recommendations. For example, encouraging the victim to keep a journal of the incidents of violence and evidence of violence including photographs. They recommend the victim leave when it is least expected, perhaps when things are going smoothly. They also give instructions on how to create a false trail. You and your clients can access the National Center for Victims of Crime for detailed safety planning tips and techniques.

Recognizing the Difference Between Delirium and Dementia

There are two keys to recognizing the difference between the two conditions.

Delirium	Dementia or Major Neurocognitive Disorder
Sudden onset	Gradual or insidious onset
Usually related to a physical condition and is often reversible if caught in time	Cause is often not known or if discovered, the condition is not reversible

Delirium

According to the *DSM 5* delirium is "(A) disturbance in attention (i.e. reduced ability to direct, focus, sustain, and shift attention) and awareness (reduced orientation to the environment)" (*DSM 5*, p. 596).

Onset It is important to recognize it when you see delirium and important that you not mistake it for depression, psychosis, or dementia. *The key to recognizing it is the sudden onset.*

It is an acute condition with three possible outcomes; The person can make a full recovery, particularly if the origins of the condition are found and corrected early in the course of the problem. It can also result, however, in permanent disability and, finally, in death, when left undiagnosed and untreated. It is your responsibility to document carefully for those who will actively seek an underlying cause for the delirium the conditions as reported by family doctors, family members, and others and to document the medications the person is taking.

Common Causes The primary reasons for delirium are declining organ function, increasing medical illness, multiple medications, sensory deprivation, physical, emotional and environmental losses. Delirium can be caused by an underlying medical condition. Many conditions have been known to cause delirium in some older people. You need to make sure the person has a good medical evaluation.

You may not see the usual signs of illness. For instance, older people often don't run a temperature when they have an infection, or they may not feel pain when experiencing a heart attack. Problems with endocrine glands are common as glands become less active with age. For example, a person may develop low thyroid later in life and complain of being cold all the time. Another person may have low blood sugar, possibly because of erratic eating habits and appear disoriented

Medication Problems Delirium can also be caused by medications or other substances. Medications can throw off the physiology of the person just enough to impair their ability to function well cognitively. In other cases a medication can exaggerate the signs of aging causing physical slowing and mental confusion. Older people often face problems with medication younger people do not. Below are some common medication problems that can lead to delirium:

An older person, because of changes in their metabolism, may require less medication than the average middle-aged adult, but be prescribed at the normal rate, without taking into account body weight and size.	An older person can be on a medication for awhile, taking it successfully, but as their metabolism changes the medication dose may need to be reduced. This leads some to overlook the real problem because the person took the medication successfully before.
Medications can throw off the physiology of the person just enough to impair his or her ability to function well cognitively.	Older people may have some trouble remembering when they last took their medication or if they took it at all. This can result in a person taking double doses or missing several doses and then taking the medication again. Taking medication in this manner can result in uneven amounts of the medication in the bloodstream and lead to confusion in some cases.
When an older person is seeing more than one physician or is being treated for more than one condition he or she may be given medications that have a negative interaction with each other. Some older people are keeping track of 10 to 12 medications a day.	A doctor may add a new drug to combat the side effects of the first drug without realizing these are side effects. In other words, the physician treats the side effects as if it is a new condition the patient is experiencing.
Sometimes finances cause the person to skip doses to save money, thus taking far less than the prescribed amount. In an effort to save money the older person may use over-the-counter medications and try to self-medicate, or they may be using over-the-counter medications that interact badly with prescribed medications.	Common problem medications are: sedatives, digitalis drugs, diuretics, and antihypertension medications. All of these can cause in some older people symptoms resembling depression or delirium. In addition, psychotropic medications can cause symptoms that look very much like the onset of Parkinson's disease. Over-the-counter antihistamines can also cause people to be unsteady or confused.
Delirium can also result when an older person stops taking a medication. For instance, the sudden withdrawal of a medication such as prednisone (a steroid) can result in confusion in some people.	Delirium can be the result of several causes such as several medical conditions or a medical condition and a drug interaction.

Neurocognitive Disorder or Dementia

According to The Alzheimer's Association in their 2013 report, "One in nine people age 65 and older (11%) has Alzheimer's disease" (p. 15) and the likelihood increases with age. For example, about one-third of people age 85 and older (32%) have Alzheimer's disease" (p. 15) and the likelihood increases with age.

Dementia is characterized by a number of problems and will involve memory disturbances, such a loss of long- or short-term memory. In fact, problems with memory may be the most noticeable sign of this condition.

> The Three A's for Neurocognitive Disorder
>
> *Aphasia* –the inability to understand what is said or express oneself
> *Apraxia*–the inability to accomplish purposeful movements required to care for oneself
> *Agnosia* –the inability to comprehend sensory data even though the senses are intact. For example, the inability to recognize common objects

Other Possible Symptoms In addition, the person may have trouble with executive functioning, that is making decisions and taking care of details of daily living. It is not unusual for people with dementia to repeat the same story or the same question. They may misplace things and get lost in places they should know well. You might find that simple math is now difficult or impossible for the person who competently balanced his own checkbook for years. You might find that the person is having trouble expressing himself, unable to think of words, or using confusing phrases.

Many times people suffering from dementia will appear depressed. The family may report that the individual has lost interest in things that used to be important or the person seems to have withdrawn. Another symptom you might hear is that the person's personality has changed. For instance, he or she may have become tactless, blurting out socially inappropriate comments or snapping at others.

Any combination of these symptoms will usually bring the person into contact with social services as her ability to function independently, socially, and occupationally is impaired.

Onset The onset for dementias is a slower, more gradual course. In the history people will describe a gradual decline in mental functioning.

Common Causes There are a number of causes for dementia and these causes can usually be diagnosed through laboratory tests and imaging. Some common underlying medical causes are:

Strokes	HIV disease
Parkinson's disease	Huntington's disease
Head trauma	Other medical conditions

To learn more about these disorders you can look at the chapter on Neurocognitive Disorders in the *DSM 5*.

Recognizing Dangers of Alcohol Withdrawal

Withdrawal

True withdrawal from alcohol depends on the duration of the drinking or the length of time the person was bingeing and on the quantity of alcohol that was consumed during that time. Serious withdrawal, the type that is potentially dangerous, could occur after the person consumed a pint of whiskey every day for ten days. These symptoms can start anywhere from several hours to several days after the person took their last drink. In most cases these symptoms tend to peak sometime around the second to fourth day with the third day commonly considered the worst. Acute withdrawal symptoms generally clear within 1 week.

The Stages of Withdrawal
Withdrawal Stage One Mental/physical symptoms: May report feeling tremulous, restless and jumpy. Sometimes feels like they are going to jump out of their skin. May tell you they feel generally apprehensive. Physical Symptoms: Increased pulse and respiration rate Primary Danger: The person will begin drinking to relieve the discomfort
Withdrawal Stage Two Mental/Physical symptoms: Person complains of the shakes inside Anxiety and dread intensify Extremely frightened and needs reassurance Usually oriented to time, place, and person Alcoholic Hallucinosis ("Audiovisuals") Try to understand the meaning of the hallucination to the person Physical Symptoms: Pulse, blood pressure, and respiration continue to elevate Tremors are more severe Grand Mal seizures may occur Primary Danger: Harm occurring during a seizure or a permanent seizure condition that will not abate and eventually leads to death
Withdrawal Stage Three Mental/Physical symptoms: Abject terror Hallucinations often tactile, visual, and auditory (may see and feel tiny bugs everywhere) Often persecutory hallucinations Orientation X three may be lost Physical Symptoms: Intense psychomotor agitation Pulse and blood pressure continue to rise Fever may develop Primary Danger: Significant mortality rate at this last stage

Detoxification

Detoxification is a *medical treatment*. It almost always involves sedative drugs that are titrated down over several days to a week. Medications to treat anxiety are often prescribed. The treatment also includes vitamins and minerals to address deficiencies that developed during the prolonged drinking. Some people will require anti-convulsant medication. During this process the person will require close supervision, good nursing care, and strong emotional support.

Detoxification Dangers How and where detoxification is carried out is a *medical decision*. It usually takes a physician, experienced in the drug and alcohol field, to make the right decision. As the case manager you would refer the person for a medical evaluation. The choices are fairly narrow: Inpatient or outpatient treatment, with or without medication. The physical debilitation of the patient will determine the length of detoxification period or the need for an inpatient medical unit.

People who have intact social support systems do better during the detoxification phase. As the intake worker it is important to explore what social support the person may have to help him or her through this process.

SUICIDE ASSESSMENT

Suicide

According to Kevin Caruso, executive director of the website Suicide.org, in 2005 there were 32,637 suicides in the United States with over 800,000 others attempting suicide (see http://suicide.org/suicide-statistics.html#2005). A popular notion is that those who talk about suicide won't actually commit suicide. It is important therefore, when you talk to a person who is depressed to explore suicide with them.

How to talk about it You can ask the person, "Have you ever felt so sad that you have thought of suicide?" or you might ask, "Do you just feel as if things are so hopeless you've thought about suicide?"

If the Answer is "Yes" If the person tells you she has considered suicide, it is important to understand how serious she might be about actually carrying out this act. What you want to know is where she stands at this point. You learn this information by talking to the person about this directly with warmth and empathy. Begin with an open question. "Can you tell me a little more about that?"

What you need to know:
Has she ever thought about how she would do it?
Does she have the means to actually carry it out? (a gun, medications)
Has she planned when to carry this out?
Has she ever tried to commit suicide before?

If you learn that she has decided how she will do it, has the means to complete it, has decided when she will carry it out, and has attempted suicide in the past, the danger or risk that she will do so now is extremely high.

You Are Not Putting Ideas in Someone's Head   Some workers, feeling uncomfortable about asking for this much information, do not ask follow-up questions like these. Some mistakenly believe that if they do, they are putting ideas in the person's head. In reality you are assessing the situation to determine the degree of risk and this assessment must be done.

Where Else to Look for Clues to Possible Suicide   There are other places to look for clues when doing a suicide assessment. Below are signals that can help you assess the degree of risk with the person you are interviewing.

The Social History

- ☐ The person sought help in the past but felt it was useless
- ☐ The person has an unstable life with numerous ups and downs and few things in his or her life that remain stable and supportive.
- ☐ There is considerable stress at this point in the person's life
- ☐ The person continually abuses alcohol
- ☐ The person indicates that he or she has no one really close; no family, close friends, significant others, or confidantes
- ☐ The person has little in the way of resources such as money, abilities, a supportive environment
- ☐ The person is withdrawn from normal interaction and activities
- ☐ The person coped with difficulties in the past in destructive, rather than constructive, ways
- ☐ The person's ability to function on a daily basis is poor
- ☐ The person tells you of previous suicide attempts (particularly note if these had a high probability of being successful and be doubly alert if you are seeing this person because he or she has just attempted suicide)
- ☐ There is a family history of suicide such as the death of a close relative or of a parent at an early age
- ☐ Others have been rejecting of the client's problems or symptoms
- ☐ There is no interest in or belief in a religious or spiritual system. (This does not mean the person has to belong to an organized religion or attend church regularly. You are looking here for a spiritual belief system that sustains the person and gives them strength or hope.)
- ☐ The person has a chronic illness, a chronically painful medical condition, a life threatening medical condition, a severely disabling medical condition
- ☐ The person has had a particularly long episode of depression
- ☐ The person was previously hospitalized for psychiatric treatment

Other considerations can increase the risk. For instance

- ☐ Older (over 45) or elderly are somewhat more likely than young adults
- ☐ Males are somewhat more likely than females to actually complete a suicide
- ☐ Unmarried are somewhat more likely than married people
- ☐ Unemployed or retired are more likely to commit suicide than those who are working

> ### What You Observed?
>
> What you observe when you are talking to the person is equally important.
>
> Here are factors that you might observe that would alert you to the fact that the risk of suicide is present.
>
> - ☐ The person came in alone
> - ☐ The person shows a severely depressed affect
> - ☐ You can see wrist scars from previous suicide attempts
> - ☐ The person has delusions that indicate persecution and external controlling factors
> - ☐ The person has hallucinations that command the person to commit some self-destructive act.
> - ☐ During the interview the person never relaxes or establishes any rapport. You feel as if there is no real communication taking place.
> - ☐ The person is extremely hostile (look for violence or rage in the history)
> - ☐ The person refuses any help

An Ironic Fact About Suicide An important point about suicide is that many people commit suicide just when you think they are improving. As the depression begins to lift and the person feels more energetic the energy to plan and carry out a suicide may be present. If the person is feeling better but still has ideas about suicide there is an increased risk that suicide may occur. Certainly the overall risk of suicide is reduced with treatment, but there can be this window of increased risk when people begin to feel better and have the energy to carry out a suicide.

Sources: Adapted from The National Center for Victims of Crime 2000 M Street, NW Suite 480, Washington DC 20036
1800 FYI-CALL http://www.marincourt.org/PDF/LethalityRisk.pdf; Maryland Network Against Domestic Volence
6911 Laurel Bowie Road
Suite 309 Bowie MD 20715
301 352-4574
http://mnadv.org/_mnadvWeb/wp-content/uploads/2011/10/LAP_Info_Packet--as_of_12-8-10.pdf

References

Adler, R. B., Rosenfeld, L. B., & Proctor, R. F. (2013). *Interplay: The process of interpersonal communication* (12th ed.). New York: Oxford University Press.

Alzheimer's Association. (2013). Alzheimer's Disease Facts and Figures. *Alzheimer's & Dementia, 9*(2). Retrieved from http://www.alz.org/downloads/facts_figures_2013.pdf

American Psychiatric Association. (1982). *Quick reference to the diagnostic criteria from DSM-III.* Washington, DC: Author.

American Psychiatric Association. (1987). *Diagnostic and statistical manual of mental disorders: DSM-III-R* (3rd ed., rev.). Washington, DC: Author.

American Psychiatric Association. (1994). *Diagnostic and statistical manual of mental disorders: DSM-IV* (4th ed.). Washington, DC: Author.

American Psychiatric Association. (2000). *Diagnostic and statistical manual of mental disorders: DSM-IV-TR* (4th ed., text rev.). Washington, DC: Author.

American Psychiatric Association. (2013). *Diagnostic and statistical manual of mental disorders: DSM-5.* Washington, DC: Author.

Bednar, R. L., Bednar, S. C., Lambert, M. J., & Waite, D. R. (1991). In G. Corey, M. S. Corey, & P. Callanan (Eds.), *Issues and ethics in the helping professions* (pp. 100–101). Pacific Grove, CA: Brooks/Cole.

Beisser, A. (1970). *In Fagan & Shepherd, Gestalt therapy now*. New York: Harper Colophon.

Bratter, T. E. (2011). Compassionate confrontation psychotherapy: An effective and humanistic alternative to biological psychiatry for adolescents in crisis. *Ethical Human Psychology and Psychiatry, 13*(2), 115–133. doi:10.1891/1559-4343.13.2.115

Brattner, T. E., Esparat, D., Kaufman, A., & Sinsheimer, L. (2008). Confrontational psychotherapy: A compassionate and potent therapeutic orientation for gifted adolescents who are self-destructive and engage in dangerous behavior. *International Journal of Reality Therapy, 27*(2), 13–25.

Bronfenbrenner, U. (1979). *The ecology of human development: Experiments by nature and design.* Cambridge, MA: Harvard University Press.

Burns, D. D. (1980). *Feeling good: The new mood therapy*. New York: William Morrow.

Butler, R. N., Lewis, M. I., & Sunderland, T. (1998). *Aging and mental health: Positive psychosocial and biomedical approaches.* Needham Heights, MA: Allyn & Bacon.

Caruso, K. (n.d.). *Suicide.org.: Suicide prevention, suicide awareness, suicide support.* Retrieved from http://suicide.org/

Caudill, O. B. (1997, February). The perils of isolation. *The Pennsylvania Psychologist, 57,* 22, 26.

Channel 27 News. (2010). *Nurses fired for Facebook posts speak out.* Retrieved from http://www.whtm.com/news/stories/0310/720899.html

Codes of ethics for the helping professions (2nd ed.). (2004). Belmont, CA: Brooks/Cole.

Corey, M. S., & Callanan, P. (1993). In G. Corey (Ed.), *Issues and ethics in the helping professions.* Pacific Grove, CA: Brooks/Cole.

Diclemente, C. C., & Valesquez, M. M. (2002). Motivational interviewing and the stages of change. In W. R. Miller & S. Rollnick (Eds.), *Motivational interviewing* (pp. 201–216). New York: Guilford Press.

Elliot, R., Bohart, A. C., Watson, J. C., & Greenberg, L. S. (2011). Empathy. *Psychotherapy, 48*(1), 43–49.

Frankel, A. J., & Gleman, S. R. (2004). *Case management: Integrating individual and community practice* (2nd ed.). Chicago, IL: Lyceum Books.

Gair, S. (2012). Feeling their stories: Contemplating empathy, insider/outsider positionings, and enriching qualitative research. *Qualitative Health Research, 22*(1), 134–143.

Gerdes, K. E., & Segal, E. (2011). Importance of empathy for social work practice: Integrating new science. *Social Work, 56*(2), 141–148.

Germain, C. B., & Gitterman, A. (1980). *The life model of social work practice*. New York: Columbia University Press.

Goldman, A. R. (1990). Special focus on basic rules of writing treatment goals and objectives. *Accreditation and Certification, 4*(3), 1–9.

Gordon, T. (1970). *Parent effectiveness training: The tested way to raise responsible children*. New York: David Mackay.

Greif, G. L., & Lynch, A. A. (1983). The eco-systems perspective. In C. H. Meyer (Ed.), *Clinical social work in the eco-systems perspective* (pp. 35–71). New York: Columbia University Press.

Gudykunst, W. B., & Kim, Y. Y. (1997). *Communicating with strangers: An approach to intercultural communication* (3rd ed.). New York: McGraw-Hill.

Hart, B. (1990). Assessing whether batterers will kill. In J. J. Parker, B. Hart, & J. Stuehling (Eds.), *Seeking justice: Legal advocacy principles and practice* (pp. 8–9). Harrisburg, PA: Pennsylvania Coalition Against Domestic Violence.

Holt, B. J. (2000). *The practice of generalist case management*. Boston, MA: Allyn and Bacon.

Jackson, S. W. (1992). The listening healer in the history of psychological healing. *American Journal of Psychiatry, 149*(12), 1623–1632.

Johnson, L. C., & Yanca, S. J. (2007). *Social work practice: A generalist approach* (9th ed.). Boston: Allyn & Bacon.

Kirst-Ashman, K. K., & Hull, G. H. (2009). Values, ethics, and the resolution of ethical dilemmas. In *Understanding generalist practice* (5th ed., p. 376). Belmont, CA: Cengage.

Kynoch, K., Wu, C., & Chang, A. M. (2010). Interventions for preventing and managing aggressive patients admitted to an acute hospital setting: A systematic review. *Worldviews on Evidence-Based Nursing*. doi:10.1111/j.1741-6787.2010.00206.x

LaBruzza, A. L. (1994). *Using DSM-IV: A clinician's guide to psychiatric diagnosis*. Northvale, NJ: Jason Aronson.

Leaman, D. R. (1978). Confrontation in counseling. *Personnel and Guidance Journal, 56*, 630–633.

Lechman, C. (2006). The development of a case load weighting tool. *Administration in Social Work, 30*(2), 25–37.

Lukas, S. (1993). *Where to start and what to ask: An assessment handbook*. New York: Norton.

Miley, K. K., O'Melia, M., & Dubois, B. (2007). *Generalist social work practice: An empowering approach* (5th ed.). Boston, MA: Allyn & Bacon.

Miller, R. M., & Rollnick, S. (2002). *Motivational interviewing: Preparing people for change*. New York: Guilford Press.

Meyer, C. H. (Ed.). (1983). *Clinical social work in the eco-systems perspective*. New York: Columbia University Press.

National Association of State Mental Health Program Directors. (2006). *Technical report on mortality and morbidity*. Washington, DC: Author.

Polcin, D. L. (2003). Rethinking confrontation in alcohol and drup treatment: Consideration of the clinical context. *Substance Abuse and Misuse, 38*(2), 165–184. doi:10.1081/JA-120017243

Polcin, D. L. (2009). Who receives confrontation in recovery houses and when is it experienced as supportive? *Addiction Research & Theory, 17*(5), 504–517. doi:10.1080/16066350801968732

Rogers, C. R. (1980). *A way of being*. Boston, MA: Houghton Mifflin.

Rothman, J., & Sager, J. S. (1998). *Case management: Integrating individual and community practice* (2nd ed.). Needham Heights, MA: Allyn and Bacon.

Siegel, M. (1993). In G. Corey, M. S. Corey, & P. Callanan (Eds.), *Issues and ethics in the helping professions* (p. 105). Pacific Grove, CA: Brooks/Cole.

Stadler, H. A. (1990). Confidentiality. In B. Herlihy & L. B. Golden (Eds.), *AACD ethical standards casebook* (4th ed., p. 102). Alexandria, VA: American Association for Counseling and Development.

Stephan, W. G. (1985). Intergroup relations. In G. Lindzey & E. Aronson (Eds.), *Handbook of Social Psychology* (Vol. III, pp. 599–658). New York: Addison-Wesley.

Strong, T., & Zeman, D. (2010). Dialogic consideration as a counseling activity: An examination of Allen Ivey's use of confronting as a microskill. *Journal of Counseling and Development, 88*, 332–339.

Summers, N. (2002). *Fundamentals for practice with high risk populations*. Pacific Grove, CA: Brooks/Cole.

Tarasoff v. Regents of University of California, 551 P.2d 334 (Cal. Sup. Ct. 1976).

Testimony of Elizabeth J. Clark, PhD, ACSW, MPH. (2002). Washington, DC: NASW. Retrieved January 16, 2014, from http://www.socialworkers.org/pressroom/events/nfcmh.asp?print=1

Three Rivers Center for Independent Living. (n.d.). *Language references*. Pittsburgh, PA: Author.

Unger, M. (n.d.). A deeper, more social ecological social work practice. *Social Service Review, 76*(3), 480–497.

Warmington, S. (2011). Practicing engagement: Infusing communication with empathy and compassion in medical students clinical encounters. *Health, 16*(3), 327–342.

Weiner, I. B. (1975). *Principles of psychotherapy*. New York: Wiley.

Yamatani, H., Engle, R., & Solveig, S. (2009). Child welfare worker caseload just right. *Social Work, 54*(4), 361–368.

Index

A

Acceptance, skills and attitudes, 265
Accessing the file, 55
Administrative case management, 381
Adversarial, 55, 258–262
Advice, 45, 164, 170, 171, 214, 259
Advice, peer support and, 7
Advocacy, 12, 13
　with collaterals, 214–215
　　macro level interventions and, 88–89
Affect, emotional state, 345
Agencies, 2, 3, 6, 11, 17, 52, 126, 127, 128, 130, 143, 150, 206, 214, 215, 227, 231, 233, 243, 250, 287, 288, 289, 290, 338, 359–360, 363–364, 443, 444. *See also* Provider agencies
　case management and, 19–20
　case manager, 379
　client and family involvement, service plans, 368–369
　collaborating with other, 416–417
　DSM, 322–323
　face sheet and, 390–392
　formal, 6
　information release from, 363
　phone intakes, 277
　provider, 25, 367–369, 395, 398, 404, 417, 421–432, 440
　service coordination and, 13–16
Alcoholics Anonymous (AA), 56
Ambivalence, 255–256
　stages of change and, 240
　trapping the client and, 258
American Journal of Psychiatry, (Jackson), 169
American Psychiatric Association (APA), 324–325
Anger, 39, 122, 128, 176, 213
　addressing, 225–235
　angry outburst, 229–231
　common reasons for, 225–226
　disarming, 160, 226–227, 229–231, 267
　mistakes, avoiding, 227–228
　what not to do, 231–233
Anxiety, 99–100
APA. *See* American Psychiatric Association (APA)
Aphasia, 344
Appearance, 18, 35, 341
Appointment form, 282–283
Assessment, 1, 4–5, 8, 9, 10, 12, 14, 18, 20, 80, 288, 379, 381, 383
Assessment forms, 297, 314–315, 369–372, 422
Athetosis movements, 343
Attention, 346
Attitudes, 40, 43, 95, 117–118, 341–342
　basic, 118–123
　changing, 110–111
　with skills, 265–267
　we-*versus*-them, 86–87

B

Background information, social history and
　behavioral health, 304
　education, 303
　employment history, 303
　family of origin, 302
　legal history, 304
　living arrangements, 303
　marriages and significant relationships, 302
　medical history, 303
　military service, 303
　religious activities, 304–305
　social and recreational interests, 304
　successes, strengths, and resources, 305
Barriers, 107, 127, 130, 170
　communication and, 152
　understanding, 375
Basic helping attitudes
　empathy and, 119–122
　genuineness and, 118–119
　warmth and, 118–119
Bednar, R. L., 60
Bednar, S. C., 60

Behavior, 21, 23, 122, 123, 124, 125, 154, 204, 206, 212, 240, 242, 243, 245, 342–343
Beisser, Arnold R., 176, 247, 248
Biological characteristics, micro level ecological model, 80
Body language, 178, 205
Boundaries, 23, 40, 96, 106, 121–122, 266
　detrimental, 129
　and power, 140–141
　understanding of, 127
Brainstorming, 258–259
Bronfenner, Urie, 82, 86
Burns, David, 229

C

Care planning meetings, 379
Case management, 19, 22, 23, 415, 422, 430, 440, 441, 447
　caseloads, 25
　generic, 26
　guidelines for, 14
　history of, 2
　level of, 381
　levels of, 16–18
　managed care and, 21–25
　process, 1–13, 18
　use of, 3–4
Case management units, 50, 89, 98, 259, 278, 359, 367–376, 387, 389, 395, 422, 429, 430, 440
Case managers
　documentation and, 396
　follow-up, 416
　generic, 26
　goals and objectives, developing manageable, 423
　for intakes, 323
　intensive, 16
　leave office, to visit clients, 418
　managed care and, 21–22
　mental status, observations, 338
　termination and, 439
Catatonic behavior, 343
Change, stages of, 240–244

Change talk, 253
Charts, 55, 390, 392
Charts, social history in, 299
Child and Adolescent Service
    System Program (CASSP),
    23, 266
Children's panel, 379
Choreic movements, 343
Clark, Elizabeth J., 25
Client, balanced view of, 82–85
Cognitive functioning, 24,
    346–349
Collaboration, 241, 266
    adversarial to, 258–262
    ambivalence, 244–247
    encouragement, 247–250
    resistance, 244–247
    stages of change, 240–244
Collateral contact, 395, 398
Collecting summaries, 260
Communication
    culture and, 95–96
    distorted, 81
    individualistic and collectivistic
        cultures, 106
    negative, 229
    oral, 53
    order or command, 152
    privileged, 56
    roadblocks to, 151–156
    thoughtless *versus* thoughtful,
        100–103
    transactional, 151
    warning of consequences,
        152–153
Concentration, 346
Confidence, 256–257
Confidentiality, 47–51
Conflict, 108, 109, 110
Confrontation, 106, 112, 159,
    203–204, 258
    I message in, 207–213
    rules for, 208–211
Confusion, 352–353
Consciousness, 346
Contact, labeling of, 398
Contact notes, 395, 440, 443
    sample of, 397
    writing, 396–397
Context, ecological model, 81–82
    personal, 80
    social, 80
Countertransference, 130
Criminal Justice Background, 299
Crisis, 12, 14, 17, 63, 64, 68, 127,
    144, 145, 205, 214, 216, 240
Crisis, responding to, 419–420
Cross-cultural communication,
    111
Cultural competence, 266
    anxiety and uncertainty, 99–100
    and communication, 94–95

ethical responsibility, 96
individualistic and collectivistic
    cultures, 106
sociological level differences,
    96–98
Cultural relativism, 109, 110
Culture, 96–97. *See also*
    Communication
    barriers, 375
    characteristics of individualistic,
        104–105
    and communication, 95–96
    dimensions of, 104–109

D

Delusions, 350–351
Depressive disorders, 332–333
Developmental transitions, 86
Diagnosis, 45, 229, 240, 323, 326,
    331, 380, 382
Diagnostic and Statistical Manual
    of Mental Disorders (DSM)
    background information,
        323–327
    cautions, 322–323
    diagnosis and, 323
    mental health tool, 322
    using, 321–333
Diagnostic labeling, 59–60
Diclemente, C. C., 240
Disabilities, 24, 81, 249, 375, 440
Discharge summaries, 443–444
Discouragement, 124, 125, 126,
    129, 250
Discrepancies, 204–205, 254–255
Disordered perceptions, 350
Disposition planning meet-
    ings. *See* service planning
    conference
Disputes, unpleasant, 416
Dix, Dorothea, 324
Documentation, 443
    best practice, 399–401
    clear and precise in, 400
    client, balanced picture of, 404
    contradictions, avoiding, 400
    disabilities, describing in, 401
    evidence of agreement, 404
    facts and impressions, distin-
        guishing between, 403
    in fee-for-service agencies, 396
    government requirements for,
        402
    hostility, avoiding, 399
    importance of, 396
    initial inquiries, 277–284
    interactions with client, 399
    judgmental words, avoiding of,
        402–403
    plan, making changes to, 404
    quotations usage in, 400

service monitoring, 398–399
significant aspects of contact,
    399–400
understandable language usage
    in, 400–401
Door-knob syndrome, 295
Drain off feeling, 176–177
DSM. *See* Diagnostic and
    Statistical Manual of Mental
    Disorders (DSM)
DSM 5, 327–328. *See also* Diagnostic
    and Statistical Manual of
    Mental Disorders (DSM)
    coding process, 330–331
    current diagnostic manual,
        328–330
    disorder, no number, 333
    V codes, 332
DSM Handbook, 380
DSM-II, 325
DSM-III, 325–326
DSM-IV, 326–327
Dual relationships, 35–39
Dysarthria, 344

E

Ecological model, 14, 77, 79
Elliot, R., 120, 121
Emotions, 107, 108, 120, 121, 122,
    127, 130, 170, 174, 177, 227,
    345–346
Empathy, 82, 119–121, 157, 164,
    170, 208, 227, 228
    and boundaries, 121–122
    and compassion, 121
    and safety, 121
Empowerment, 250
Encouragement, 37, 62, 81, 84,
    247–250
Encouragement, *vs.* discourage-
    ment, 250
Environment, 5, 77, 78, 79, 82, 84,
    86, 87, 88, 101, 102, 103, 130,
    141, 143, 157, 170, 176, 195,
    266, 355–356
Ethical behavior, 34
Ethical code, 34–35
Ethical principles, 33, 34
    to colleagues, 65–67
    competence and, 65
    to the profession, 65–67
    violations of, 67
Ethical responsibilities, 61–62
Ethnic group, 97
Ethnocentrism, 109–110
Evaluation forms, 422
Everyday Lives, 24
Exchanging views, 204
    discrepancies, 204–205
    reasons to, 205–206
Exploitation, 13, 33

F

Facebook, 55
Face sheet, 390–392
Feedback, 14, 124, 162–163, 203, 208, 213, 214, 233, 240
Feeling Good (Burns), 229
Final interview, in case termination, 442
First interview, 26, 179, 287–295
　ask for clarification, 292
　beginning, with introduction, 290–291
　end of, 295
　expectations, of client, 293
　information to collect in, 292–293
　note-taking during, 291
　open questions in, 291–292
　preparing for, 288–289
　purpose of, 287
　tasks, 294
　wrapping up, 294
　your office for, 290
　your role in, 288
First sentence, of presenting problem, 300
Follow-up, 16, 39, 80, 164, 217, 233, 385, 416
Funding, target date and, 388

G

Gair, S., 119, 121
Generalist approach, 88
Generalized Anxiety Disorder, 332
General release form, 360–363
Gerdes, K. E., 122
Goal plan, 370–371, 375–376, 385, 404
Goals, 7, 9, 11, 12, 13, 15, 20, 22, 43, 44, 45, 80, 83, 104, 125, 126, 205, 209, 245, 247, 248, 249, 252, 266, 368, 369, 372, 373, 417
　client focus, 424
　client participation/collaboration, 422
　combining goals and treatment objectives, 426–428
　defined, 430
　developing, 421–432
　elements in, 424
　finishing touches in writing, 428–429
　long term and short term, 430
　for meeting, 380–381
　numbering system for, 428
　objectives of, 425
　positive outcomes, 423–425
　proper endings, 428
　review date and, 388
　review dates for, 429–430
　target date and, 388
　target dates for, 428
　treatment intervention, 429–430
　writing, 424, 425
Goodwill Industries, 414
Gordon, Thomas, 151, 207
Greif, G. L., 78
Greisinger, Wilhelm, 324
Group contact, 398
Gudykunst, W. B., 97, 98, 99, 103, 104, 110

H

Health Insurance Portability and Accountability Act (HIPAA), 52–55, 291, 360, 414
HIPAA. *See* Health Insurance Portability and Accountability Act (HIPAA)
HIV/AIDS-related release forms, 361–363
Home visit contact, 398
Homocidality, 353–354
Hull, G. H., 40, 51
Human service directory, 380

I

I-messages, 266–267
　in confrontation, 207–213
　examples of, 207
　firmer, 159
　overbearing, 215–216
　rules for confrontation, 209
　ways to start, 159
Impressions and recommendations, 305, 312–314, 383
Impulse control, 354
Individualized planning, for clients, 374
Information, 2, 5, 7, 19, 81, 82, 87, 88, 108–109, 127, 143, 150, 162, 174, 177, 212, 213, 259, 349
　HIV/AIDS, 48
　issues related to, 365
　from other agency, 363–364
　privacy, 51
　protected health information (PHI), 52
　receiving and releasing, 359–365
　in release forms, 360–363
　releasing, 47–48
　sending for, 359
　useful, 233
Informed consent, 46, 265
Initial Assessment Form, 315
Initial inquiries, 18
Initial inquiries, documenting, 277–284
Inquiry form, 277
　guidelines for filling out, 278
　steps for filling out, 278–282
Insight, 118, 139, 140, 354
Intake Evaluation Form, 315
Intakes, 4, 10, 100, 169
　phone, 277
　in social history, 310–311
Intellectual disability, as barrier, 375
Intelligence, 349
Intensive case management, 381
Intention to harm, 57–58
International Classification of Diseases (ICD-10), 326
Interpersonal style, 341–342
Interventions, 18, 19, 26, 37
　crisis, 12, 16
　developing a, 87–88
　macro level, 87–89
　treatment, 429–430
Involuntary commitment, 60

J

Jackson, Stanley W., 169
Judgment, 56, 110, 117, 118, 122–123, 152, 154, 176, 212, 213, 240, 241, 252, 266, 354
Judgmental words, in case notes, 402–403

K

Kim, Y. Y., 97, 98, 99, 103, 104, 110
Kirst-Ashman, K. K., 40, 51
Kraepelin, Emil, 324

L

LaBruzza, Anthony L., 323, 325, 339
Lack of resources, as barrier, 375
Lambert, M. J., 60
Language, 343–344
　as barrier, 375
　in social services, 2–3
Leaman, D. R., 208, 212
Linking, 11–12
Linking summaries, 260
Listening, 19, 25, 47, 101, 119, 120, 121, 128, 130, 140, 155, 156, 163, 164, 231, 233, 234, 241, 242, 252, 256, 257, 266
　to content, 170, 174–176
　defining, 169, 170
　to feelings, 170–176
　open questions, 194–195
　positive reasons for, 176–177
Long term goals, 430
Lynch, A. A., 78

M

Managed care, 439
Managed care, case management and, 21–25
Managed care organization (MCO), 21–22
Mandated reporting, 58–59
Medically life threatening, addictions, 419
Memory, 347–348
Memory testing, 348
Mental disorders
    name of, 330
    number of, 330
    psychiatry attempts to classify, 324–325
    severity of, 331
Mental illness, as barrier, 375
Mental status examination (MSE), 5, 337–356, 399
    observations, 337–339
    outline of, 339–356
Meyer, Adolph, 324
Miller, William, 239, 244, 245, 246, 253, 259
Monitoring, 12–13
    advocating, 417–418
    case managers in, 413, 418
    documentation, 398–399
    financial purpose of, 414–416
    follow-up in, 416
    services, 417
Mood, 282, 345. *See also* Emotions
Motivation, 79, 83, 156, 282
Motivational Interviewing (Miller and Rollnick), 239, 252, 259
Multiple diagnoses, 331
The Myth of Mental Illness (Szasz), 325

N

NASW. *See* National Association of Social Workers (NASW)
National Association of Social Workers (NASW), 25
Neurological language disturbances, 344
Neurovegetative signs of depression, 345–346
New Referral form. *See* Inquiry form

O

Obligations, 293–294
Observations, 5, 212, 337–339
Observations, documenting, 338–339
Office visit contact, 398
One Flew Over the Cuckoo's Nest (film), 325
Open questions. *See also* Questions
    formula for, 192–195
    reflective listening, 194–195
    tips for, 192
Outcomes, 5, 22, 23, 84, 104, 124
Outcomes, of goals, 423–425

P

Paranoid delusions, 351
Parent Effectiveness Training (Gordon), 151
Peer support, 7, 251, 252
Perception, 8, 80, 103, 124, 194, 205, 230, 232
Perls, Fritz, 176
Perseveration, 344
Personal context, 80
Phone contact, 398
Planning, 44–45, 123, 243, 245, 249, 262, 266
    change, stages of, 241
    continued, 10
    follow-up to, 80
    individualized, 8–10, 266, 374
Polcin, D. L., 203, 213
Practical Communications (Goldman), 422
Practice, 267
Praise, false, 162–163
Prejudice, 43, 81, 82, 110
Preoccupations, 353
Presentation, 383–385
    making, 383–384
    preparing for, 383
    sample, 384–385
Presenting problem, 4, 293, 298
President's New Freedom Commission on Mental Health, 25
Privacy, 51, 52, 53, 142
Privacy, and self-disclosure, 106
Protected health information (PHI), 52
Proverbs, 349
Provider agencies, 25, 367–369, 395, 398, 404, 417, 421–432, 440
Provider agencies, defined, 367
Psychological characteristics, micro level ecological model, 80
Psychomotor activity, 342–343

Q

Questions, 4, 5, 23, 55, 96, 144, 145, 155–156, 233, 244, 252, 256, 261
    acceptable answer, 191
    in advocating, 417
    assumptions, 192
    to case manager, diagnosis, 382
    change the subject, 191
    closed, 158, 188
    in final interview and letter, 442
    important, 187
    multiple questions, 191
    open, 158, 189–190, 192–195, 266
    why questions, 190
Quotations, in case notes, 400

R

Records, 9, 88, 126, 359–360, 365, 395–404. *See also* Documentation
    case termination and, 439
    legal documentation of work, 396
    social history in, 299
Recovery, 21–23, 128
Recovery Model, 250
Recovery tools, 250–252
Referral notification form, 389
Referrals, 12, 20, 56, 322, 385, 387–392
Reflective listening, 130, 170, 176–177, 194–195, 231, 233, 234, 266. *See also* Listening
    drain off feeling, 176–177
    responses time, 177
    self-acceptance, 176
    solution phase of, 177
Relapse, 242
Release forms, 360–362
    examples of, 362–363
    general form, 360–363
    HIV/AIDS–related, 361–363
Reliability, 355
Resiliency Model, 23, 266
Resistance, 47, 101, 214, 228, 244–247
Resource coordination, 17, 381
Resources, 37, 251
    community, 25
    coordination, 17
    Generic, 6
    informal, 7–8
Responses, 50, 124, 130, 149–165, 171, 175, 254. *See also* Listening
    to content, 174–176
    to feelings, 170–174
Responsibilities, 293–294
Review dates, 372, 388
    defined, 431
    for goals, 429–430
Reviews, 414
Rogers, Carl, 121
Rollnick, Stephen, 239, 244, 245, 246, 253, 259

S

Schizophrenia, 43, 47, 87, 125, 126, 144, 214, 341, 343, 344, 345, 351, 352
Segal, E., 122
Self-acceptance, 176
Self-determination, 24, 45–46, 142, 143, 251, 253
Self-direction, 250
Service coordination, 13–16
Service interventions, 423
Service planning conference
   benefits of, 381–382
   collaborative activity, 382
   DSM Handbook to, 380
   follow-up to, 385
   goals for, 380–381
   human service directory to, 380
   preparing for, 379–385
   presentation, 383–385
   tentative service plan to, 380
Service plans
   client and family involvement, 368–369
   creating, 372–373
   developing, at case management unit, 367–376
Services, 2–3, 77, 84, 85, 95, 98, 99, 120, 123, 124, 141, 143, 233, 381
Services, case termination and, 439, 440, 441, 442, 444
Severe akathisia, 343
Short term goals, 430
Siegel, M., 51
Social context, 80
Social history, 4, 20, 26, 293, 369
   agencies and, 297
   brief, 311–314
   in chart, 317
   client's appraisal and, 301
   client's personal background, 302–305, 311–312
   on computer, 316
   description of, 298
   details of clients, 305–306
   impressions and recommendations of, 305, 312–314
   intakes in, 310–311
   layout of, 298–299
   in other settings, 310–311
   outline for, 298
   presenting problem in, 299–301, 311
   questions, open and closed, 299
   taking in home, 316–317
   who took, 243, 306–310
Social network, 55–56
Social service agencies, 322
Social services, language in, 2–3
Speech, 1, 24, 343–344
Stadler, H., 51
Staff behavior, 234–235
Stephan, W., 110
Stereotypes, 101, 109, 129, 131, 342, 344
Strengths identification, of clients, 373–374
Subculture, 97, 111
Substance abuse, 3, 6, 7, 15, 21, 22, 36, 40, 87, 368
Substance Abuse and Mental Health Services Administration (SAMHSA), 3
Suicidal case, emergency, 419
Suicide, 51, 57, 353–354
Summarizing, 259–261
Supervision, 16, 17, 48, 51, 88
Szasz, Thomas, 325

T

Tarasoff, 57, 58
Tardive dyskinesia, 343
Target dates, 388
   defined, 431
   for goals, 428
Targeted case management, 381
Tentative service plan, 380
Termination, of case, 439–447
   case manager and, 439
   dies or moves away, individual, 439
   disappearing of, individual, 440
   dissatisfaction of services, individual, 439–440
   documentation, 443
   examples, 444–447
   feelings about, 440–441
   final interview, 442
   finance services and, 439
   successful, 440–441
Third ear, 120
Thought content, 350–351
Thought processes, 351–352
Transference, 129–130
Transition summaries, 260–261
Trapping the client, 258
Treatment, 3–6, 11, 12, 16, 17, 20, 21, 23, 86, 105, 119, 123, 214
   goals, 368
   interventions, 429–430
   plan, 372–373
Treatment planning conference. *See* Service planning conference

U

Uncertainty, 99–100
Understanding, of client, 288
Unger, M., 89
Universal precautions, 48
Unspecified Disorders, 333
U.S. Department of Health and Human Services, 21, 23
Using DSM-IV: A Clinician's Guide to Psychiatric Diagnosis (LaBruzza), 323

V

Valesquez, M. M., 240
Value conflicts, 40–44
Value conflicts, self-assessment exercise, 42
V codes, 332
Verification form, 282–283

W

Waite, D. R., 60
Warmington, S., 121
Who owns the problem, clarifying, 139–145
Workplace, safety in, 233–234